DEDICATION

To all the boatbuilders who have done me
the great honor of building my boats; and
to my mother

CONTENTS

Ultrasimple
BOATBUILDING

17 PLYWOOD BOATS ANYONE CAN BUILD

Gavin Atkin

International Marine / McGraw-Hill

Camden, Maine • New York • Chicago • San Francisco • Lisbon • London • Madrid • Mexico City
Milan • New Delhi • San Juan • Seoul • Singapore • Sydney • Toronto

The McGraw·Hill Companies

2 3 4 5 6 8 9 10 CCW CCW 0 9 8

Library of Congress Cataloging-in-Publication Data
Atkin, Gavin.
   Ultrasimple boatbuilding : 17 plywood boats anyone can build / Gavin Atkin.
         p.   cm.
   Includes index.
   ISBN-13: 978-0-07-147792-5 (pbk. : alk. paper)
   1. Boatbuilding—Amateurs' manuals.   I. Title.   II. Title: Ultrasimple boatbuilding.
   VM351.A88 2008
   623.82'023—dc22   2007023935

Questions regarding the content of this book should be addressed to
International Marine
P.O. Box 220
Camden, ME 04843
www.internationalmarine.com

Questions regarding the ordering of this book should be addressed to
The McGraw-Hill Companies
Customer Service Department
P.O. Box 547
Blacklick, OH 43004
Retail customers: 1-800-262-4729
Bookstores: 1-800-722-4726

Unless otherwise noted, photos and illustrations by Gavin Atkin.

A note on the tables of coordinates: Readers may notice what appear to be slight inconsistencies between corresponding millimeter and inch measurements within the tables of coordinates. For every ⅛ inch there are almost 3.2 millimeters; therefore, because there are three millimeters within the smallest measurement we're using on the inch scale, the millimeter measurements give us higher-resolution measurements. Small differences that can't be detected when measuring to eighths may be detectable when working in millimeters, hence the discrepancies in the tables. Please note, however, these discrepancies are way too small to matter when building these boats.

PART TWO
The Boats

ACKNOWLEDGMENTS

Huge thanks to my editor at International Marine, Bob Holtzman, and my friends Chuck Leinweber and John Welsford for help, ideas, and encouragement; to Ed Bachmann for his help with visuals; to those many builders of my boats who were kind enough to let me use their photographs; to the members of the Yahoogroup Mouseboats for their wonderful support and interest; to Jim and Eileen Van den Bos, Rosalind Riley, and Tim Bull for letting me use their lakes; and to my family for their tolerance, and especially to Julie for providing essential life support without which this book would never have been written.

THE VIRTUES OF SMALL AND SIMPLE BOATS

I adore secondhand bookshops—the smell, the atmosphere, and the low prices. In my time, I've found lots of great books about building small boats on their dusty shelves. Because many aspects of boating remain the same over generations and even centuries, older books on the subject can often be as relevant today as newer editions, and there's much indeed to learn from them.

However, I have come to dislike certain types of older books, particularly the ones on boat carpentry that are more discouraging than helpful. They usually start with an anecdote about how the author tried building a boat at a young age and failed. In some versions, the author's father, in a fit of misplaced and misguided rage, destroys the original boat with an ax because it's both shameful and unsafe; in others he burns it. Usually the experience leaves the fledgling boatbuilder in tears but determined to win the father's approval. Stories like this make me angry because when I see them I know it's more than likely that the author, deliberately or not, is replaying the part of the angry father and making readers feel inadequate, as

if they were children unable to build a proper small boat.

In contrast, this book is meant to make you feel confident about building small boats. It will explain how to make the project go as smoothly and efficiently as possible, bearing in mind that mistakes aren't the end of the world. You can correct many with the help of good old cousin Poly Urethane, Uncle Epoxy, and Auntie Filler, so there's no need to fear potential blunders along the way. Thankfully, the bad old boat-chopping and boat-burning days are long gone.

The methods I'm presenting here are tried, tested, and known to work. Over the years I've noticed beginners often wonder whether alternative cheaper materials and methods might work just as well as the ones I have included in this book. The answer is some will, some won't, and some will only to a point. If you have a slightly wacky idea about building a boat from expanded foam or using water-resistant (rather than waterproof) glue, ask about it on the Internet discussion groups. Someone will almost certainly have tried the same thing and will be only too happy to give you the benefit of his or her experience.

This book is based on the principle that boating is fun, and that the fun can be enhanced if we don't make it too expensive, too much like hard work, or too serious. On all these points, simple and small home-built boats are often the best way to go, and that's what my own boating journey has been all about.

Let me explain some of the advantages of small boats:

■ Small home-built boats have low costs in every way, and they are also environmentally friendly. Building them consumes small amounts of precious wood and small amounts of hydrocarbons in glues and paints. If powered by engines, they use far less fuel than larger, faster craft and put far fewer pollutants into the air. They also don't require antifouling paint because you typically haul them out after each use, and so they don't release toxins from the paint into the water.

■ Handling a small boat is easier and takes much less hard work than a big boat. Small boats require less strength to manage and, if you're sailing, the lines are shorter and easier to manipulate. One person can often launch and haul the boat or even transport it on the roof of a car. Novices frequently feel more comfortable gaining experience on a small boat since the forces involved in sailing it are much more manageable.

■ Building a small plywood boat is easy, quick, and satisfying, and owning and maintaining it is has less of a financial impact on other family members than buying and owning a large boat. This is an important point: if a small boat has less of an impact on family life in terms of money and time spent on maintenance, it is much less likely to cause domestic strife.

■ On the fun side, in a small boat you're in such close touch with the water you feel each puff of wind and every ripple.

Being close to the elements can clear a tiring, stressful week out of a tired brain in a matter of minutes.

■ Another advantage of being so close to the elements is that it hones your boating skills. In sailing, sitting just inches from the water in a boat light enough to react to every ripple and puff of wind forces you to react quickly. Even slight changes in conditions will require you to make adjustments regarding crew placement and boat balance, sail trim, helm, or oar or paddle strokes. The same is true even when you're steering an outboard motor.

■ Most children enjoy being captain of their own ship far more than watching their parents have all the fun. Kids who have done any amount of summer boating will readily say that falling in the water with a crowd of other kids is the funniest and most popular part of the experience, and that's only possible in the smallest of boats. Introduce them to small boats at an age as early as six years old (with proper oversight, of course) and let them play with friends in their own tiny craft. Almost certainly they'll quickly come to love boating on their own and with you.

I mentioned my own boating journey, and I should explain where it has taken me. I did a small amount of sailing as a boy and even built one of the famous Mirror stitch-and-glue dinghies with my father. My father was often busy and stressed, and as a youngster it felt good to have a joint project with him. More than that, though, by the time the boat was completed, I was pretty sure I'd acquired a lifelong interest in small boats and would be quite confident about building one if ever I felt inspired. Sadly, the Mirror got very little use. We lived a good distance from any body of water ideal for sailing and my father hated driving. Alas, boating slipped below my horizons, and for years I concentrated on traditional music, dancing, and hiking for entertainment.

But that sad, boatless condition changed two and a half decades after my father and I built and briefly sailed the Mirror. I was persuaded to sign up for some dinghy sailing classes, and these proved to be the start of something really big. The boating bug bit hard in the way it sometimes can, and it took over my imagination, my bookshelves, and finally my computer. As an incurable romantic, I became fascinated by the old working boats, partly for their looks, partly because many of them were associated with heroic, half-desperate fishing communities living on the knife edge of survival (as well as on the edge of the land), and partly for what they can still tell us about what makes a good boat for a particular stretch of water.

Also, as someone who'd studied physics in college out of an inherent curiosity about the world, I soon wanted to understand how the dinghies I was sailing worked in greater detail, with a more profound understanding than simply knowing what line to pull and when. It seemed to me that my boats quite often didn't do what I had been taught to expect by my sailing teachers, and that fanned my curiosity. I also thought boating would be fun for me and my family, but as a practical parent with a big monthly mortgage, I wanted to discover how to get the most fun out of it for the least amount of cash.

So, I spent a lot of small sums of money in secondhand bookshops and took the books home for study. I steadily learned more and more, and began to see that not only could I build small boats, I could go much further and involve myself in designing and building a new generation of cheap and simple small craft. They would deliver a huge amount of fun with a small investment of money and time, and they could be built with the basic skills within the reach of most ordinary people.

I struck gold when two things came together. I was asked to go to New York to play fiddle for a sword dancing team. I was pleased to be invited, but the really great news was that the invitation finally afforded me an opportunity to visit Mystic Seaport: The Museum of America and the Sea in Mystic,

Connecticut. It also just so happened that I'd received an e-mail from David Colpitts, a teacher and paddler who lived in Hartford, not far from Mystic. David asked for a boat design so simple and basic that the kids he was teaching during summer vacation would be able to build it on a tiny budget, using cheap luaun plywood, polyurethane adhesive, and drywall tape. We arranged to meet.

After some memorable and possibly mad midwinter boating, we went to a bar to talk about the new boat. I remember the excitement as I told him my idea for a little pram-bowed craft with a shallow V-bottom and vertical sides, which would require just one sheet of plywood (if you didn't count the decks). I drew the Mystic Mouse on a beer coaster, David said he was pleased, and I went home, where I formalized the design and e-mailed it to him. He promptly built the boat in ⅛-inch plywood, using polyurethane glue and tape to hold the joints together, and pronounced it an instant success.

What do you do with something that makes a great toy that you want to share, but has little or no monetary value? My answer was to make it available for free over the Internet, and very soon people were downloading the plans and building the Mouse. I was delighted.

Having started down this track, I set up the Mouseboats Yahoogroup (http://groups.yahoo.com/group/mouseboats) to promote the idea in a more substantial way and to provide a forum for related original ideas. At the time, I had no idea what fun I would have, what friends I would make, or how many people would build the little boats. As I write, more than 180 boats have been built (which doesn't include the ones not reported to the Yahoogroup).

Owners and builders of Mouseboats have become a wonderful community, who merrily swap Mouseboat experiences and offer new builders the best support and advice available anywhere. These people know everything there is to know about building the boats and their many uses. Some give them to their kids to play with, some use them for fishing or hunting birds, while others use them for daysailing or even for extended camping trips. Most seem to use their little boats for far more than I envisaged that day when I drew a smudgy sketch on a beer coaster for David Colpitts.

I strongly recommend anyone building a Mouse to join and register their boat. Not all online communities are friendly places, even those dedicated to building and using boats. But Mouseboats has always been the gentlest and most helpful of discussion groups, and I intend to ensure it remains so. Where my other boats are concerned, I'd love to hear from you. I'm particularly keen to receive pictures and reports of how the build went and how the boat worked on the water. I will also help builders whenever I can, often through another Yahoogroup: http://groups. yahoo.com/group/gmaboatbuilders.

Through it all, I'm half-ashamed and half-proud to say that I remain the roughest and quickest of boatbuilders. I'm not going to tell you how to gild your plywood boat, for I'd never do it myself. I'm not going to suggest difficult and expensive boatbuilding techniques or designs, nor confuse you with information you don't need to get the job done quickly and at a very low cost. If perfectionism and glitter is your game, this book probably isn't for you.

But if you're looking for a book that offers plans and instructions for a collection of very simple small boats, capable of being built (though possibly not painted) in a weekend, you've found what you're looking for. In many cases, you can be afloat in a half-painted boat in a week or so and have your boat fully finished in two weeks tops, which is enough time for your paint to fully harden.

To make this possible, I've focused on boats that involve making small numbers of easily shaped frames, as most beginners with few woodworking skills are happier and work faster with goo and tape, or with my version of the chine log method. The boats come from a small range of designers with a strong interest in small craft, and every design has a particular set of attributes that make it especially suitable for a particular task. My own designs here include:

- Minimouse and Micromouse: flat-bottomed paddling prams with perfect stability for kids new to small boats, and sufficient carrying capacity for small to medium-sized adults
- Lilypad: an extremely (wonderfully) basic punt
- Mouse (the original): similar to the Minimouse but with a more capable V-shaped hull, requiring just a little more building effort
- Rowing Mouse: similar to the Mouse, but set up for rowing
- Cruising Mouse: designed for paddling but can also be rowed, with enough displacement and volume for more comfort and longer voyages; requires two sheets of plywood
- Dogsbody: an outboard-powered skiff or garvey; suitable as a workboat or for fishing
- Jiggity: the simplest possible adaptation of the traditional pram-bowed Auray punt; makes a fine tender
- Cinderella: an elegant (if I do say so) and somewhat round-bottomed, double-paddle canoe; perhaps a bit more involved than most of the others, but still simple as boats go
- Flying Mouse: an 8-foot sailing version designed to teach young kids how to sail single-handed boats with just the right combination of confidence and excitement
- Eek!: an 11-footer with the same purpose as the above for teens and adults
- Doris: a light, 16-foot, straight-sided dory for oar or sail

As previously mentioned, I've also included several fine boats from other designers:

- David Beede's Summer Breeze: a small plywood version of the traditional American flat-bottomed rowing and sailing skiff
- John Wright's version of the Puddle Duck Racer: a sailing box that is the world's simplest, lowest-cost racing

class. (As this is a development class and subject to continuous improvements, I've redrawn and made some changes to John's original design.)

■ Steve Lewis's Poorboy: a simple, two-sheet, low-horsepower skiff that can be built to 10 feet or 11 feet 6 inches overall

■ Jim Michalak's Piragua: an extremely popular design for a pirogue—a kind of flat-bottomed canoe

■ Murray Isles's Aurette: an interpretation of the Auray punt. It is a bit more sophisticated than my Jiggity and adapted for sail

■ Finally, I have included a gallery of additional designs from these and other designers for your consideration, but without full plans or building instructions

All of these boats are in this book because they meet the key criteria for happy small boating: keep it cheap, make it easy, and have fun.

I hope these little boats will give you, your families, and your friends more fun and more value for your money than you imagine possible. I'd certainly love to hear about it. Please send me your stories and pictures (gmatkin@ clara.net)!

I'd like to make a few important points about these designs. Many are by amateur designers. No claims are made for their performance or safety. The designers and author accept no responsibility for any accident or loss that may take place during building or in use. No individual may build more than two or three boats to these drawings without the permission of the designer.

Finally, a word about terminology. Since U.S. and U.K. English differ somewhat—especially in regard to technical items such as lumber (timber) and latex (emulsion) paint—I've tried to include both forms in the first instance of use, with the U.K. version appearing in parentheses immediately following the U.S. version. Thereafter, I've used just the U.S. form for the sake of brevity.

PART ONE

How to Build Them

CHAPTER I

THREE SIMPLE METHODS

A flat-bottomed Micromouse built with external chine logs, using the simplified chine log method. (Henry Massenburg)

The boats in this book can be built using a combination of three ultrasimple methods—simplified chine log, polyurethane stitch-and-glue, and epoxy stitch-and-glue. Let's examine and compare them to show you just how simple it is to build these boats and to give you the information you need to decide which method best suits you. Then as you read the following chapters, you'll be able to focus on the specifics for the method you plan to use and learn more about how it can be applied to a given design.

THE SIMPLIFIED CHINE LOG METHOD

For many amateur home boatbuilders, the best-known method of plywood boatbuilding is the one described in Dynamite Payson's *Instant Boatbuilding.* This book caused a revolution in this segment of the boatbuilding

market and was one of the main reasons for the resurgence of home boatbuilding's popularity in recent years.

My simplified chine log method is a further simplification of Payson's concept for the "instant boat." It is in essence the same as Payson's, but made simpler by the fact that many of the boats in this book have vertical sides, which cuts out much of the beveling and shaping work usually involved in building small boats. It will appeal to those who favor conventional woodworking procedures over slathering adhesive and tape to make glass-and-resin fillets, which although having the same function as internal chine logs, are constructed quite differently. Briefly, a chine log is a long piece of lumber (timber) used to make a joint between two pieces of plywood running the length of the boat; for example, between the bottom of a flat-bottomed boat and its sides. A fillet is made of filled epoxy or polyurethane and shaped using a tool such as the back of a large spoon. It is covered on the inside with glass tape.

While all three building methods share some of the same techniques and procedures, it's all a matter of degree. The simplified chine log method relies the least upon glues and the most upon mechanical fasteners (i.e., nails or screws).

In the simplified chine log method, the plywood bulkheads and transoms have pieces of lumber called "cleats" nailed all around their perimeters to hold fasteners. Most of these cleats must be beveled so that the sides of the hull will lay flush against them. Plywood panels are cut to shape and bent around the bulkheads and transoms to make the sides of the hull. Fasteners are driven through the plywood into the cleats.

Next, chine logs are bent and then nailed and glued to the bottom outside edge of the side panels. They hold the fasteners for the bottom, performing the same function that the frame cleats do for the sides. The bottom is then attached to the cleats and bulkheads with glue, nails, or screws. With "ultrasimplified" designs where the boat's sides are at right angles to the bottom, this operation could not be simpler.

TWO STITCH-AND-GLUE METHODS

The second and third building methods are both called "stitch-and-glue." They differ a little, depending on whether you choose to work with polyurethane glue or epoxy, but the principle remains the same. Stitch-and-glue boatbuilding relies as much on the shape of the hull panels as on the frames or bulkheads, to define the shape of the hull. When you join the curved edges of the hull panels to each other, they almost magically form themselves into a boat.

Because the frames and bulkheads define the shape of the hull in the simplified chine log method described earlier, it is not necessary to carefully measure the bottom panel before cutting it to shape. You can simply trace the shape of the boat's sides onto the bottom panel of plywood, cut roughly along the pencil lines, and trim it flush after it's fastened down. However, when using either stitch-and-glue method, it is all of the hull panels that define the shape of the hull. Therefore, you must also measure and cut the bottom panel to create a hull with a good shape.

As I've said, the other major difference in construction methods is the way parts are put together: stitch-and-glue is more adhesive-intensive, while the simplified chine log method relies more on nails or screws.

In stitch-and-glue construction, it's normal to drill lots of little holes along the edges of the hull panels, and stitch the panels together with plastic cable ties or bits of wire to hold them in place while you apply glue and fiberglass tape to the seams. Most of the boats in this book have such easy shapes that you'll probably be able to use duct tape or something similar to hold things in place. (Since this is stitch-and-glue without the stitches, you may prefer to call it "taped-seam construction.") However, some designs will require at least some stitching, perhaps because they have tighter curves or you're using thicker, stiffer plywood.

Once the panels are in place next to each other, they are joined on the inside with fillets. As noted earlier, these are beads of thickened adhesive goop that run the entire length of the seams, then are smoothed and hollowed to a nice concave (or coved) shape. When the goop is half-hardened but still a little soft, a strip of fiberglass tape is laid onto the fillet, then another coat of adhesive is laid on it and smoothed to finish the seam. After the fillets are made, the outside seams are covered with glue and fiberglass.

Polyurethane Stitch-and-Glue

In concept, the stitch-and-glue methods are pretty much the same whether you use polyurethane glue or epoxy, but these two materials are different enough in how you work with them that I think we should view them as distinct methods.

9

A stitch-and-glue Cruising Mouse hull made by Anthony
Smith is nearly complete. The seams between all the hull
panels and bulkheads have been filleted with epoxy-based
putty and taped over with fiberglass tape and more epoxy.
(Anthony Smith)

Generally, polyurethane glue is easier to work with than epoxy, and it is often regarded as less toxic. Most polyurethane glue comes out of a cartridge and is applied with a caulking gun. About the same consistency as bathtub caulk, it's not runny and so stays where you put it. It is not quite as strong as epoxy, and it doesn't have epoxy's long track record of success in "serious" boatbuilding, but it's strong enough for our purposes. Several of my boats have been built using this method and, as reported on the Mouseboats group, the builders have been very satisfied with the results.

Epoxy Stitch-and-Glue

Epoxy, on the other hand, is a famously effective material that is proven in boats ranging from dinghies to ultralight ocean racers and huge motoryachts. It's more expensive and more difficult to work with than polyurethane glue since you must *carefully* measure and mix

two components (resin and hardener) before using it. Once you've mixed the two, depending on the particular recipe of resin you have bought or the temperature in your backyard, you may have a very limited amount of time to use it before it "kicks," or begins to harden. Furthermore, as it comes out of the container, it is quite runny.

To make the fillets, you have to mix in some kind of powdered filler material. The filler helps to bulk the expensive epoxy and add strength, although the real strength of a fillet comes from the glass tape that lies over it. It's best to use only the very finely ground, purpose-made fillers supplied for mixing with epoxy, as nothing else I've tried or heard about is as good. In my experience, even fine sawdust from machine tools is rough stuff when mixed with epoxy, resulting in a rough surface, and is much less effective in making the epoxy go a long way.

Epoxy is also somewhat dangerous to work with. Ignore the health warnings, and you could injure yourself!

In spite of its drawbacks, however, epoxy is certainly better than polyurethane, from both a structural and aesthetic point of view. It will produce a stronger, more durable, and better looking boat than polyurethane.

Some boatbuilders recommend polyester resin as an economical alternative to epoxy as an adhesive in stitch-and-glue boats. Polyester is the plastic from which many "fiberglass" boats are made (the proper name for fiberglass boat construction is glass-reinforced plastic, or GRP). It is similar to epoxy in that it is a two-part material that requires mixing a resin and a hardener. Although there is plenty of evidence that it works well, it is no longer widely used by home boatbuilders, and I can't personally vouch for it since I haven't used it.

All three of these methods are good, and deciding between them should not be a cause for anxiety—none represent a "wrong" choice. They're all easy and fun and will result in a strongly constructed boat that won't fall apart. They all have the potential to produce a good-looking boat, depending on how much work you put into the final product. (As I said earlier,

Making a Stitch-and-Glue Seam

A Cable tie or wire stitches in place.
B Epoxy fillet in place. It covers the tie on the inside and is shaped in a soft curve to accept glass tape.
C Epoxy glass tape in place on top of the shaped epoxy fillet.
D When the inner seam has hardened, trim off the cable tie on the outside and round off the outside of the seam, filling with epoxy where necessary.
E Brush epoxy to the external seam, then apply epoxy and glass tape to the seam.
F As an alternative to the external tape, in small boats and canoes, the external taping may be omitted if the boat is covered with glass cloth.

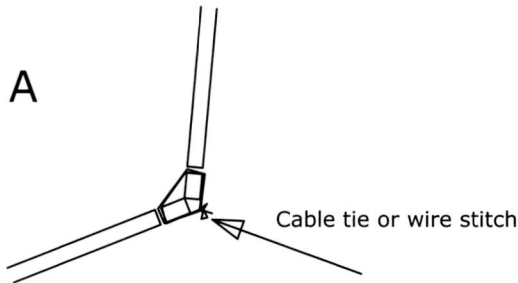

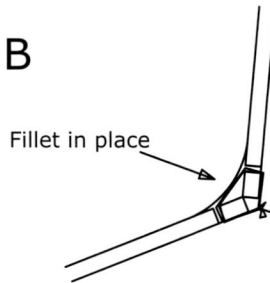

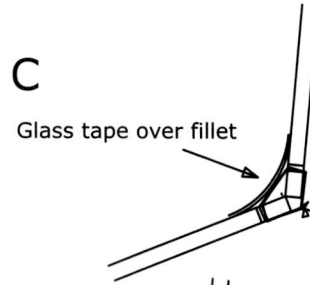

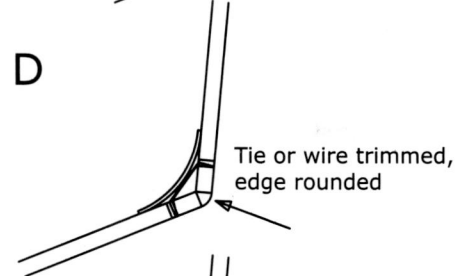

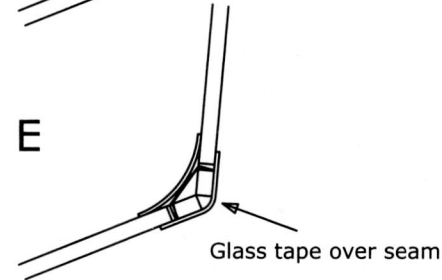

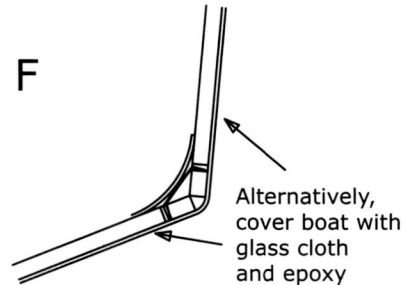

C

Glass tape over fillet

D

Tie or wire trimmed,
edge rounded

A

Cable tie or wire stitch

E

Glass tape over seam

B

Fillet in place

F

Alternatively,
cover boat with
glass cloth
and epoxy

The steps to making stitch-and-glue joints.

it's by no means essential that your boat look good. You're perfectly entitled to build an ugly boat if getting out on the water without delay is more important to you than the compliments of your loved ones.) Even the most expensive choice—epoxy stitch-and-glue—will be downright cheap since the amount of epoxy required for these tiny boats is correspondingly small. So just pick the method that sounds like the most fun. The boats are so simple and inexpensive that you can (and probably will) try all three methods before too long. The main thing at this point is to make your decision and get started on the next step of the process.

CHAPTER 2

MATERIALS AND TOOLS

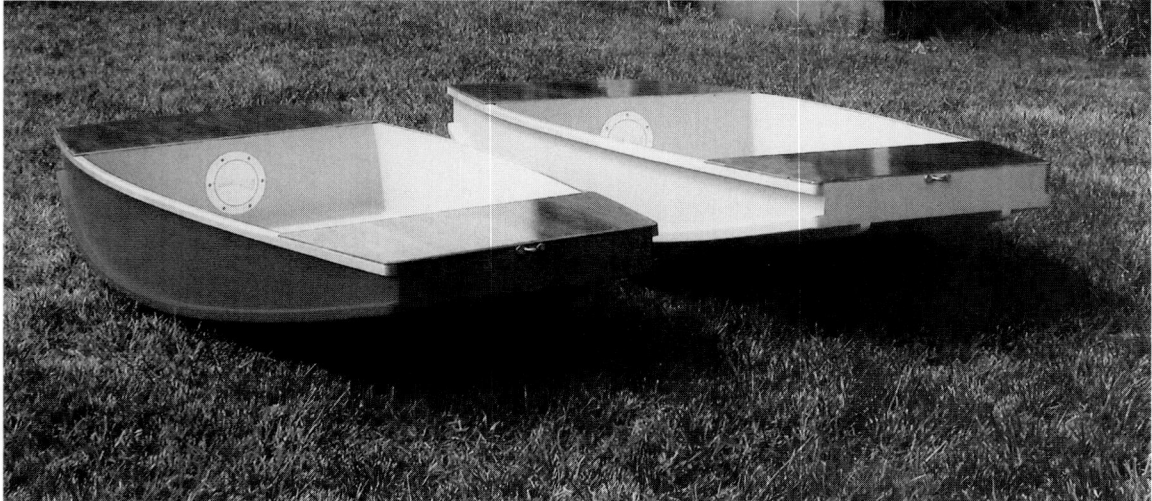

A brace of Micromice built by Henry Massenburg for a pair of lucky kids. (Henry Massenburg)

Having made my case that small boats are fun, useful, and well within your ability to build, and having tantalized you with a few pictures, it's time to talk a little more about materials and tools. Let's think of a simple flat-bottomed Mouseboat like Minimouse or Micromouse. There's no doubt that flat-bottomed versions can take as little as a day, or even half a day, to assemble. However, anyone new to boatbuilding or working with wood will naturally take longer, so please bear that in mind. If you're unfamiliar with the materials and tools needed to do the job, it'll take you some time to feel comfortable and to work at optimal efficiency.

I'll obviously go into extensive detail on building the various boats in this book later, but let's use the Micromouse as an example here to illustrate the tools and materials that you'll typically need. The following list will give you a good idea of how inexpensive it is to build these boats.

To build the 7-foot 10-inch flat-bottomed Minimouse using the polyurethane stitch-and-glue method, you will need:

- 1½ sheets of waterproof plywood
- 30 feet of 1-by-1-inch pine or similar
- 3 pints of epoxy
- 1 or 2 tubes of polyurethane glue
- 50 feet of carpet tape
- A few dozen stainless steel screws
- 1 or 2 quarts of paint
- A little varnish
- 4 drain plugs
- 1 galvanized steel cleat for attaching the painter
- 3 to 4 yards of cheap polypropylene line for the painter

To make a double paddle, you'll need:

- A pine or similar closet pole (make sure there's no joint; if there is, it may not be well engineered and could have been made using non-waterproof glue)
- Some scraps of plywood to make blades
- 8 stainless steel screws

You may well have quite a few of these things in your garage. Even if you don't, when you add up the cost of this little list, it's unlikely to come to much, particularly if you use the Internet to find the best prices on some of the pricier items, such as stainless steel screws, glue, and epoxy.

I live in the United Kingdom, one of the most expensive countries in the world, but if I shop around, I swear that it's hard to see how this shopping list can add up to much more than, say, £80 to £100 (roughly $160 to $200), provided that you plan to use epoxy only for the seams and not to cover the entire hull. What's more, if you're prepared to use the very cheapest waterproof plywood and polyurethane glue, and you can find some unused exterior paint from another project, it's possible to bring the price down a lot lower than this figure, maybe as little as £50 ($100 or so) or even less.

The tools and skills can also be minimal, if you're happy working that way. There's always room to add more effort and patience in boatbuilding, but the boats covered in this book are probably best suited to the kind of rough and ready building jobs most of us can complete in a couple of weekends.

PLYWOOD

Home boatbuilding today is not at all like it used to be. At one time, it was almost impossible to make a boat that didn't leak when you first put it in the water. You had to wait until the hull planking soaked up water and swelled to form a seal that stopped most of the leaks.

Fortunately for the home boatbuilder, traditional methods of boat construction are no longer necessary because of a single breakthrough that changed all the rules. In 1934, a Harbor Plywood Corporation chemist named James Nevin developed a fully waterproof formaldehyde adhesive that could be used to create the wonder material we now call waterproof plywood. Waterproof plywood is made up of sheets of wood peeled from the

trunk of a tree. These veneers of wood are then stacked and glued at right angles to each other. Picture a compass: the grain in the first layer of wood runs from east to west, the grain in the second layer runs from north to south, and the grain in the third layer runs from east to west again, and so on.

Plywood is easy to take for granted in the modern age, but for an amateur boatbuilder and user, it really is a wonder material. It's very dimensionally stable—it swells and shrinks very little in any direction with changes in temperature or humidity. This means that boat structures made from waterproof plywood can be glued together. Unlike the old boats, modern plywood boats don't depend on soaking to close up any leaks between the strakes of the hull. Therefore, they can be stored all year in a garage or shed, and then thrown into the water and used immediately. What's more, compared with dimensional lumber, plywood is very uniform, easy to mark out and use, and strong for its weight. It also takes glue well and provides a good substrate for painting.

Plywood's weak point, though, is that half the layers on any edge consist of exposed end-grain. Unlike its flat surfaces, the edges of plywood *will* soak up water, which can lead to rot. This can be avoided by sealing the edges before assembly, preferably with epoxy.

Some people don't like plywood. Many plywood detractors are ultra-traditionalists who appreciate only the best old-fashioned workmanship, and perhaps also derive some odd pleasure from having wet feet and sloppy bilges from time to time. Lots of us appreciate high-quality workmanship, pretty designs, and lovely materials but while we admire traditional boats, many of us find we have to choose the conveniences that plywood provides: easy trailering and launching, light weight, low cost, and easy maintenance. Without these conveniences there might be no boating or boatbuilding for many of us.

Buying Plywood

In order to use plywood effectively, there are a few things you have to know and a few decisions you have to make.

First, the plywood must be made with genuine waterproof glue. The best stuff is real marine grade plywood made to British Standard BS 1088. This is available in the United States as well as the United Kingdom, is usually quite expensive, and according to the standard, should have no voids in the inner layers.

Exterior grade plywood, or what we call water- and boil-proof (WBP) plywood in the United Kingdom, may also be acceptable—if you can find some good stuff. Considerably cheaper, WBP plywood usually has at least a few voids, which are undesirable because they weaken the structure invisibly and, worse, can hold water and cause rot. Many builders try to plug them with epoxy or polyurethane glue. Having tried this myself, I'm skeptical about how effectively this can be done, so in addition to plugging the hole, I have at times used a doubler, or sister. This is a second piece of plywood glued to the surface of the offending section, on the inside of the boat.

I wouldn't normally use WBP here in the United Kingdom because you rarely find any that isn't dreadful, although the situation varies here, and I gather also in the United States. I do know some people in the United States who have, at different times, found sources of luaun underlayment plywood that wasn't too bad. A useful message here, though, is that the quality of plywood can vary greatly from place to place and from country to country.

Interior grade plywood is made with glue that softens when wet and should not be used, no matter how well made.

Second, the material of the faces is important. In three-layer plywood, it's clear that the grain in both of the outer layers runs in the same direction, which has the consequence that it is less easy to bend in the direction of the grain than across it. This matters when we come to cut material for different parts of the boat. If we're going to

Voids in a sample of water- and boil-proof (WBP) plywood. Voids represent a serious inherent weakness of their own, but they also tend to hold water, which in turn causes rot. Voids should be eliminated by filling with epoxy or glue, and should also be backed with a void-free piece of scrap plywood. Notice that the central plywood is made from poor quality wood, which will also tend to soak up water and rot over time.

clamp an engine on the stern, for example, in cutting out the transom across the back of the boat, the grain on the outer faces must run across the boat to provide the strongest possible support for the engine. Cut it the wrong way and the transom may flex.

The difference in flexibility with direction becomes less important as the number of layers increases, but even with five-ply plywood it's still significant. I'd also add that where the grain of the wood will show, say, through varnish, it will always look better if it runs along the longest direction.

Third, even superficially similar plywoods are not equal. It's difficult when writing a book for an international audience to be very specific about plywood because the materials available vary greatly from country to country, and, in large countries, from region to region. So I have tried to make some specific comments that apply in both the United Kingdom and the United States, and may make sense in other areas of the world.

There is a fundamental, inescapable choice to be made right at the beginning of a plywood-and-glue boatbuilding project concerning the level of quality of the build, which should be consistent. For example, if the materials are cheap in one area, say the hull, they should be cheap everywhere. This issue applies to most of the projects in this book and to many other designs as well.

In the case of home boatbuilding, the choice starts with the quality of plywood, how much you are prepared to pay for it, and what is available where you live. The level of quality in plywood that is used or is acceptable in building small boats is a matter of constant debate on Internet boatbuilding forums. It will almost certainly horrify some people to hear me say so, but I'd argue that it can be rational in some cases to build a boat using scrappy exterior grade construction plywood sealed with paint. You might be a person who hasn't the skills, inclination, or time to do anything but work roughly. Yet you still enjoy building and using small boats, do your boating in shallow, sheltered water, and wear your PFD at all times. If so, who has the right to tell you that you shouldn't use the least expensive materials?

Keep in mind, though, that plywood with paper-thin outer layers will most likely require hand-sanding when it comes time to finish and paint your boat. Whereas, with good quality marine-grade plywood, it is just about possible, with care, to use an orbital sander. The marine-grade plywood has thicker faces (outer layers) that are less likely to be seriously damaged (see page 57 for more details on sanding). You can also get by with just paint to seal the hull. Yes, this also goes for exterior grade construction plywood in a very low-end build, but there is always the risk that the boat won't hold up well over time.

As a general rule, hulls built from plywood of any quality will benefit from being sealed with a "spit coat" of epoxy (a thin layer of epoxy without any fillers added), and they will benefit still more if covered with epoxy and fiberglass cloth. How far you go in this direction is up to you. Some boatbuilders take one or the other of these steps, even when building with marine grade plywood, to protect their investment.

Adding epoxy and glass will significantly increase the weight of the boat, but they also will make it stiffer, stronger, and harder wearing. There are many advantages over using paint alone to seal the lumber and plywood.

Further, epoxy spitcoats, and epoxy and fiberglass coverings, will likely require hand-sanding, as opposed to doing the job with an orbital sander, so that should factor into your strategic thinking early on in the planning stages of the project. See Chapter 4, page 49 for details on covering hulls with epoxy, or epoxy and fiberglass.

The interior of the hull should be sealed as well. The basic approach is to use paint only, and quite often people use a spit coat of epoxy covered in paint. If you go this route, you'll have to rely on more careful application of paint on the inside and pay greater attention to maintaining the interior and keeping the bilge free of water when the boat is not in use.

Where do I stand on all this? On balance, I've come to the conclusion that I'd rather use better quality marine plywood, even if it means I can only afford a spit coat of epoxy over most of the boat, and perhaps some epoxy and glass cloth on the bottom. My reasoning is that because the material is relatively thin, weak spots due to voids, bad cores, and flawed outer layers can't be tolerated. In addition, it seems to make sense to make the bottom of a boat hard-wearing.

As I said earlier, look for the BS 1088 designation and make sure the outer layers are not paper-thin. However, it has also been my experience that the BS designation is no absolute guarantee of quality, even in the home country of the British Standard. You should look closely before you part with any money!

LUMBER

Old-time boatbuilders built their boats with the expectation that they would live in the open air and often on the water for years at a time, and so chose wood varieties—cypress, certain kinds of oak and pine, teak, and so on—that could withstand that kind of treatment without rotting. Generally, these trees had grown slowly in the wild, and consequently the wood material cut from them was slow to rot. They didn't know how lucky they were, or maybe their customers didn't know how lucky *they* were, for their boats would sometimes go for years without any care at all.

Over the decades and centuries, however, our resources of slow-grown stands of these types of timber have generally been cut down and used, which means that unless we have a lot of money at our disposal, we don't have easy access to these materials. Even if we do buy the same varieties of lumber, they're generally cultivated, fast-growing varieties that don't contain as much of the protective resins as the naturally-grown lumber.

In our time, this means that if we want to make cheap boats, we usually have to make compromises that would have horrified our ancestors. That's just a fact of life, but there are things we can do to minimize the problems of rot and fungal infection. One option is to choose the best materials we can find. In the United Kingdom, I tend to buy pine, Douglas fir, and spruce rather than the very cheapest no-name softwood, and I've found that the largest building suppliers sometimes have stocks of meranti from the Philippines at reasonable prices. The term "meranti" includes four related species, but much more important it is nice to work with: free of knots, lightweight, strong, and a nice color. It's only moderately resistant to rot, so it should be sealed and the boat kept off the ground when not in use. Oak is available from specialists but at a price that can't easily be justified at my level of craftsmanship. In similar establishments in the United States, I've been impressed to see light, beautifully clear cedar at prices that make me green with envy when I compare it to what I see in the United Kingdom, and I understand that cypress, fir, and long-leaf yellow pine are often available. I've also learned from my American contacts that hemlock and poplar should be avoided.

Buying Lumber

While we might hope to build some of our boats using wood from some pretty cheap sources, we should still choose the pieces we want with some care. In almost any pile of lumber of a suitable variety, there's usually something that's usable. In my experience, most lumberyards will let you pick and choose without taking offense, though they may occasionally point out that if you want better lumber you could make it easier on yourself by going to a more expensive yard and paying more for it.

As you go through the pile, look for straight grain, no knots or small knots in preference to large ones, and material that isn't wildly bent and twisted. Avoid material with obvious flaws. Unless you're making spars, a little twist isn't too much of a problem in building small plywood boats. The shape of the panels defines the shape of the boat, and the rigidity of the boat's structure will almost always pull a twisted 1-by-1-inch piece of softwood into line without difficulty.

If you are making spars, a little curve is fine, so long as you laminate your mast together from two or more pieces of material with opposing and roughly equal amounts of bend.

How to get your lumber home is an issue that you should consider, as you may find the yard's prices for delivery are a little steep. I usually carry it on top of the car on a roof rack, securing it in the middle and at the ends and tying the lines off at the towing points many car manufacturers place just below each bumper. If the boards are particularly long, it can be useful to tie a brightly colored cloth to the back end to warn the driver or motorcyclist behind you that you've got a long load on board.

If you're worried about scratching your paint, you might take the precaution of wedging a soft cloth between the cord and the bumper. I don't trouble about this myself as I reckon that cars, unlike boats, have no souls, feel no pain, and try harder to please if you let them think you don't love them.

ADHESIVES

As previously mentioned, the great leap forward in plywood technology came with the invention of a truly waterproof adhesive in 1934, and the original formaldehyde glue that made it possible was followed by a series of better, stronger formaldehyde types that could be used in making plywood in factories. Formaldehyde glues are cheap, but I don't favor them because they require that my joints fit more tightly than I tend to achieve, and I generally don't have the patience to start again when they aren't tight. Instead, when dealing with joints, I tend to use epoxies, which are more expensive and require careful mixing and completely dry lumber. But no matter how bad my carpentry, they can always be mixed with a filler that enables them to fill any gap and produce a strong bond.

Epoxy

After waterproof plywood, epoxy is the second wonder material at the disposal of the modern amateur builder of small boats. The credit for its invention by synthesis of bisphenol-A-based epoxy resins is shared by Dr. Pierre Castan of Ciba Geigy in Switzerland and Dr. S. O. Greenlee of Devoe & Reynolds in the United States in 1936.

Epoxy became popular in the aerospace industry in the early 1950s. It was then used to build canoes, and, in 1963, it was used in the construction of the popular Mirror dinghy, which introduced thousands of amateur boatbuilders to the stitch-and-glue building technique described in Chapter 1.

Epoxy is a good adhesive in many applications, and it pays to know how to use it. See page 34 in Chapter 3 for more information about working with epoxy.

Polyurethane Glues

Epoxy may be a miracle material, but not everything in a small boat is best glued with it. The reason for this is you have to mix up more glop and filler every time you wish to use a few drops to assemble a bit more of your boat, which might just be a simple seat, an oarlock support, or a breasthook.

Designed by Jack Holt and Barry Bucknell, the Mirror sailing dinghy was the first dinghy to be built in any numbers using the stitch-and-glue method, although canoes had previously been built using the technique. It's still a popular boat, and kits, hulls, and parts can be obtained from Trident in the United Kingdom at www.trident-uk.com.

For this kind of carpentry, I go for polyurethane glues almost every time. Polyurethanes were developed as a kind of imitation rubber adhesive for the shoe industry, but they're a gift for boatbuilders. They're relatively inexpensive and available in various forms, including fast- and slow-setting, and filled versions suitable for slightly wider joints. They work in more humid atmospheres where epoxy would combine with the moisture to form a useless slimy mess called amine blush or bloom. The secret of polyurethane glue is that it's actually the moisture in the air that makes it react and harden—so a little moisture in lumber or even a little light fog is unlikely to cause problems. See page 33 in Chapter 3 for more information on working with polyurethane glue.

Warning: One of the great temptations for inexperienced boatbuilders is to try to save money by using water-resistant adhesives and non-waterproof plywood, thinking that if the boat is well covered and sealed with paint the glue will hold up. If you're considering doing something like this, please don't. You'll always be wondering if the glue or plywood will fail while you're using the boat,

17

and that will take the fun out of being on the water quite quickly indeed.

FASTENERS

Stainless steel screws are ideal for building small boats. In boats that are in and out of the water on a regular basis, they never seem to rust. And because they have a tighter thread than most other screws, the resulting fastening is wonderfully strong, although you have to put in a few more turns when driving them home. You often can obtain stainless steel screws quite cheaply by mail order. If you prefer bronze screws, you can buy them from specialist boatbuilding suppliers.

Generally, it's best to avoid brass and bright zinc-coated screws, as saltwater quickly destroys brass screws, and zinc coatings don't usually last very long.

The nails used in boatbuilding also have to be resistant to rusting, and the preferred ones are bronze boat nails or ring nails, which won't rust even in seawater and have rings around their lengths that stop them from working out of the wood. However, in very low-cost boatbuilding, if we use nails at all, it seems reasonable to use the galvanized type—because they're cheap, come in various sizes, and will often outlast a boat built from low-quality plywood.

PAINT AND VARNISH

Varnishes are made up of a resin that is capable of sticking to wood and other materials mixed with an evaporating solvent and various additives that create a shine. Paints are similar but contain additives that produce color and make them opaque.

The resins vary, as do the solvents, but one result of varnish's lack of opacity seems to be that it succumbs to the effects of weather and the sun's ultraviolet rays more quickly than paints. In my experience, a varnished piece of wood that lives in the open air and unprotected from the sun will need to be refinished two or three times for every one time that a painted piece of wood will need to be recovered. What's more, both paint and varnish are expensive, so I think

it's right to decide which we use with a little care to keep both costs and maintenance under control, and that starts with using paint, not varnish, wherever possible.

A few small areas of bright varnish can be very appealing and may be justified. Spars, rudders, seats, breasthooks, outer gunwales, and so on, are easy to coat and can look really good—but a little is all you need. Paint the large majority of the boat, or you may find yourself spending more time on maintenance than you wish.

Now, a word about paint. Most of the boats in this book are frankly basic. They're simple, cheap, and fun to build and use, and don't require much effort to carry and launch. However, there's no special pretence that they're going to turn heads while you're out on the water. Further, most will spend their lives in a garage or in the backyard under a tarp. Perfectionists will go their own way (and for the sake of their nerves and blood pressure should probably not be reading this

Large areas of varnish may be very attractive, but they can be hard to maintain, as the owner of this Solo racing dinghy would surely tell you. Another approach is to paint the boat and varnish just a few small areas, which can look very handsome and requires considerably less maintenance. See David Beede's Summer Breeze (page 184).

book), but for the rest of us, the good news is that some of the cheapest finishing options are perfectly acceptable, including ordinary exterior house paint.

The cheapest sources for these things can be garage sales and clearance sales at your local building supplier. Just hope the color you're able to get isn't too gruesome, as it's likely to be the stuff no one else wants!

I'm no painter, just as I'm no boat carpenter, but I've had my best results with either:

■ A layer of spirit-based paint primer, followed by sanding with a medium grit; followed by a couple of layers of undercoat, followed by sanding with a finer grit; and a couple of layers of a top-coat, OR,

■ A lightly sanded single or double coat of epoxy or fiberglass-and-epoxy, painted with water-based exterior gloss latex (emulsion) paint.

Don't underestimate gloss or enamel latex paint: if you leave it alone for a week or two after application, it hardens to a tough, long-lasting coating. That may come as surprising news to boatbuilders in countries such as the United Kingdom, where these paints are usually only applied to masonry and interior walls. However, in the United States, the epoxy-latex emulsion approach is very popular in amateur boatbuilding and widely regarded as a good way to go because it is cheap, environmentally friendly (it won't release clouds of invisible volatile organic solvent into the atmosphere), and far less likely to irritate the user's throat and lungs. It is also effectively resists UV rays for several years at a time, even in the hottest and sunniest climates.

Chandlers, dealers, and magazines will try to sell you a variety of high-tech paints as well as synthetic stains and finishes, but I generally avoid them because they're expensive and aimed at upscale boatowners. The boats I build usually don't justify that kind of expense, even if they may get more use than the average yacht.

TOOLS

If you're one of the many people unaccustomed to using tools, don't despair. Everything you need to know will be explained. The only thing I ask is that you double-check what you read and double-check what you measure before you cut a piece of wood. You may have heard the experienced woodworker's mantra that you should measure twice and cut once to avoid wasting wood. I'd add that you should also read twice to avoid potentially costly mistakes. While there are plenty of ways to correct errors, it's better to steer clear of them from the outset.

You will need very few tools to build the boats in this book. Clamps (or in the United Kingdom, cramps) are the exception: you can never have too many. If you can't afford them, there are ways to work around the problem using screws or by making your own. To make your own clamps, buy a section of 3- to 4-inch diameter PVC pipe and cut it into sections 3 inches long to make a series of rings. Then cut straight through one side of each ring to create a 90-degree slot. Now you have a bunch of small, cheap, and useful clamps that have the added advantage that they can be doubled-up, one over the other, wherever necessary.

The only power tool you'll absolutely need is an electric screwdriver. You may wish to obtain an orbital sander and a power planer, but neither are essential. If you are building with heavier plywood, a saber saw

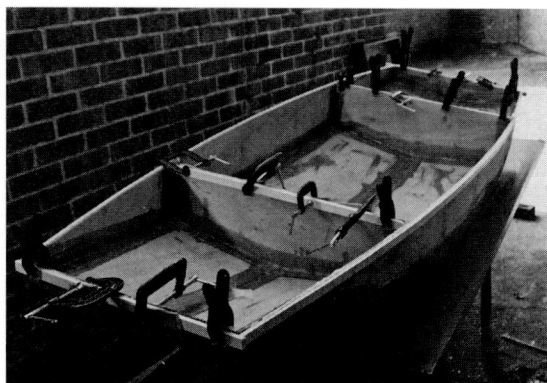

Whenever you take a trip to the hardware store, always buy at least one clamp. They soon accumulate, and it's almost impossible to have too many, as you can see illustrated here with this Mouse.

(also known as a hand jigsaw) might also be useful.

Generally, though, I'm not enthusiastic about power tools: leaving aside the cost and danger of buying and using power where you don't need to, perhaps the worst thing about them is that power transforms a small mistake that could be fixed with judicious use of a little extra Uncle Epoxy mixed with Auntie Filler into a big mistake that requires you to buy another sheet of material or length of lumber. Even the orbital sander can quickly take off critically important areas of surface veneer and weaken your nice rigid structure.

The tools you buy *can* be the cheapest kind, but it's almost always worth spending just a little more for something that's strongly made and comfortable to hold. With cheap screwdrivers in particular, you can find the soft metal breaks or rounds off very quickly in use, and comfort is particularly important for saws. Don't forget that many tools, including clamps, vises, chisels, and planes, can often be picked up for a song at yard or garage sales.

The essential woodworking list should look something like this:

- A workbench fitted with a vise—one of the cheap foldaway Workmate-style devices with a long clamp on top works fine—but in a pinch a sturdy wooden table or chair and a good big clamp will also work.

- Something to put your boat on so that you don't have to bend over all the time while working. Sawhorses that are solidly built are good; I've found cheap plastic ones are just about bearable, but they wobble and will probably only last for one or two small projects.

- Some soft pencils and soft erasers (rubbers). Get at least three, so if you lose one you won't have to stop working.

- A tape measure. Some designs are in the traditional inches and fractions, while others are in inches and tenths. To measure tenths of an inch, use an engineer's scale rule, available in most office supply stores.

Raising your work off the floor will make construction and painting easier. If good sawhorses aren't available, you can make do with a kitchen table, a pair of 5-gallon pails, or almost anything that's stable and strong enough. Anthony Smith put together some homemade stands using expanded polystyrene for the purposes of painting his boat. (Anthony Smith)

- A small woodworker's square (aka tri-square or set-square) or a combination square. A draftsman's T-square or even a large envelope will also work.

- A long straightedge (a 3- or 4-foot rule marked in inches and fractions), while not essential, would be a bonus.

- A flexible batten for drawing curves. I'd suggest one of those 8-foot lengths of plastic molding you can buy in a hardware store.

- An 18- to 20-inch handsaw (jacksaw)

- A smaller tenon saw with finer teeth. You could also use a Japanese-style pull-saw.

- A small- to medium-sized hammer. (There are no big nails in these boats.)

- A hand drill and a set of wood bits

- A utility knife

- A nice big sanding block: 10 inches to a foot long

- A small collection of chisels, and a block and oil for sharpening them

- As many clamps as possible

- A collection of small finishing nails (panel pins), screws, and so on
- A Stanley Surform and some spare blades
- A plane (if you can handle it and sharpen the blades)

When working with epoxy and other glues and fillers you will need:

- The manufacturer's instructions (read them, particularly the safety information)
- Several pairs of disposable rubber gloves
- Face masks or, better, a respirator
- Shallow, flat disposable trays for mixing epoxy. Trays used to package meat are fine as long as they're clean and dry. Do *not* use yogurt cups. Their deep shape tends to retain heat, which can cause the epoxy mix to become volatile and produce toxic vapor, not to mention a distorted pot full of useless hardened epoxy.
- Vinegar for cleaning up epoxy
- Disposable brushes for epoxy
- A sponge for wiping "bloom," or blush, off epoxy surfaces before painting or applying more epoxy. Bloom is slippery stuff that forms when water vapor meets partly hardened epoxy.
- A selection of sandpapers and wet-and-dry papers in various grades

- The usual stuff for painting: brushes, a roller handle and covers, stirring sticks, mineral spirits, etc.

A list of desirable and justifiable tools to look for when you start on your second or third boat might include:

- A 4-foot T-square. These are expensive but wonderful. Sold for marking off drywall, they're also perfect for marking out the component parts of plywood boats.

Finally, you'll need a space to work in. I've found I need an area no less than about 12 feet by 12 feet to build a standard Mouseboat. I know this, because this is the size of my small British living room. I have built a boat in it and can't imagine successfully building a Mouse in anything smaller. Even though the Mouse measures just 7 feet 10 inches by 30 inches, you need extra room to maneuver the components and work on all sides of the hull. I'd add that you should not seek to share this space with anyone else, even as a corridor, unless they are either helping you or are *very* understanding. You may have to wait until your housemate or spouse goes away for the weekend. It's also a good idea to make sure you have a way of getting your boat out of your working space without sawing it up or making holes in the wall.

CHAPTER 3

BASIC SKILLS

Depending on what you may know or not know about basic carpentry, this chapter, which is about working with tools, glues, wood, and plywood, either will be important to you or won't. If you already feel confident about what you're doing in the workshop, skim or skip this chapter. But if you're in the least bit doubtful about your abilities, take a few minutes to see what's here. It may help!

WORKING WITH LUMBER

Cutting starts with selecting and marking out your piece of material for a particular purpose. As always in boat construction, it's important to avoid using lumber embedded with large knots that can badly weaken the structure of a boat. Make the grain run along the length of the board, not across it. If the grain will be obvious, say, in a varnished gunwale or a breasthook, it's well worth choosing lumber with an attractive fine, even, and straight grain.

Give a little thought to how the board may want to warp over time, particularly if it becomes wet. Lumber tends to "cup" in the same direction as the curve of the grain when viewed from the end of the board. Where possible, the cupping should be turned toward whatever the piece of lumber is being fastened to, so that the edges of the board will remain flush against its neighbor.

Marking

The next step is marking out. For this you need a few basic tools, including a pencil and eraser (rubber), a 12-inch or 18-inch ruler or scale, a tape measure, a long straightedge, and a square. The eraser is essential, of course, for if you follow the marking out equivalent of the old carpenter's rule of "measure twice and cut once" by measuring and marking twice before cutting, you're bound to find mistakes

and have to remake some of your marks in new places. If this happens and you don't thoroughly rub out the wrong marks, the resulting mess will mislead you, and you'll end with a cut that's slightly but significantly out of place. I've fallen foul of that one and been forced to start again more than once.

Rulers and tape measures can provide some challenges of their own, not least because many of the boat designs on the market use a variety of units. Both in the United States and the (allegedly metric) United Kingdom, rulers and tape measures are regularly marked in inches and fractions—halves, quarters, eighths, and sixteenths. These correspond well with the traditional boatbuilding distance unit of the foot-inch-eighth (you'll often see these used in what are known as tables of offsets, which are often used to define hull shapes). But you will also find boat designs that are dimensioned in inches and tenths and even in meters and millimeters. Rulers and scales can be obtained for all of these, although in the United States it might be necessary to obtain an engineer's scale for working in inches and tenths.

Tape measures have a small trick up their sleeves that's worth mentioning in case you don't know about it. In using some tape measures, you may have noticed that the little hook on the end moves, and thought, as I once did, that it was just the result of some strangely shoddy workmanship. It isn't—it's a device to account for the thickness of the hook itself. When making an external measurement, the hook pulls outward to measure to its inside edge, but when making an internal measurement it pushes slightly inward to measure to its outside edge.

Straightedges can of course be bought at the hardware store, but I'd suggest that the straight factory edge of a piece of plywood will generally serve, and if it's scrap from an earlier project it will not cost anything. If your lumber supplier is very friendly, he might even give you a suitable scrap. When marking out small pieces of lumber, a proper carpenter's tri-square is well worth having, but I've certainly used a plastic drafting square without problems, and even an old envelope.

On the subject of pencil marks, some expert woodworkers use marking knives instead of pencils. This is a great technique for the expert because it helps to produce some nice clean edges, but it is probably not helpful to less experienced woodworkers. You should work in pencil to draw a line that can be rubbed out and replaced, and still seen and followed even after several corrections have been made.

Marking a length of lumber to cut is reasonably straightforward, but if you haven't done it before there are a couple of pitfalls to know about and avoid. Let's say that we want to take a piece of machine-planed lumber straight from the supplier and cut it to, say, $12\frac{3}{8}$ inches in length, and let's imagine that the designer has specified that it should be a piece of 1 inch by $1\frac{1}{2}$ inch. The width and depth measurements won't be the true size, unless you have personally taken a larger piece of lumber and cut and planed it down to this size. This is because for some historical reason lumber is generally sold by what it measured *before* it was planed smooth or dressed by the sawmill. So, in truth, your 1-by-$1\frac{1}{2}$-inch piece of lumber is more likely to measure $\frac{7}{8}$-by-$1\frac{3}{8}$-inch, or slightly less.

Cutting

Having taken this esoteric and needlessly awkward piece of information on board, the next thing to ask is does the piece of lumber need to be cut to this exact length at this point, or can it be left slightly long and trimmed later? Try to leave it slightly long and trim it later as a precaution against errors in some other area. It may also give you a useful handle to lift your half-built boat if you happen to need to move it.

Next, check to make sure that the end you're starting from is cut true and square. If it

isn't, you can't hope to measure accurately from it or make marks that you can reliably use for sawing. If necessary, use your square and mark out how the end should be cut to create a reliable square-cut end. Just mark a line a little way from the end and use your square to guide the pencil all the way around the board before cutting along the waste-wood side of the line. Obviously, you want the board firmly secured in a clamp or vise and your saw carefully held at right angles so that it follows all the lines you have drawn on all sides of the board.

When cutting to any measured line, always cut on the waste-wood side of the line and err on the generous side. It's relatively easy to trim just a little off if you've left too much, but it's wasteful and annoying to have to start again with a new piece of wood because you've tried to cut your board finely to the line and failed.

Working with Saws

You'll see a range of saws in your hardware store, but a general purpose handsaw with a blade about 18 inches in length with teeth of 14 points to the inch will cover everything from cutting plywood to the kind of rough and ready carpentry that's needed in these boats. If you're planning to build more than one boat, you might consider buying two saws. One saw, for cutting plywood, should be a carpenter's saw about 22 inches long, with no less than 12 points to the inch, or your plywood will have ragged edges. The other saw, for fine work, can be a much shorter tenon saw with small teeth and a slim blade supported along its top edge by a thicker spine of steel. A Japanese-style pull saw has an even slimmer blade than a tenon saw and cuts on the pull stroke rather than the push stroke, the norm for all Western-style saws. Pull saws are a matter of taste. I'm too rough and ready a boatbuilder to find that their advantages justify the extra cost, but some people like them and feel that they can be more accurate and do finer work.

Regardless of the type of saw you use, the critical factor in obtaining the best results is keeping the lumber secured while you saw.

It's essential to clamp the work securely in order to make a clean saw cut.

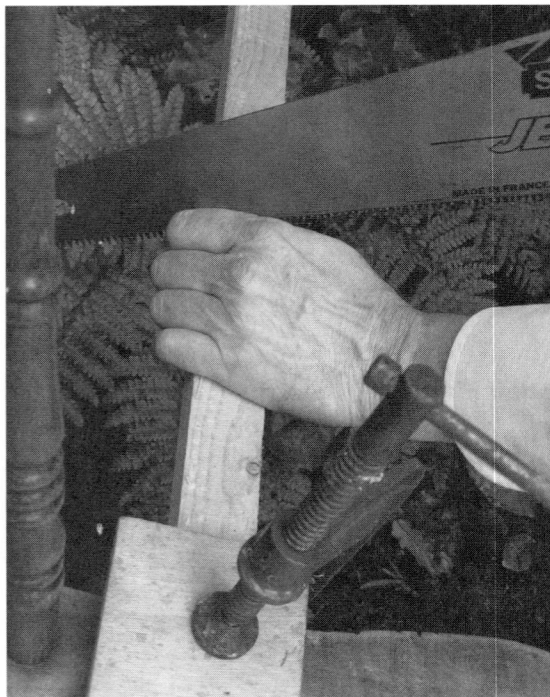

When starting a saw cut, line the last joint of your thumb up parallel to the cutting line, and rest the saw blade against it as a guide. Make the first few passes very gentle, with no downward pressure on the saw.

A proper workbench equipped with a vise or clamps is ideal for this purpose.

When working with a conventional Western push-saw, start the cut very gently, using the last joint of your thumb as a guide for the blade. Without pressing down, draw the saw toward you, then push away, and then draw toward you again once or twice, just letting the weight of the saw do the work of beginning the cut. Doing this very gently helps to make sure that the blade doesn't bounce and that you are completely in control of the exact position at which the cut begins to deepen. It also gives you a chance to check that you're consistently holding the blade at the right angle to the work before going too far. Practice on some scrap material and concentrate on a straight, easy action.

When it's clear that the cut is well placed, slowly increase the pressure and the length of each stroke, taking care that the saw remains at right angles to the material and that the cut is following the pencil line until you are almost finished cutting. Just before the saw breaks through the last of the material, change the angle of the saw so that the handle points downward, and reduce the pressure to a minimum so that your cut ends with a nice rectangular end rather than a split corner. It helps to get someone else, or another clamp, or use your spare hand, to hold the loose end carefully as you cut through the last bit.

Once the end is true and square, you can usually use a ruler, scale, or tape measure to mark your required length. As always, measure, then mark, then measure again before getting the saw out of the box. It's astonishing how often little checks like these can prevent you from wasting expensive materials. And then, of course, carefully make your cut on the waste-wood side of the line.

Joining Lengths of Lumber

Because almost all of the boats in this book are quite short, you can cut most of the long pieces that you'll need from a single, long piece of lumber. There may be times, however, when you want to join two lengths of lumber running in the same direction to make a longer

Butt joint   Scarf joint

Two ways to join lengths of lumber or plywood, viewed from the edge. The length of the sister piece or butt strap on the butt joint should be eight times the thickness of the material. Likewise, the length of the scarf should be eight times the material thickness. The butt joint is easier to make, while the scarf joint is "cleaner" and allows the piece to take a more natural bend.

board, perhaps because you bought a bunch of 8-foot boards on sale, or you were put off by the high cost of clear (i.e., knot-free) lumber in long lengths. The joint between these two pieces can be made with a *butt joint* in which the joint is "sistered," or backed up and reinforced by a piece of the same material. The sistering should be the same width and thickness as the material you're using it to reinforce, while its length should be eight times its thickness. For example, if you're joining two pieces of 1-by-2-inch lumber end-to-end, the sister should be an 8-inch 1-by-2.

An alternative method for joining boards end-to-end is the *scarf joint*, in which each piece is tapered, as shown, over a distance of eight times the thickness of the material to be joined, and then glued and fastened. This is obviously a much more difficult joint to make, but it produces a clean and smart result that's compatible with the smooth flow of water around the outside of the boat. This means that it can be used for gunwales and inwales, and for both internal and external chine logs. It also leaves no vulnerable end-grain exposed to dampness and possible rot.

WORKING WITH PLYWOOD

In most of the plans presented in this book, the shape of the boat is established by cutting plywood into defined shapes. These shapes can be shown in a variety of ways, but my favorite is to use coordinates, which I'll explain below. After you've marked the shapes on the plywood, you'll cut them out with a saw. For boats longer than a single sheet of plywood (4 by 8 feet), you'll have to join sections end-to-end into longer hull panels, and for this you'll make simple butt joints with either butt blocks or fiberglass tape and epoxy. Then

you'll join the edges of panels together in long seams to create the hull.

Watching the hull take shape as you bend and glue together these flat sections of plywood can be amazing and impressive the first time you see it. But what's even better is the fact that it happens time after time on many different boats.

Marking Out Using Coordinates

From looking at the chapters in this book that refer to my boat plans in particular, you'll notice that I've included tables of pairs of numbers or coordinates for marking out your plywood. One way to think of these is as lines that you draw in parallel first to the left-hand edge (the X coordinate), and then to the bottom edge (the Y coordinate). The point where the lines cross is the point you mark with a cross, preferably in soft pencil.

Another way to think of plotting these X,Y coordinates is as measuring a distance (X) along the bottom edge of the plywood sheet followed by a second distance (Y) directly upward at a right angle to the bottom edge. In practice, this is the way you'll most likely plot them out.

From the Mouseboats group of boatbuilders, I know that just hearing the word *coordinates* can worry some people, as it makes them think they're about to be drawn into some area of higher mathematics where they'll be uncomfortable and may even become puzzled. If you're one of these people, relax. All you're going to be asked to do is to mark a length along the bottom edge and mark a height using a tape measure, a 4-foot straightedge, and pencils.

You'll also need a cushion to protect your knees, as you'll be kneeling on a piece of plywood for an hour or two. Unless you lay floorboards or fit carpets for a living, I doubt you'll be exactly used to this sort of thing!

Because it helps me to see where my mistakes are (most of us make at least a few) and because it's nice to use a short ruler or scale for the small-scale stuff, I always start by marking out my plywood in 10-inch squares.

I do this using a tape measure and pencil, and begin by laying out the material crosswise with the short ends on my right and left and the long bottom edge in front of me.

Then I make pencil ticks along the bottom of the sheet at 10-inch intervals. Starting at the left-hand side, I mark at 10, 20, and 30 inches, and so on. I do the same thing along the long top edge, starting from the left-hand side. Then I lay the straightedge against both ticks so that I can draw a straight, clear line between the two. I do the same thing along the vertical (shorter) sides, again starting from the bottom left-hand corner—depending on the length of your straightedge. You may need to add some more 10-inch ticks along one or more of the lines you've just drawn around the middle of your sheet of plywood.

There's a good argument for marking these squares using a colored pencil and then marking the actual panels in a normal pencil. If the drawn lines are all exactly the same color, it can be too easy when sawing to follow a squaring-off line rather than the panel line you meant to follow.

Further, in order to keep the straightedge steady and prevent it from sliding while drawing, I often find it useful to put a couple of heavy old books behind the ruler on the plywood. They make good weights and can often be made to slide smoothly until the straightedge is precisely in place.

Now that you have your plywood squared off in, say, a nice shade of blue, it's time to think about marking out all those points that make up the shapes of the panels, the coordinates. Before you begin I strongly suggest you check each measurement after you make your mark just to make sure it's correct. Because these are small boats, I have chosen to mark the measurements in inches and fractions on the grounds that they're rather more familiar and easier than the feet-inches-eighths that wooden boatbuilders in the English-speaking world traditionally use.

Before you actually make any real marks, think a little more about the meaning of the X/Y measurements by making some trial marks. Do this in normal pencil so that you can rub out the marks you make.

First think of a point 1,1 in the accompanying drawings. What this means is a point 1 inch from the left-hand edge of the plywood and 1 inch up from the bottom edge. Using your tape measure, here's how to do it: Mark a tick 1 inch to the right of the left-hand corner along the bottom edge, then use a set-square (or some other reliably rectangular object) to guide you in making the tick just a little longer up and down through your measured point (see the diagram). Then along this line mark a height of 1 inch up from the bottom edge. Where they cross is the point 1,1. Check it.

Now, using the same method, mark 6,6: this means 6 inches along the bottom or X direction, then 6 inches directly upward in the Y direction. Check this new mark.

Now try some odd numbers and fractions: 3, 6; then 1½, 6; then 4, 10⅛.

Plywood squared off in 10-by-10-inch squares

4-by-8-foot plywood is often slightly longer than shops say it is.

Working with coordinates #1: Start by drawing a grid of 10-inch squares over the plywood panel.

Plot coordinate 1,1 here

Y axis

Along 1 inch

Up 1 inch

X axis

Plot coordinate 6,6 here

Y axis

Along 6 inches

Up 6 inches

X axis

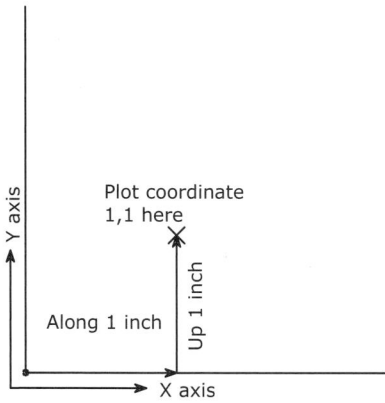

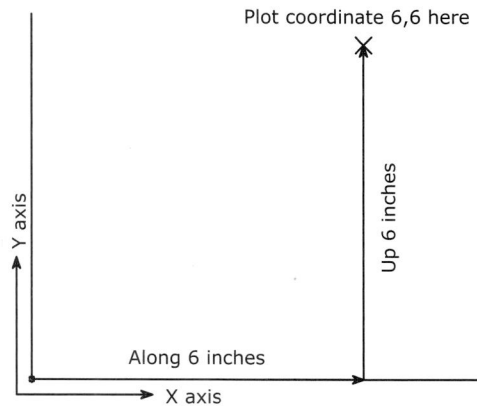

If the coordinate reads 1,1, that means 1 inch to the right along the X (horizontal) axis, and 1 inch up along the Y (vertical) axis (left). A coordinate of 6,6 means 6 inches along both axes (right).

Now, to take you out of the bottom left-hand section of your plywood sheet, try marking the point at 12,4¼! By now I think you're probably getting the idea pretty well, and I guess you also now see why squaring off the plywood is such a good idea. It means you only ever have to measure a few inches from the bottom and left-hand outlines of your 10-inch boxes!

Rub out your experimental pencil marks and start making out panels in earnest.

You may have noticed that the designs in this book vary in the number of coordinates that you have to plot. There's a trade-off here. With plans with many coordinates, you simply have to plot the coordinates and join them up with a pencil and a straightedge; you can be sure of getting a good shape. This works well with small boats, while with larger boats it would produce panel profiles with visible flat spots.

With larger boats, mark the coordinates and then drive finishing nails (panel pins) into them. This allows you to bend a long flexible object around the nails, which will produce a curve you can trace with a pencil. This long, flexible item is properly called a batten and in the past would have been made of wood: in our time, plastic moldings bought from a hardware store work very well.

Don't forget to double-check the dimensions before moving on to the next part of the process.

Moving on to the long panels, begin with the panel at the bottom of the plywood sheet. With these long, curved lines it's well worth plotting the points along each of the long lines one at a time, taking time to double-check your measurements and to stand back and look at the curved pencil lines that slowly emerge. If you see any sign that a particular point is beginning to look like an awkward

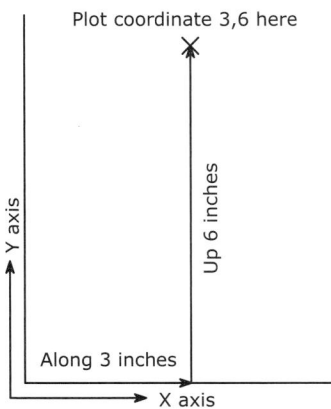

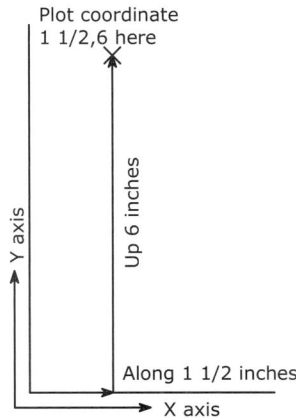

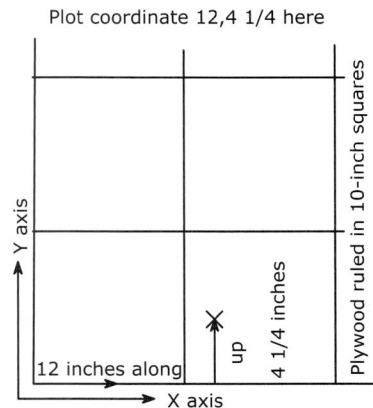

Plot coordinate 3,6 here

Y axis

Up 6 inches

Along 3 inches

X axis

Plot coordinate 1 1/2,6 here

Y axis

Up 6 inches

Along 1 1/2 inches

X axis

Plot coordinate 12,4 1/4 here

Y axis

12 inches along

up

4 1/4 inches

Plywood ruled in 10-inch squares

X axis

More examples of coordinate plotting: 3,6 (left); 1¹/₂,6 (middle); and 12,4¹/₄ (right).

kink, triple-check the point itself and a couple of points on either side; when correct these long curves are gradual and even graceful along their length.

In marking out, it's particularly important to mark a few final lines to show where the bulkheads or frames are meant to go. If you forget to do this, you'll kick yourself later because this is the last time you'll see the coordinates showing where they should be. In my designs in this book, the bulkhead and frame positions generally coincide with a pair of the coordinates you have been plotting. Don't rub out these lines, as you'll need them to guide you when you place the internal frames.

My favorite layout tool is the drywall square. This measures a couple of feet across, and 4 feet from top to bottom, making it perfect for measuring and marking out plywood. If you find yourself building more than one boat from this book, I think you'll find the cost of a drywall square is justified by the ease with which it allows you to mark out your plywood and the ease with which you can double-check each mark before cutting.

Finally, before cutting, it's a really good idea to write the name of each of the plywood parts of the boat onto the material itself so that there's never any doubt which part goes where. In larger boats, this is absolutely essential, but even in a small boat it'll save you time. All you need to write in this case are "forward transom," "aft transom," "forward frame," "aft frame," "bottom," "left side," and "right side." It also might help to indicate which is the forward end of each side and which the aft.

Cutting the Panels

Once you're happy that the markings on your plywood are looking just like those on the drawing and that everything has some kind of identification written on it, it's time to start thinking about cutting. For most of us, the big issue in cutting panels drawn on a 4-by-8-foot sheet of plywood is how to hold such a big piece of thin material steady so that it can be cut smoothly without wobbling and binding in every direction.

An ideal solution might be to have two or three waist-high benches placed close together so you can cut your material in the space between them. However, few of us have one workbench, let alone two, and in any case, it can be difficult to reach across the width of the material. My approach is to find four or more props about a foot high, preferably made of plastic or wood so that my saw won't be ruined when I inevitably cut into them. At home, I usually use toolboxes made of polypropylene. I place two of my props on either side of the line to be cut and two more in the general area where I'm going to kneel. (I should mention that an old cushion will help as well.)

When everything's ready and steady, you can start cutting. Start with the long cuts, and always work along the waste-wood side of the line, leaving little or no space between the line and the saw. There's no need to hurry. In the grand scheme of things, this part of the project won't take very long (no more than 30 minutes, or less than half the time for marking out) and you do *not* want to make a mistake that could lead to having to buy another piece of plywood. Stop at regular intervals to move your supports so that your saw won't hit them as the cut progresses, and to make sure your own position is well supported, so that you won't fall or break the material. Check to be sure that the curves you have marked out really are emerging from the plywood.

Use the first side panel as a template to trace out the shape for the second one as allowed for in the drawings. If the boat's over 8 feet long, you'll have sections for the front and back of both sides. Make sure that the port and starboard sections of both side panels are mirror images of each other, which they *must* be. Amateur boatbuilders rarely say anything about the issue, but I'd be prepared to bet that quite a few of us have made up identical panels with the butt joints on the same side.

The last part of the process is to make the sides identical by clamping them together and fairing a little along the curved edge using a Stanley Surform or a rasp. Either will do the

A rasp and a Stanley Surform tool make rounding corners, smoothing curves, and relieving edges easy.

job well. Aim for smooth, identical curves along both side pieces, but please don't overdo it. Each sweep of the cutter removes a little of the boat's buoyancy, and that's definitely something you should try to preserve.

Butt Joints

Making a butt joint is one of the first steps in many boatbuilding projects over 8 feet or so, and several of the designs in this book require one or more of them, usually one for each side, and a third for the bottom.

Butt joints may be made before or after the plywood is marked out, depending on the design: in principle, if the boat is made up of wide panels with long butt joints, the panels can be safely assembled after the panel components have been cut out, but if the butt joints are narrow, it's safer to join the material before cutting out. However, when you do this you may be surprised to find that the material you are working with is actually more than 8 feet in length.

This may seem like another odd lumber trade eccentricity, but I'd go further and say that to boatbuilders working from a set of coordinates, it's a real nuisance because it creates some issues that you need to find a way of working around when squaring off the plywood (see the section on marking out). One possible way to deal with it is to carefully remove the surplus over 8 feet with a saw and then plot the coordinates, but this solution is probably best left to those with

access to a big power saw as it can be difficult to accurately saw the long edges by hand.

Instead, I think the easiest approach is to lay a thin scrap of plywood under the sheets you're marking out, making sure they are end-to-end and flush against each other. Temporarily join the sheets together by driving small nails through them into the thin scrap of plywood. The plywood scrap is important. When it's time to remove the nails, all you have to do is insert a screwdriver or a knife under the scrap and carefully pry it upward. The heads will catch on the scrap and lift the nails out of the plywood sheets. You will use this technique often when building small boats, so it's a good one to know.

Now, mark out the material with coordinates, panel outlines, frame positions, and so on, as if the joint isn't there. After the marking is done, you'll separate the 4-by-8-foot sheets from each other, cut out the panel sections to the marks, and make the butt joints to hold the fore and aft sections of each hull panel together later.

In fact, the joints can be executed in several ways. Many home boatbuilders choose to screw and glue a solid wood butt block to the inside of the joint. This is simple and effective, but it can be ugly in certain kinds of designs, particularly if the boat you are building has multiple chines or the butts do not line up with some other internal feature, such as a frame.

Where a wood block is used, its thickness should be two or three times the thickness of the plywood used for the hull. It should cover the whole joint and its width should be twelve times the thickness of the hull. Some might say you can use a minimum width of eight times the thickness of the hull, but I prefer to err on the safe side in this area. So if the hull plywood is ¼-inch thick, I'd suggest the width of a solid butt block should be 3 inches.

Although plywood straps perform the same function as the wood block, they can usually be thinner. They must still be twelve times wider than the thickness of the hull material is thick, but they are typically the same thickness as the plywood from which the hull is made. They should be cut so that the exterior grain lies across the joint rather than along the joint. Some of the designs in this book have a strap designed into the plywood cutting plan.

In the past, many people have made these joints using glue and bent copper nails, or even very short stainless steel screws, to hold the parts together, but I find I can successfully glue them together without nails. I do this by carefully laying out the components on pieces of scrap plywood to ensure the resulting panel will be straight, and then I apply glue or slightly filled epoxy to the plywood strap. I butt the panels I'm joining against each other and place the butt strap over the joint, making sure it is centered over the joint, and I cover the external seam with glass tape and epoxy.

Finally, I cover the whole thing with a piece of polyethylene (polythene) or plastic grocery bag material and weight it down with a toolbox for a night. I find this produces a good strong joint, and in the kind of warm dry weather that epoxy likes, I can get straight to work the following morning.

However, my favorite way to make butt joints involves no butt-block or butt strap at all; instead use epoxy and a couple pieces of 3-inch wide fiberglass tape over the joint on both sides of the panel. You will also need a sheet of reasonably thick polyethylene or similar material to prevent the wet epoxy from sticking to the floor while you're working.

Lay the polyethylene flat, stretching it slightly so that it really is flat and without wrinkles. Mix the epoxy in small quantities in a flat, open tray. Push the two sheets of plywood together over the plastic sheet and brush the epoxy onto the border area—2 inches on either side of the joint is about right. Then paste down the tape, making sure everything's nicely wetted. Leave it overnight. In the morning, gingerly, and probably with some help, turn the whole 16-by-4-foot panel over so that the taped side is now lying on the plastic sheet. This operation needs to be done carefully to avoid cracking the half-taped joint.

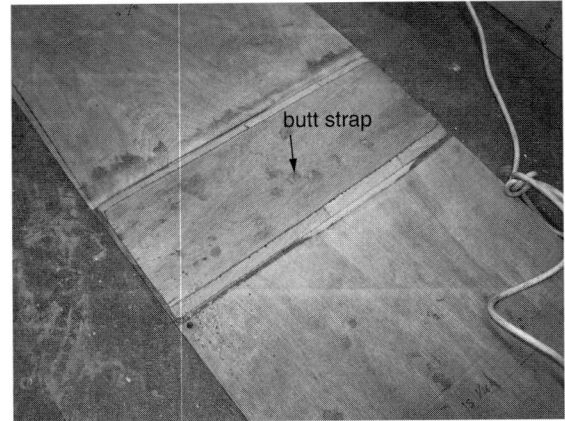

A plywood butt strap glued in place. Although not essential, this one's edges have been tapered for a slightly cleaner appearance. This photo shows a Cruising Mouse under construction by Anthony Smith, a first-time boatbuilder whose excellent work indicates how easy it can be to build one of these boats without prior experience. (Anthony Smith)

Once the panel is safely back on the plastic sheet, look at the butt joint. If the gap is very small, four hundredths of an inch or less, fill it with unfilled epoxy, but if it's any larger some filled epoxy of about the consistency of tomato ketchup is better. Then epoxy another strip of glass tape to cover the joint. If you've mixed your epoxy well and the weather is favorable, the whole thing will probably be good and solid by morning. It's true that the taped edges require some filling and sanding before they can be painted, but they can be made almost invisible, and doing so is a lot easier than traditional scarfing and a lot less unsightly than butt blocks.

When I'm in a hurry, I sometimes make these joints in a single overnight operation by applying epoxy and tape to *each side*, making the butt gap as small as possible and filling it with epoxy, covering the upperside with plastic and a plank, and weighting down the whole thing as evenly as possible with a few heavy items from around the garage. This might work for you, too, but I'd suggest you don't try it with your first boatbuilding project. It's easy for a beginner to mess up the job, and that can be stressful.

USING FASTENERS
One of the most important differences between a lot of household joinery and boat carpentry is that in boatbuilding any splits

around fasteners are bad. They cause weak points in the structural integrity of the boat, and they may encourage water to creep into the lumber, possibly causing rot.

In household do-it-yourself jobs, you have probably seen how a nail or screw can cause a split, perhaps because it was too near to the end of the wood or because it was put in without being drilled. Around the house, there may not be too much to worry about when this happens. Houses are deliberately over-engineered, which is why they last for generations rather than falling down in ten years. But because we don't generally over-build when making small boats, we place our screws and nails carefully and, apart from the very smallest temporary pins, holes for nails and screws must always be pre-drilled before the fasteners can be driven home.

To prevent splitting I always make the distance from the end to the site of the screw or nail at least equal to the maximum thickness of the material. If the material is 1-by-2-inch softwood, I'll place the last screw 2 inches from each end at a minimum.

Pre-Drilling Holes

How large the hole should be depends on the material and on the fastener. When driving a nail into a relatively soft wood of the kind I usually use, I make a hole that's about half the width of the nail and find that it'll almost always drive home neatly and straight.

When working with screws, it's a bit more involved because the screw has three parts: the head, the threaded portion, and the shank, which is the unthreaded section just below the head. First, drill a hole equal in diameter and length to the shank. Then, using a smaller drill bit equal to the diameter of the shaft of the screw inside the threads, drill deeper, to the length of the screw less a quarter inch or so. When driving flathead screws directly into lumber, you may have to countersink the surface of the wood with a special countersinking bit. This is not usually necessary with plywood, however, since the cone-shaped bottom of the head will often embed itself in the surface of soft plywood.

Switching drill bits two or three times for every screw hole can be a bit of a pain, but there are two ways that you can simplify the process. The easier but more expensive way is to buy two or three drills, and set up each one with the right bit. The cheaper way is to drill all the holes that you need with one bit before switching to the next one. This is not always feasible, however, because in some cases you will have to fasten one screw down tight before you can know where the next one should go. This is especially so when bending a long piece around a curve, as when fastening gunwales or chine logs.

When driving fasteners into harder, more expensive woods like oak or mahogany, drill slightly more generous holes. It's worth experimenting with scraps of the lumber you're using to determine the optimal size of the drilled holes for the nails and screws you're working with.

There's a neat trick to ensure that you drill the holes to the right depth. Place the drill bit and the nail or screw side-by-side, and mark the necessary length on the drill bit by wrapping a little colored tape around it at the point where the top of the fastener should go. Once you've drilled down to the tape, you've gone far enough. This is a really good technique, and you'll be particularly glad you know it when drilling a lot of holes, such as when fitting a boat bottom or a gunwale.

Working with Hand Drills

Hand drills may be more complicated to look at than a saw or a hammer, but they're still relatively straightforward tools to use. I find that if I hold the rotating handle with my left hand I can usually unscrew the chuck with my right hand to open it to insert or remove the drill bit. Many people keep their drill bits in a little compartment in the long handle, by the way, so if you've bought yours second-hand somewhere, look inside and you may find some useful little treasures.

As with any cutting tool, the main issues are to ensure that the work is held absolutely steady, partly for reasons of safety, and partly to ensure that the cut that you're making starts in the right place, goes straight, and finishes in the right spot. Starting can be made just a little easier if you use a bradawl or a nail to make a small dent in the material at just the right location. You only need a small hole to prevent the point of the drill bit from wandering. As with using a saw, it's best to start carefully and gently until you're sure all is well and the cut is going straight. Also, it's good to practice using some scrap material until you're confident about holding the hand drill at the right angle, which is usually 90 degrees to the piece of material you're working on.

Using Hammers and Driving Nails

Hammers seem simple enough at first glance, but we all know that things can go wrong. Thumbs can get bashed and nails can bend, forcing you to yank them out, which will leave your boat pockmarked with nail holes that must be filled and sealed to prevent water from seeping into the lumber or plywood. Here's some simple advice for anyone who hasn't been formally taught to use a hammer:

- Make sure the hammer you're using is a reasonable size for working with the nails you've got: you wouldn't use a 5-pound club hammer for driving small slender nails because you'd bruise your thumb and bend a lot of good nails.

- Hold the handle at the end farthest from its head, and start the hammering gently, making sure that each strike is in the middle of the head and in the direction the nail is meant to go. Until the nail's direction is well established, use the weight of the tool to drive the nail, not muscle power.

- For safety's sake, never use a hammer with a damaged or loose head.

Once the nail is driven all the way in, it's worth using a nail set (a little, hardened steel punch) to drive it just a little farther so that its head is just below rather than just above the surface of the wood. This can then be filled with a dab of filled epoxy or another filler appropriate for the job.

Using Screwdrivers and Driving Screws

I'm sure that everyone has used a screwdriver at some point, but there are a few essential pieces of advice that are well worth repeating:

- As with nails and hammers, it's important to use a screwdriver of the right size for the screw. You'll find it's nearly impossible to drive a large screw with a small screwdriver or to persuade a large and wide screwdriver to fit in the slot of a small screw. However, there may be times when you'll find yourself driving a screw with screwdriver that's just a little too wide. The result will be that at the place where the screwdriver tip sticks out of the side of the screw, it will leave a nasty gouge in the surface of the wood.

- Don't use a screwdriver with a worn head because it will damage the screw. Don't use the cheapest screwdrivers because they are made with soft and easily damaged metal, making them useless almost immediately. It's worth buying screwdrivers with a hardened tip.

- Most operations with screwdrivers are best carried out two-handed, with one hand on the handle and the other steadying the shaft.

WORKING WITH CHISELS

Chisels are not really necessary in some of the simpler projects described in this book, but they're such a widely used and useful tool a few words about how to handle them are in order.

In some ways they're the opposite of the twist drill bit. While most people buy a new drill bit when an old one wears out, it's not only normal to sharpen a blunt chisel but essential if you're going to use it safely. A chisel may look simple compared to a drill bit. Yet they require more thought and care in use, not least because they're so dangerous.

Chisels should always be treated with the utmost respect. If you buy one, get the necessary paraphernalia to sharpen it, including a stone and a device to hold it at the required angle.

When using chisels, always keep the following in mind:

- The work must be held rock-steady in a vise or clamps.

- The tool must cut away from you, not toward you or anyone else. You must always be in control of the cutting motion.

- Don't try to cut off too much in a single motion or you may find that either the chisel gets stuck and then breaks through the material, or that it'll get stuck and may even break itself (if it's a finely made one).

Chisels are very likely to get stuck if you're getting into the kind of operation where you're chopping the grain of the wood with the flat of the chisel, or when you're cutting a slot into a piece of solid lumber. Tackle this by first cutting a small trench within the marked out slot area with the bevel-side downward. (It's easier to control in this position.) Once the job is nine-tenths done, turn the chisel around and clean up the dead vertical cuts required on the sides of the slot. This can be challenging, but the trick is to go easy, be patient, and to try to cut just a little at a time.

If you find you can't cut a little at a time under good control, it may be a sign that you need to sharpen the chisel!

WORKING WITH GLUES

There's no getting away from the fact that modern synthetic waterproof glues like epoxy and polyurethanes are sticky and generally nasty. While working with these materials, you need a well-ventilated area, rubber gloves to keep the glue off your hands, a mask or respirator to guard against inhaling harmful vapors, and goggles to protect your eyes. It can take many days to remove these glues from your skin, and they'll ruin your clothes, which is why you should always wear old ones.

Using Polyurethane Glue

Polyurethane (PU) glues and sealants work very well both as glues and sealants in general boat carpentry and in boats built with chine logs,

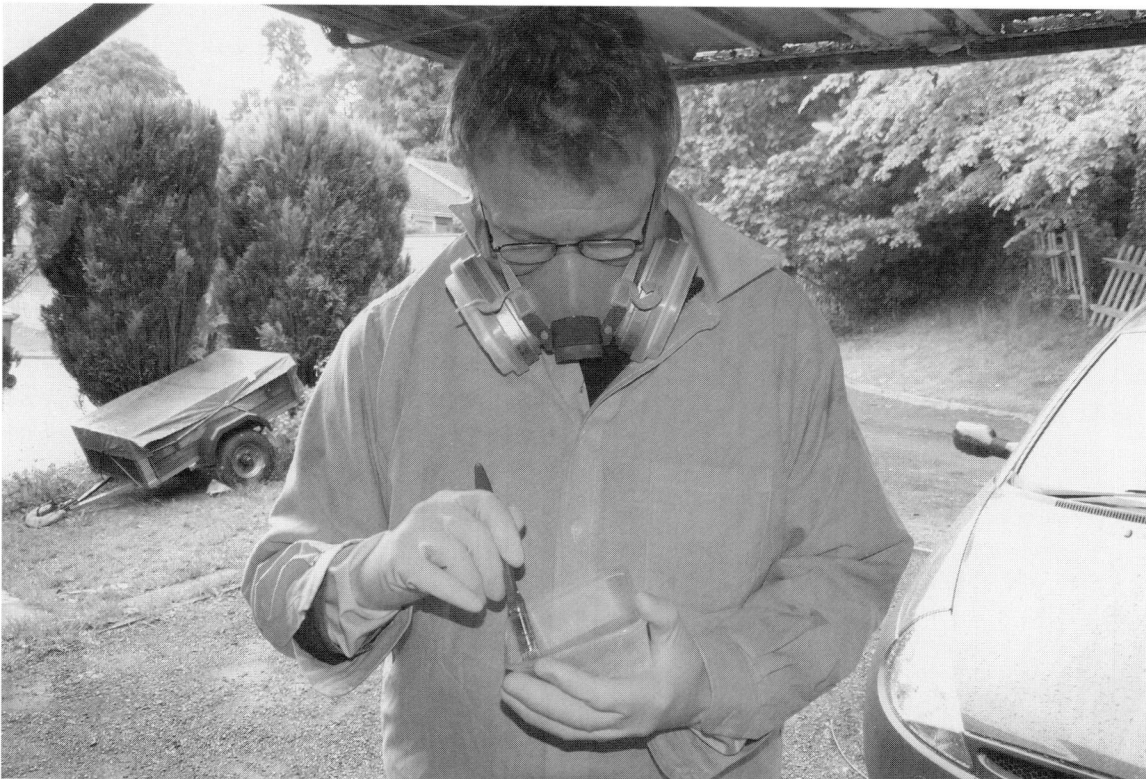

Be sure to wear protective clothing when working with glues and epoxies.

- Work in a well-ventilated area. The fumes can be harmful to your health.

- Polyurethane adhesives like moisture. Wiping the glue area with a damp cloth and giving it a few minutes to soak in improves adhesion.

- Polyurethane adhesives are applied with a caulking gun. Compared to caulking, it takes more effort to get polyurethane adhesives to squeeze out of the tube, and it has a tendency to "run on" after you finish covering your gluing area and take pressure off the gun. Keep a rag handy to catch the runoff before it drips where you don't want it, and be sure to keep something under the nozzle when you put it down.

- Using just the right amount is one of the trickier aspects of working with polyurethane adhesives. With tight-fitting situations, such as gluing on a gunwale, it's best to spread polyurethane adhesives thinly with a putty knife or equivalent on one surface, wet-fit the pieces, remove them, re-fit them, and then clamp them together. Be prepared to remove the excess *immediately*. The stuff will creep out as the glue starts to set, so you'll probably have to scrape it off again. You may find that you still use too much, but that's preferable to using too little.

- Remember that polyurethane adhesives foam and expand. To prevent the joint from expanding, you must clamp firmly and evenly, but without so much pressure that you squeeze all the glue out of the joint. In addition to clamps, you can anchor the joint with small screws or nails, to be removed later.

- Scrape off as much of the polyurethane adhesive as you can while it's still soft to avoid having to scrape and sand it off after it dries. I've had great luck painting over polyurethane adhesives with most paints, but I've had problems with some paints not adhering.

- To clean tools while the polyurethane adhesive is still soft, use WD-40 followed by mineral spirits.

- Avoid contact with skin by wearing latex or polyethylene gloves. Polyurethane adhesives melt vinyl gloves. It can be helpful to double-glove if you're a messy worker, or if you'll be using your fingers to smooth a fillet.

- Polyurethane adhesives really do stick to (almost) everything, especially clothing. Work in clothes that you don't mind ruining.

- Polyurethane adhesives are extremely strong and virtually permanent. Think twice about using them if you might want to remove the glued part at some time in the future.

To this list, I would add that as always it's essential to read the label. For one thing, not all polyurethane glues and adhesives are fully waterproof, and those that are not cannot be recommended for building boats.

and they can also be used to make quick, low-cost stitch-and-glue boats. One of the reasons is that many of them harden in 30 minutes or some even in 5 minutes, and this is what makes the one-weekend boat possible.

However, PU glues have one very important characteristic that should be borne in mind: they expand as they set. As a consequence, it's crucial to clamp parts you're working with together or the foaming of the adhesive will push them apart. A joint that's been pushed apart in this way may not look too bad, but it will have a weak, foamy interior. Nevertheless, when you adjust your clamps, don't over tighten them because you may squeeze so much glue out of the joint it will be weak.

Even in a properly clamped joint, the foaming of PU glue will usually result in a good amount of foamy excess around the edges. This can usually be cut off after it sets, but it is better to scoop it up with a putty knife before it hardens completely. Finally, the tendency of PU glues to foam means that they can't be used for laminating larger areas of, say, more than 4 inches across. Otherwise the stuff will foam up, and the result will be an unevenly glued joint across the surface.

If you're using PU glue such as PL Premium in the United States or Balcotan in the United Kingdom, please see the sidebar for what experienced polyurethane user and serial amateur boatbuilder Bryant Owen has to say about these materials. I couldn't have put it better myself.

Using Epoxy

Epoxy is a wonderful material, but it can be exacting. The two parts must be carefully measured according to the manufacturer's instructions and then thoroughly mixed for at least a couple of minutes. Another essential to success is that the work has to be absolutely dry, or the mix will break down and make a slimy, slippery mess that will have to be thoroughly removed before you can start again. Yet another essential is getting the temperature right. There are minimum and maximum temperatures for working with epoxy, but even within the specified limits, I've found it can either harden so fast that it's difficult to work with or so slowly that it can be days before it sets, if then.

Get these conditions right, and you'll find epoxy is also a fascinating material to work with. In boatbuilding, we use epoxy in several ways:

■ As a coating, without filler

■ As a glue, with filler

■ For encapsulating large areas of plywood with sheets of glass cloth

■ For creating neat, easy butt joints supported on both sides by glass tape and epoxy

■ As a structural member, with filler and glass cloth tape in the form of the stitch-and-glue "fillet" joint

It's also possible to buy two-part epoxy paints, but while they're great, they're also very expensive and almost certainly overkill for most of the boats included in this book.

Before I go any further, I feel I should mention that epoxy that hasn't yet set *can* be removed before it destroys a floor or an item of clothing, using either a cleaning fluid available from your epoxy supplier, acetone, or, perhaps surprisingly, plain old cheap vinegar. In truth, vinegar isn't the best of these three, but it is at least cheap and widely available. In the end, however, it's always best to take precautions and work carefully and cleanly, taking time over everything and preventing spills and messes from happening in the first place.

Tips on Using Epoxy

As I've said, the best containers when working with epoxy are open, flat trays made of plastic (like the ones used for packaged meat) or aluminum, and that getting the two-part epoxy mixture right are among the most important considerations in successful epoxy work. Measuring out the two epoxy components need not be too difficult.

Ignore the typical sales pitches that say you need to buy expensive plastic measuring pumps. You can mix epoxy by weight as easily as by volume, and you can do it with the help of an inexpensive electronic household scale with the tray covered with plastic wrap. This is what you do:

■ Put your plastic wrap over the scale and the mixing tray on top of that. Then zero the scale.

■ How much you need to make up at a time will vary. Six ounces is a reasonable average, though you might mix more when gluing a large sheet of glass cloth in place, or a lot less for a small and well-defined item of woodwork, such as the foot of a mast.

■ If you are indeed mixing up about 6 ounces of epoxy, calculate the proper amounts of resin and hardener according to the manufacturer's recipe. Pour the resin into the tray, and add the hardener on top to make a total of 6 ounces.

■ Start mixing. I often do this with a disposable brush.

■ I've heard it said that two minutes is long enough for mixing, but I'd argue it's safer to mix a little longer.

■ Generally, a brush up to 3 inches wide is fine for applying epoxy to a small boat. One of those neat little rollers they sell for applying gloss paint can be useful when working on large surfaces, but it is far from being essential. Brush the stuff on rather than pour it—pouring can result in an uneven thickness, particularly in hot weather in which the mix gels quickly.

■ The epoxy companies will also sell you a range of fillers to mix with your epoxy. The fillers will give it structural qualities when used as a glue and help make the relatively expensive resin go a bit further. The high-density filler is the stuff to use, as it creates a strong material when embedded in epoxy. Don't use lightweight fillers such as microballoons, as they're not structural and are meant to be employed as an easy-to-sand surface filler rather than a structural element in a boat.

■ On the health and safety front, I've mentioned several times that these resins and glues are not the nicest materials— many of them are capable of causing some nasty allergic reactions, and they have a way of sticking ferociously to anything they come into contact with, including clothes and skin. In fact, epoxy is generally thought to be rather worse than polyurethane, so whenever you are working with epoxy, wear old clothes or overalls, including long-sleeved shirts and long trousers. Essentially, you want no skin exposed. Gloves and masks are an absolute must, and a respirator and goggles are highly recommended.

■ Wherever I find myself working with epoxy and glass cloth together, I pay a little extra attention to the glass cloth's edges. The fibers are fine enough to puncture the skin and cause dermatitis on their own, but what's even worse is they can help to drive epoxy into your body, making an allergic reaction more likely.

■ Finally, have everything set up before you begin work so you only have to wear your safety gear for a short time.

CHAPTER 4

BUILDING THE HULL

We've had lots of talk about the individual techniques of building these small, cheap boats. Now it's time to cut to the exciting part—taking the flat pieces of cut-out material and joining them to make a real, boat-shaped boat. Even seasoned backyard boatbuilders find this part thrilling, and I hope you will, too.

SIMPLIFIED CHINE LOG METHOD

Several boats in this book have flat bottoms and vertical sides and can be built using the simplified chine log method described on page 8 in Chapter 1. After the hull panels, transoms, and frames (which are often also called bulkheads) have been cut from the plywood, the next step is to fit the cleats—the framing that goes around the perimeter of these parts. Remember that the shape of the transoms and bulkheads, or frames, is the key to this method of boatbuilding.

Cleats for Frames and Transoms

Frames and transoms generally start with a cut-out plywood shape, which is then reinforced on one side with cleats around the edge. You'll need to bevel the edges of the cleats so that there are no gaps between the frame and the hull panels greater than about $\frac{1}{16}$ inch.

The bulkheads, or frames, (but not the transoms, in this case) will be fitted so that the cleats face the ends of the boat—in other words, the cleats on the forward bulkhead should face forward, and the cleats on the aft bulkhead should face aft. Take a look at the plans for one of the flat-bottomed boats, such as Micromouse, Minimouse, or Flying Mouse, and picture how the bulkheads fit relative to the curving bottom and sides of the hull. You

can see how, if you position the bulkheads facing the right direction, you'll have to trim the outer edges of the cleats just a little, so that the gluing surface of the cleat will be parallel to the bottom and side panels of the hull. Bevel the edges of the cleats so that there are no gaps greater than about $\frac{1}{16}$".

But while you're thinking of the way the sides fit these frames, no doubt you've just spotted a small complication. On the bow and stern transoms, the framing faces the other direction: the cleats on the stern transom face forward, and the cleats on the bow transom face aft. Picture this in context with the way the bottom and sides curve, and you'll see that the plywood will be bowing outward, away from the cleats, not cutting into them.

We deal with this by fastening the bottom and side cleats on the transoms so that the long edges of the cleats stick out beyond the edges of the plywood by about a quarter of an inch. This will give us enough wood to bevel down to the edge of the transom's plywood surface at an angle, so that the side and bottom panels will have a flush surface for gluing. Because the decks are flat, there's no need to add any extra here; the cleat at the top of the transom should be attached flush with the edge of the plywood.

Begin by cutting a cleat from a piece of 1-by-1-inch framing lumber slightly longer than the top edge of the bulkhead. Using either polyurethane glue or epoxy mixed to the consistency of ketchup, glue the cleat in the required position along the edge of the plywood and hold it in place with small nails or clamps until the glue sets. You can screw the cleat to the plywood if you wish, but if you're using good waterproof glue, it's by no means essential and will add to the labor involved, as well as to the cost and weight of the boat.

These transoms and bulkhead frames all have their top cleats glued in place flush to the top edge of the plywood. For chine log construction, additional cleats would be added to the bottoms and sides. (Anthony Smith)

lumber. Consider whether this is an outward-curving or inward-curving joint, and then apply adhesive and fasten it in place.

Do the same for the other short side.

With these shorter pieces, I always cut slightly long and then trim until it fits exactly. While cutting, make sure the saw is at a right angle from the work. You can mark nice straight and square lines all the way around the piece of wood before you start to cut or, if you've caught the tool-buying bug, you might buy a miter box to help you saw straight.

If, in spite of all your care, there is a gap of more than $\frac{1}{16}$ inch or so between the ends of a side cleat and the top or bottom cleats, polyurethane glue won't fill it with sufficient strength, so cut a new piece to ensure a tight fit. (However, a little epoxy and filler will fill almost any gap.)

After considering the type of joint, fit and fasten a cleat on the opposite (lower) edge of the bulkhead, on the same side of the plywood.

To complete the first bulkhead, take some more of the 1-by-1-inch framing lumber, hold it along one of the short sides of the bulkhead, and mark and cut it to exactly fit the space between the two previous pieces of

Assembling the Bulkheads, Transoms, and Sides

Now, we can go to three dimensions! I've developed a quick method for assembling Mouse-type boats that works well and can save a lot of tweaking and fiddling. It goes like this:

Unless they happen to be precisely at the widest point in a boat (where the sides of the

A Lilypad hull with transoms, bulkheads, and open-topped frame in place. The builder combined aspects of two construction methods, using the more carpentry-intensive simplified chine log method for bulkheads, frame, and transoms, and the stitch-and-glue method for the long side-to-bottom seams. (The outside seams have been taped; the inside seams have yet to be filleted and taped.) (Robert Holtzman)

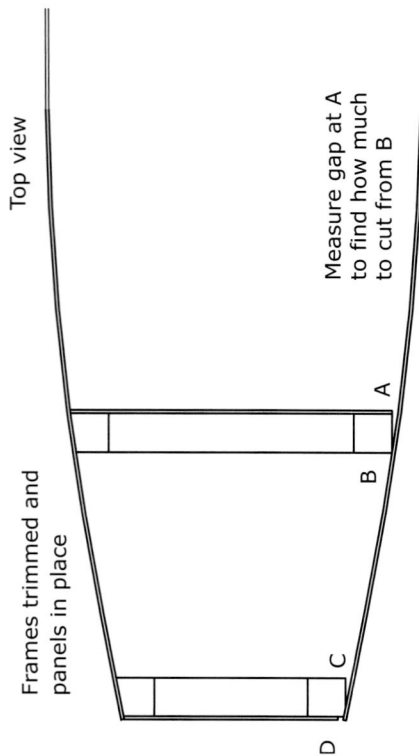

Plywood side of transom; side and bottom cleats are proud of the plywood edge so they can be trimmed

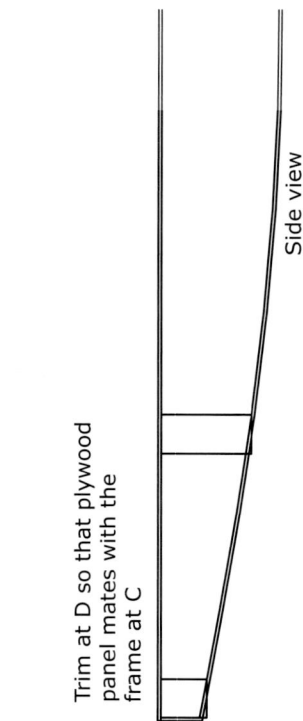

Cleat side of frame; side and bottom cleats line up with the plywood edge and can be trimmed to accept inward curve of sides and bottom

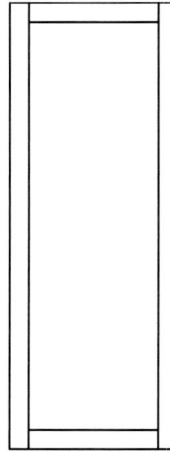

Top view

Frames trimmed and panels in place

Measure gap at A to find how much to cut from B

A

B

C

D

Side view

Trim at D so that plywood panel mates with the frame at C

The exaggerated drawing on the bottom left shows how the cleats must be trimmed to accept the curve of the sides and bottom. To make this required shape possible, the cleats around the tran-soms must be attached so that they are $1/4$ to $1/3$ of an inch or so proud of the cut-out plywood transom at the bottom and sides. The cleats of the frames need only to line up with the cut plywood edge. Where the frames meet the deck the cleat does not need to be trimmed and should be lined up with the cut edge of the plywood.

hull are parallel), the bulkheads, transoms, and frames have to be trimmed with a plane or Stanley Surform to conform to the curve of the hull. Some designers give precise angles for cutting the framing, but I find a good alternative is to partially assemble the hull using clamps very loosely tightened to find out what the angles should be. Once you've clamped everything roughly in place, it's not too difficult to see how much material you need to remove. Mark it with a pencil on the frames, remove the clamps, and trim as required using a rasp or Stanley Surform.

Instead of clamps, at the ends, we might drill small holes near the ends of the hull panels, and stitch them to the transoms with cable ties or bits of wire. Or we could use a "Spanish windlass," which is created by tying a loop of rope around both hull panels, passing a stick through the loop, and winding the line around so that the cord progressively tightens to hold the sides in position. Small nails can be used to prevent the rope from falling off. Simple and effective, a Spanish windlass is a handy trick to know when building a small boat.

My technique may not be elegant and purists might well scoff, but for rough and ready boatbuilders like us (who probably haven't a clue how to go about measuring and cutting a 27.5-degree bevel in any case), it's easier, more straightforward, and in my experience, can be every bit as effective as it needs to be.

In assembling the hull temporarily, we can use clamps against the cleats to join the bulkheads to the hull panels. Here is the procedure:

■ Temporarily clamp or nail the hull side panels to the forward and aft bulkheads, making sure that the cleats are facing the right direction, as described above. If you use small thin nails, drive them through a piece of scrap plywood before tapping them in; it'll make them much easier to remove. Remember, this assembly is temporary and must be disassembled, so *don't glue anything at this stage.*

■ Now we'll turn to the bow and stern transoms. Use duct tape to temporarily attach the transoms to the sides of the boat, with the cleats facing inward. This will give you a clear idea of how much you need to bevel the cleats so that they will lay flush against the hull's side panels. Thinking about the bow transom for starters, notice the gap between the hull's side panels and the aft edges of the cleats. This is the amount that must be removed from the forward edges of the cleats to obtain the correct angle. One way to measure this is to place a piece of paper across the gap that appears on the outside of the boat-to-be and mark it with a pencil. After removing the transom, transfer the mark to the opposite edge of the cleat to show how much needs to be removed. Cut the bevel with a Stanley Surform, a rasp, or a plane. Do the same on the opposite side, then give the transom another trial fit between the side panels, holding them in again with tape or clamps. If your bevel doesn't extend right down to the edge of the plywood face of the transom, remove the transom again and continue planing the whole bevel down at the same angle, until the hull's side panels come flush against the front edges of the transom with no significant gap in between.

■ Now you can permanently attach the transoms to the sides with the glue of your choice. Don't worry if the rasp has left a rough surface on the cleats, as it will help the glue to stick. In fact, I generally quickly go over any surface I'm about to glue with coarse sandpaper because I believe it helps the glue do its job. While the glue is drying, you can hold things in place with temporary nails or clamps, or you can add permanent nails or screws if you wish. Just make sure these are good quality stainless steel, copper, or bronze fasteners, or else they'll rust.

■ You'll feel better if you know your boat is straight. I use a tape measure or a piece of non-stretchy plastic packaging string to measure the hull's two diagonals (from the left front corner to the right rear, and vice versa): they will be the same if the boat is truly straight. If they're not, jiggle the corners gently until they are.

■ The bevels on the bulkhead cleats are treated much the same way as the transoms. Measure the gap, or mark it off on a "tick strip" as before. If you're eager to move forward on this step and the glue at the transoms has not yet had a chance to firmly set, cut some bits of scrap lumber to prop out the sides of your boat and retain the correct shape before removing the bulkheads. Once the required shaping has been done, glue and attach the sides to the bulkheads. It's worthwhile getting out your tape measure or length of string to check that the boat's two diagonals are still the same.

Adding Chine Logs

External chine logs were one of many innovative ideas introduced in Dynamite Payson's *Instant Boatbuilding* (see page 8 in Chapter 1, for a discussion of chine logs). When a boat has a flat bottom and vertical sides, these are the easiest way for beginning boatbuilders to make long plywood-to-plywood joints, and it is for this very reason that several of the boats in this book use external chine logs in the design. They require quite a lot less work than internal chine logs, and you more quickly reach the hugely satisfying moment when the boat goes "three-dimensional"; with bulkheads, sides, transoms, and bottom in place, it really starts to look like a boat.

In making an external chine joint, assemble the sides and frames, and attach and glue pieces of 1½-by-1-inch lumber along the bottom edge of the hull sides. These are the chine logs. Be sure the chine logs are nicely flush to the edges of the plywood so that the bottom will lie tightly against them, and then attach and glue the bottom. If the sides are 90 degrees to the bottom, this is easy. However, if the sides of the boat are not vertical or the bottom is V-shaped, not flat, the chine logs must be beveled

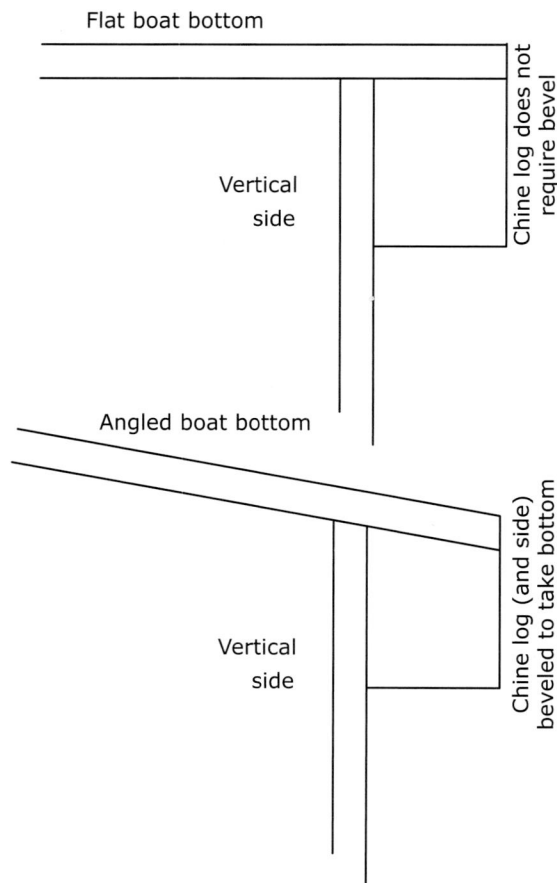

Bevel the chine logs so that the bottom panel will fit flush against them. No action is needed for a boat with a flat bottom and vertical sides, but boats with V-bottoms or angled sides require some work with a rasp, Surform, or plane after the logs are in place, or the angle may be cut into the log as the lumber is ripped on a table saw.

with a rasp, Surform, or plane to provide a flat gluing surface that conforms to the bottom.

There's no need to measure out the boat's bottom panel on your plywood, because you can use the actual boat to trace around with a pencil to show where your cuts should be. This is best done with the boat upside-down, and it's useful to ask a couple of pals or family members to hold the bottom material in place so that you can draw around the boat underneath.

Having drawn this line, leave a margin of maybe half an inch all round as you cut out the bottom. This allows a little jiggle room in case a mistake has occurred in the tracing, and it's fairly easy to remove it later with a saw or a Stanley Surform. Keep the excess within reason, however. A mere ½-inch overlap won't interfere with the use of clamps, but a couple inches all around could cause problems.

Apply glue to the surfaces to be joined and progressively clamp the plywood to the chine log so that the drawn line representing the edge of the plywood coincides with the edge of the log as exactly as possible. As the glue sets, you may or may not choose to drive a series of stainless steel flathead screws through the plywood and into the chine logs, having first drilled the holes as described in Chapter 3. The screws should be very small, as the material you're working with may be no more than 1 inch deep and ¾ inch wide.

With good quality waterproof polyurethane (PU) glue or epoxy applied liberally along the length of the logs, you don't really need screws; consider them an option. It is possible to produce a hull with almost no metal fasteners. However, if you don't have access to a lot of clamps (for example, 20–30 clamps for a 12-foot boat), or you haven't made a set of the drainpipe clamps I described in Chapter 2, you would be better advised to go with the screws.

After the glue has set, remove all the clamps and trim off the excess of the bottom panel using a saw, plane, or Stanley Surform. It's good to seal the edges with epoxy or epoxy and glass tape to protect the edges of the bottom from water ingress before you paint. As I said in Chapter 2, and it bears repeating here, the great weakness of plywood is that its edges always present endgrain to the elements, and end-grain can soak up rot-inducing moisture like blotting paper.

That, then, completes the hull of your simplified chine log boat. The installation of seats, gunwales, and other parts is identical for all three methods, so we'll deal with them later in this chapter.

STITCH-AND-GLUE
The stitch-and-glue joint consists of a fillet of filled epoxy or PU glue covered by glass tape and embedded in still more epoxy or PU. Stitch-and-glue is one of the most popular methods among beginning boatbuilders, and it's a great choice whenever you want to make long joints where the angle between the panels isn't 90 degrees—even more so if the angle varies along the length of the joint. Several of the projects in this book, including the original V-bottom Mouse, use this method.

In addition to being easy to make, stitch-and-glue seams have other advantages: they are light, don't suffer from rot, and don't rely on an external chine log, which, to many observers, looks odd and awkward. However, it can take practice and great skill to produce a flawless taped seam capable of satisfying the most exacting eye. If you're willing to settle for less than aesthetic perfection, however, stitch-and-glue joints, no matter how ugly, are generally very strong. For more information on stitch-and-glue methods, see Chapter 1.

Bulkheads and Frames
As stated previously, stitch-and-glue boats derive their shape as much from the hull panels as from the bulkheads and intermediate frames. Bulkheads or frames have an important role in

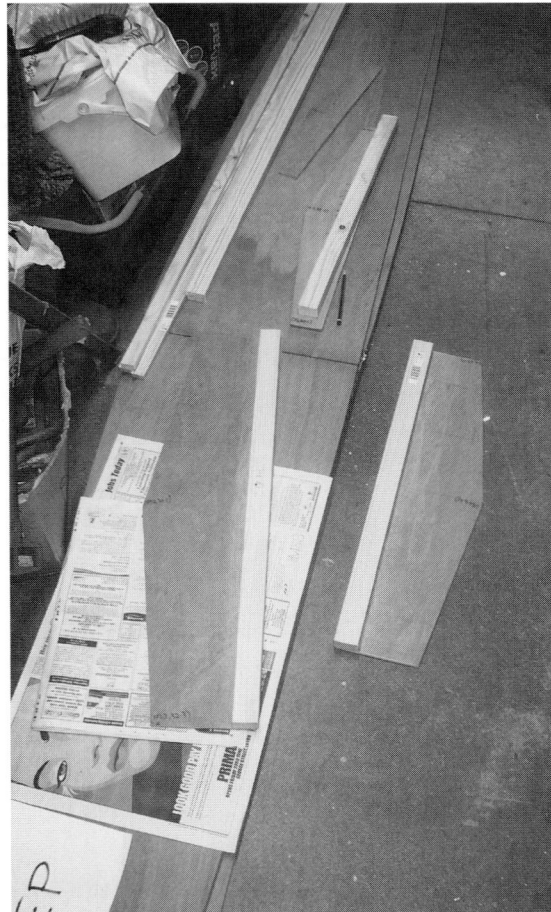

Transoms and bulkheads being assembled for stitch-and-glue construction require cleats only along their top edges.
(Anthony Smith)

holding the panels in place during construction, and also in strengthening the structure after they have been permanently installed.

After marking and cutting out the major plywood components, stiffen the frames and transoms by gluing a single 1-by-1-inch (or similar) lumber cleat along the upper edge so that they remain straight and stable as you assemble the boat and the glue cures. You do not need to add cleats around the entire perimeter of these components.

In some designs, the frames are somewhat U-shaped rather than solid bulkheads. In these cases, the stiffening cleat is a cross-brace that spans across the legs of the "U."

Initial Assembly

Once the glue you have used in the frame reinforcements has hardened, the next task is to set up the boat. With the goal of assembling the boat in an upright position, begin working with just the sides and the bow and stern transoms. These must be firmly taped together with duct tape in such a way that the edges of the plywood kiss all the way

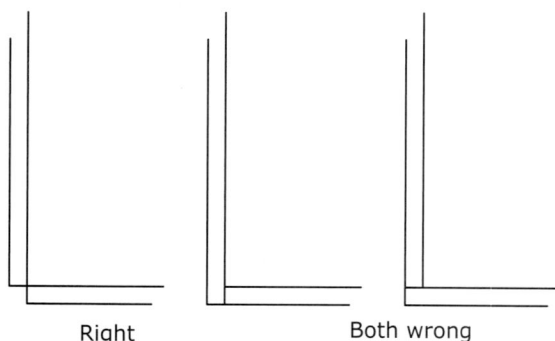

Right   Both wrong

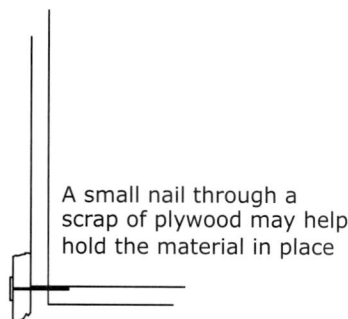

A small nail through a
scrap of plywood may help
hold the material in place

The long edges of the plywood joints should meet so that the corners kiss along the entire length. Be sure the flat of the edge does not lie on the surface of the plywood. Driving a nail through a scrap of plywood may help keep the material in place while you make the inside fillet.

from the top to the bottom. You want each panel to be in full contact with the adjacent one all along an edge, but not along the butt-end surface—in fact, it's better if there is a little space that allows resin and filler to penetrate into the joint. Getting the edges of the side panels and the transoms lined up perfectly is often awkward and irritating, but that's just how it is. Even when you think you have everything just right, the blasted plywood has a habit of slipping when you're not looking.

Where the plywood panels seem stable, there's no need for concern. Where they are trying to move into the wrong place, I drive small nails through scrap plywood into the panel edge to secure them. (Recall that the scrap plywood is used to easily lever the nails out when it's time to remove them.) Here and there you might need to use several to achieve your aim of making the long edges kiss in just the right way.

Once the sides and transoms are in position, place the frames or bulkheads so they correspond to the lines you marked for them on the sides of the boat. They will make the sides bend outward to form the boat's final shape but will tend to slide about. To prevent this, drive a couple of panel nails or small screws through the side of the boat and into the frames. Again, drive them through pieces of scrap plywood first.

If you are working on a flat-bottomed boat like Minimouse, add the bottom next. Slip the bottom panel beneath the partially assembled hull and line up the sides and the bottom so that the edges are flush all around. Take your cloth-backed tape and cover the seams all around the bottom. You may find that it helps to temporary flip the boat over at this stage.

If you are working with a V-bottom boat, like the original Mouse, there are two bottom panels to attach, and a bit more shoving and bending to get all the edges to align with each other. Start taping where panels come together at their widest points, and simply "roll" the plywood panels together toward the ends, taping as you go. Again, you will probably find this easier to do if the boat is upside-down.

Taping vs. Stitching

In small boats with gentle curves and made from thin plywood (¼ inch or less), the seaming is simplicity itself, as duct tape is quite sufficient to hold the panels together. This is great news, in part because it means you don't have to do any drilling or stitching! In addition, the tape will seal the outsides of the seams against drips while we make our fillets on the insides, minimizing the mess and the waste of good adhesive.

However, where the curves are sharper, and particularly in larger boats made from thicker plywood, tape won't have the strength to hold the panels in place and will have to be helped along using small nylon cable ties, like those from Zip-Ty, that you can buy in the electrical department in most hardware stores. Some people use twists of copper wire, but I like the way the nylon cable ties can be neatly sliced off with a craft knife once they have served their purpose.

Each cable tie requires a pair of holes large enough to pass it through, and the holes must be sited opposite each other on the edges of adjacent panels. Typically, the holes are drilled about ⅓ inch from the edge of the plywood, but in areas of higher tension, where the panels must take a hard bend (such as around the stem of a small dinghy), you may want to increase that to ½ inch or more to prevent the cable tie from tearing through the plywood.

The cable tie goes through both holes before being tightened up on the *outside* of the boat. You place the clasp of the tie on the outside of the hull because you generally tape the inside of each seam first, and if the clasp were on the inside, it would make an awkward lump in the seam unless it was cut out. Dynamite Payson wrote about taping the outside first, and there are some people who still prefer the method, but I think it's now unusual. I've tried it and found that it's

Nylon cable ties are used to "stitch" together the hull panels of Anthony Smith's Cruising Mouse. Since you'll be fiberglass-taping the inside of the hull first, place the clasps of the cable ties on the outside of the hull. (Anthony Smith)

The inside seams of the Cruising Mouse's hull have been prepared for fiberglass-taping. The bulkheads, transoms, and hull panels are cable-tied and masking-taped in place. Large gaps between the bottom and side panels are backed up with masking tape to prevent the glue from falling through. The puss is about to be banished from the shop, as cats and epoxy do not go well together. (Anthony Smith)

those gaps nicely. Once they are covered with fiberglass tape and another layer of adhesive, the joints will be more than adequately strong, and no one will know about the gaps. Stitch-and-glue *likes* sloppy workmanship!

We will begin making the permanent seams on the inside of the hull, so the boat must be upright. If necessary, find a couple blocks of wood or bricks to keep the boat stable on your work surface: it's annoying if it rocks or shifts as you work. Before you begin making the fillets, check that the tape you applied a short while ago is still firmly in place. You don't want it to slip off the plywood while you've got a mass of wet glue in the joint because it probably won't stick again and will make a mess. If you're building with epoxy or polyurethane glue and are inexperienced in using either adhesive, refer to the section beginning on page 9 for more details before proceeding to make your fillets.

Making Fillets with Polyurethane

The best polyurethane (PU) glue for filleting is the kind formulated for use with masonry, as it seems to contain filler and it bubbles up less than the other kinds. You will also need some mesh-type fiberglass drywall tape.

Dampen the plywood adjacent to both edges to be joined with a wet sponge. The presence of moisture makes PU adhesives "go off," and wetting the plywood is a particularly good idea if you're working in dry, sunny weather. Using your caulking gun, run a bead the width of your finger along the length of the joint. Using a rounded tool like a tongue depressor, the rounded end of an old tablespoon, or a Ping-Pong ball, smooth the bead out to a hollowed cross section. Let it harden a bit, then lay a length of the mesh-type drywall tape on top, and then apply another couple of beads of PU, spreading it neatly so that the weave of the scrim is completely hidden.

Make fillets along all the internal joints of the hull: between the sides and the bottom; between the frames or bulkheads and the bottom and sides (on both sides of the frames);

difficult to make sure that the seam is properly filled when pasting the filled epoxy into the inside of the boat. I could see whether the internal fillet was complete, but what was behind it, between the internal fillet and the taped outside seam, remained in question. There was a good possibility that a void might exist.

Even where ties are used, it's a very good idea to cover all the seams with tape, because it prevents the glue from dribbling through the gaps.

Speaking of gaps, if you've measured and cut your panels accurately, and pushed and shoved and generally adjusted the placement of the panels when taping or stitching them together, those gaps will be fairly small. If you've been somewhat less accurate in your work, the gaps may be somewhat larger. In traditional boatbuilding, this would be bad and you'd be called incompetent and told to throw the boat out and start again. Not here! One of the great beauties of the stitch-and-glue method is its tolerance for loose joints and (frankly) fairly sloppy workmanship. The thickened epoxy or naturally thick PU adhesive you are using will fill and strengthen between the two bottom panels in V-bottom designs; and between the transoms and the hull panels. It'll take a little time to do them all, but it's well worth making sure your fillets are smooth before cutting the scrim neatly to length and gently laying it into the adhesive and gently bedding it with whatever you used to shape your fillet. If there are points where the tape hasn't properly bedded, apply a little more PU glop until all are securely wetted—not just adhering to the surface, but securely part of the fillet.

Stop for an hour or two to let the PU harden. Once the seams are hard, you can get down to the serious business of preparing the outside of the hull for the next stage. Turn the hull upside down and remove the tape that was holding the panels together. If you used cable ties, cut these off flush to the surface of the plywood.

Now gently round all the outer seams so that the drywall tape will lie neatly on the plywood: it won't lie flat across a sharp corner. Once again the radius should be about the same as a Ping-Pong ball. I do this job with a rasp or a Stanley Surform, both of which are quite adequate for the task. Don't use sandpaper on it at this stage because a silky-smooth finish is not required. A rasp or Stanley Surform will take off material more quickly and will leave a good rough gluing surface.

Taping the outside seams of the hull is just like taping the inside, only easier, since you don't have to form fillets. Lightly moisten the area to be glued, cut the lengths of tape required, apply the glue by spreading it with a putty knife or brush, and roll on the tape and cover it with another layer of glue until it is well bedded. The tape must be thoroughly covered, so that it will not be damaged and cut at the sanding and painting stage. At the ends of the runs, where two or more seams meet, let the tape double up. These will be vulnerable points and a double-thickness of material will add strength.

Making Fillets with Epoxy

Before making the fillets, mix up a small batch of unfilled epoxy and apply it with a brush along the edges of the plywood, making sure the edges are well saturated, and also along the surfaces to be taped. The unfilled epoxy will seal the edges and keep water out of the end-grain, thus helping to prevent rot, while wetting the surfaces that will be covered in tape and fillet will result in stronger bonds.

I use disposable brushes for this sealing job, and because I don't like throwing away brushes that can be reused I wash them in vinegar. I rinse and thoroughly dry the brushes, and I find they can be used again without much trouble. This saves on costs and that's important.

Now it's time to make up the fillets. Check the manufacturer's instructions to get the right quantities of resin and hardener, but you'll need to measure out a total of about 8 ounces and stir it really well. To make the fillet material, add filler to the mixed epoxy until you've got something that looks and handles just like peanut butter: it needs to be just soft enough to spread, and just thick enough to stay where it's put. Press it into the joint with a putty knife and smooth it out in a hollow cross section. People use a variety of tools for this purpose: tongue depressors, Ping-Pong balls, and the backs of old tablespoons are some examples I've come across.

The fact that people use tongue depressors and other objects of a similar size should give you a good idea of the size of the fillet needed for a small boat. Don't make your fillets too big because they will quickly use up a huge amount of epoxy and filler. Also, in a boat the size we're making, bigger fillets won't make it any stronger, but they will certainly make it heavier and more expensive. Another indicator of fillet size is the width of your fiberglass tape, which is typically 3 to 4 inches. When you apply it, the tape should overlap the edges of the fillet and extend onto the plywood on both sides by ½ to 1 inch or so.

After you've applied the filled epoxy, take some time out and let the epoxy harden. If you've done a good, bump-free job, it's much easier to neatly apply the fiberglass cloth to a fillet that's taken on the consistency of, say, a piece of fudge than when it's at the peanut

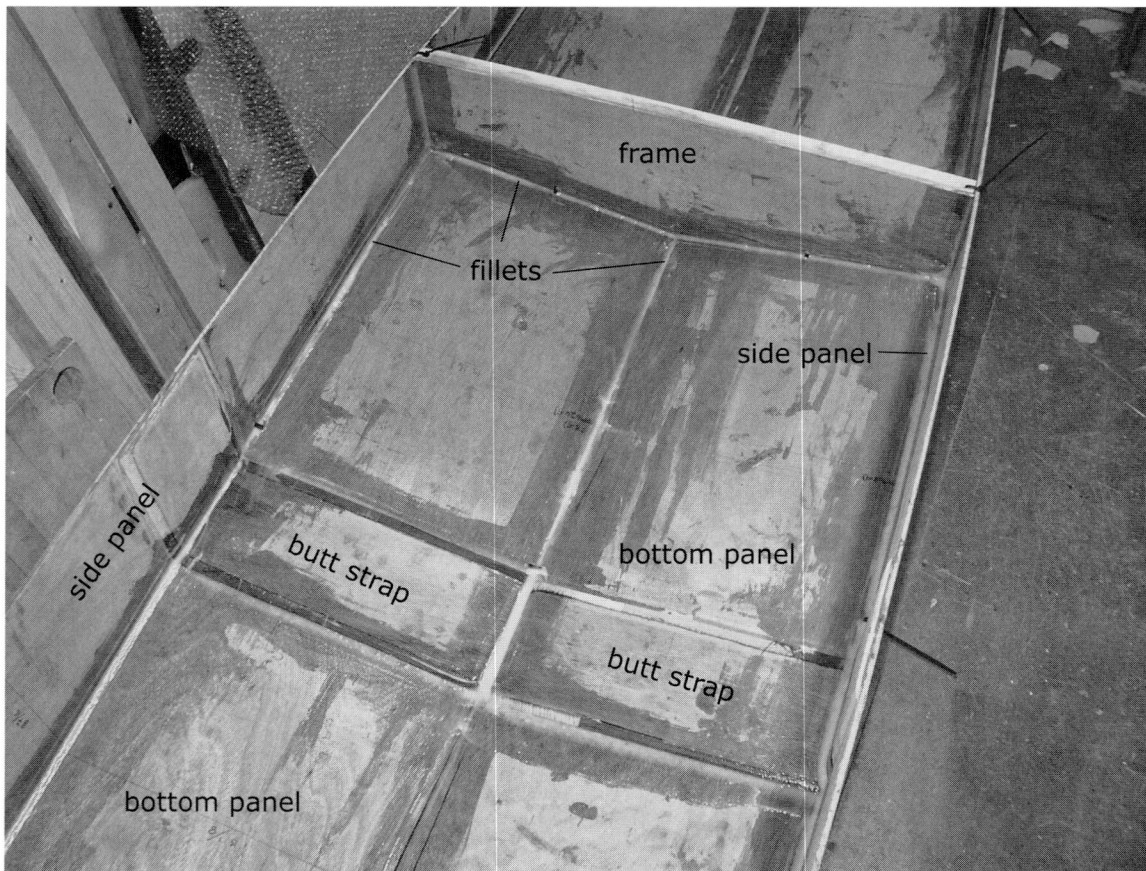

The fillets have been neatly applied to the inside seams and to the joints between the hull panels and the bulkhead frames of Anthony Smith's Cruising Mouse. Note the fillets' small size, which is sufficient for such a small, light boat. Any bigger would only waste epoxy and add weight. (Anthony Smith)

butter stage. The problem with the peanut butter consistency is that your paintbrush full of epoxy can easily be stiff enough to make a mess of your fillets and ruin all your previous neat work. Once the fillets have reached the cooled-down fudge stage, you're ready to continue, but before you do take a last look around for any unwanted lumps and messes that you should scrape up before the epoxy fully cures.

It's a good idea to measure and cut a number of sections of fiberglass tape before you mix the epoxy for the next step, because epoxy has a limited "pot life" and once you've mixed it up you want to get the most out of it before it starts to cure. You'll need to make fillets along all the internal seams: between the transoms and the sides and bottom, between the side and bottom panels themselves, and along the perimeters of the

bulkheads and frames on both sides where they contact the hull panels. Cut lengths of fiberglass to suit and lay them out in an orderly way so that you can grab the right one at the right time. If you're not of an orderly frame of mind, mix smaller batches of epoxy and do the glassing step one section of fillet at a time.

Pasting the glass tape onto your fillets should be a piece of cake. Wet the surface of the fillet by brushing it with unfilled epoxy, roll on the tape, and then brush on more epoxy, until the cloth wets through and becomes invisible, at which point it has properly wetted out. You'll need to dab at parts of the tape with the brush hairs to eliminate bubbles. It'll take a little time, but it's well worth making sure your fillets are smooth so that your glass tape will adhere without bubbles or other weak spots.

fiberglass tape

fiberglass tape

gunwale will be fitted later

Fiberglass tape has been applied over the fillets and brushed over with another coat or two of epoxy. The gunwale and shavings lying loose in the bottom indicate that another construction step has already begun on this Cruising Mouse. (Anthony Smith)

Leave the work overnight and cover it with a plastic tarp or a polyethythene sheet to prevent dew from breaking down the surface of the epoxy to form the slime known as *bloom*. Bloom can be a real nuisance because it can weaken the epoxy and prevent subsequent epoxy and paints from adhering. Once hardened, epoxy is quite waterproof, of course.

Unless you're working in a cold environment or have failed to mix your epoxy adequately, by morning your fillets should be hardening up nicely. They won't be fully cured yet—that takes a few days at least. Tear off the duct tape from the exterior of the hull (and remove any nails you used to make the plywood align properly) and reveal the elegant hull beneath. And it will be elegant, for the cloth-backed tape has prevented ugly runs and lumps from dripping all over the exterior surfaces.

The structure of the boat should soon be solid enough to gently turn the boat over. If so, get your friends and family to help you turn it together, remembering to take photos of what most home builders regard as a historic event. Once all the fuss has been made and all the photographs taken, you can get down to the serious business of preparing the outside of the hull for the next stage.

What the next stage should be is your choice: you can simply apply glass tape and epoxy to the external seams of the boat, or, if you want to make what might be a substantial investment, you can cover the whole hull exterior with glass cloth and epoxy. A halfway measure that might appeal to some is to glass and epoxy the bottom of the boat and up over the edges to cover an inch or so of the sides or chine logs, on the grounds that this is the only part that needs protection from abrasion. This is what I often choose, particularly for kids' boats that they're likely to grow out of within a few years.

Whichever way you choose to go, the next task is to gently round all the seams you are about to glass so that the fiberglass cloth or tape will lie smoothly against the curve. Once again, the radius should be about the same as a Ping-Pong ball. I do this job with a

The exterior of this Cruising Mouse hull has been prepared for glassing the seams. The cable ties have been cut off flush, and the edges have been gently rounded with a rasp. (Anthony Smith)

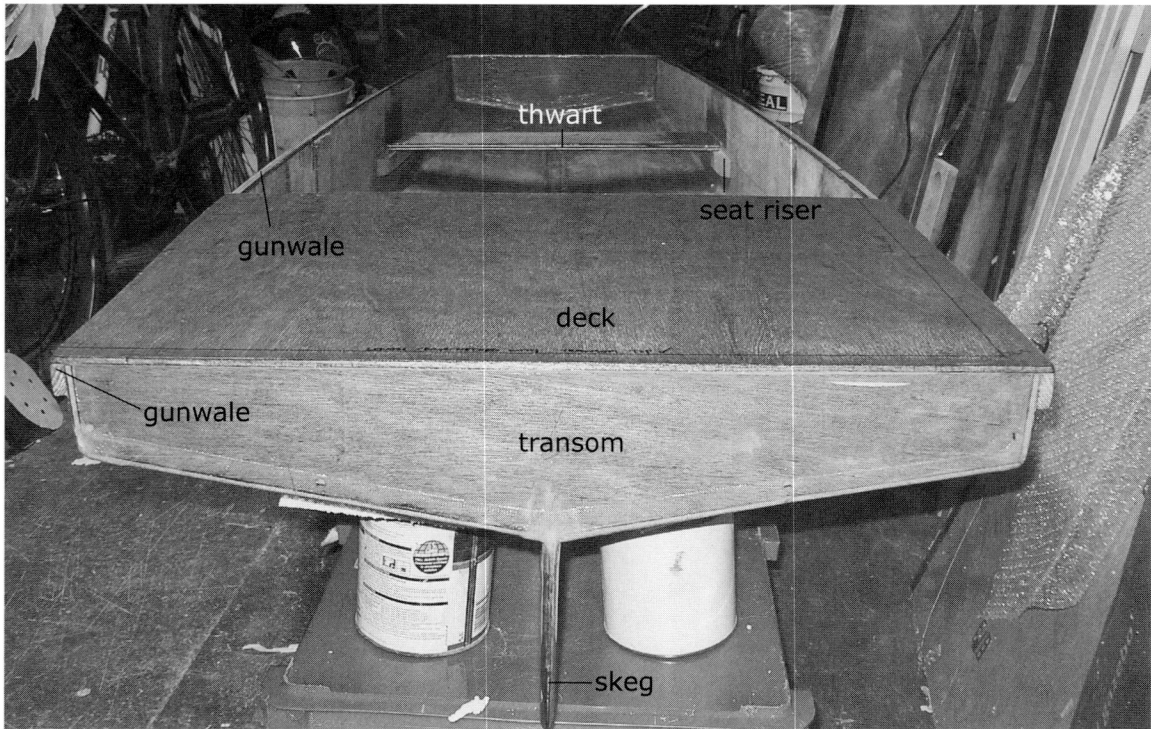

Construction is complete and Smith's Cruising Mouse is ready for finishing. The inside and outside seams have been taped; the skeg fastened; the gunwales and decks are on; and the seats (i.e., "thwarts") are in place, supported by seat risers in the form of cleats that have been epoxy-glued to the hull sides. (Anthony Smith)

rasp or a Stanley Surform, both of which are quite adequate for the task. Don't use sandpaper at this stage. The tools I've mentioned will produce the kind of fairly rough surface that gives epoxy a good grip. Also, you do not want to sand epoxy that is not fully cured for health reasons.

The outside seams are much simpler to make than the inside ones, because no fillets are required. Fill any remaining gaps with thickened epoxy, cut the tape, and wet the area to be in contact with the tape by brushing on the epoxy. Next, roll on the tape and wet it again until it has become translucent and is thoroughly saturated. You'll need to dab parts of the tape with the brush hairs to eliminate bubbles. At the end of the runs, double up the tape. These will be vulnerable points and a double-thickness of material will add strength. I find the look of three layers of tape so thick that it spoils the overall effect, so where three seams come together (like at a lower corner of a transom), I carefully snip away one layer of cloth in such a way that the third layer lies neatly over the top.

COVERING THE HULL

The hull may now be covered with a spit coat of epoxy or with a sheathing of fiberglass cloth and epoxy.

Unfilled epoxy makes a good coating for the hull. It helps prevent rot, and it makes a good surface for painting or varnishing. It's common to see boats that have been coated with epoxy and then varnished to create a beautiful deep gloss finish. These finishes are easy to create by smoothing and sanding the plywood, applying a couple coats of epoxy, allowing it to set thoroughly, sanding again, and then applying a few coats of varnish in the usual way.

Epoxy, however, has a serious vulnerability: it must not be exposed to sunlight because ultraviolet rays break it down over time. Varnish will protect epoxy so long as it remains intact, but varnish too breaks down due to the action of sunlight.

Epoxy is also often used as an undercoat to water-based exterior latex (emulsion) paint, and this too must be kept in good condition. But since exterior emulsion is much more resistant to sun damage than varnish, a paint-over-epoxy finish will usually require much less maintenance.

The other option is to sheath the entire hull (or just the bottom) with epoxy and fiberglass cloth. The solidity of the epoxy material, combined with the glass cloth's resistance to stretching, adds a great deal of strength and rigidity to the hull. It also provides excellent protection from abrasion. It does all this without greatly adding to the weight of the boat.

With the exception of the outboard-powered boats, all of the designs in this book could probably be built without exterior glass taping at all. Simply covering the bottom with a lightweight sheet of fiberglass and epoxy (combined with the usual internal glassed-and-filled epoxy fillets) should provide sufficient strength to hold everything together. This technique is widely used in building canoes.

You will need some gloves, preferably a pair that covers your forearms as well as your hands. Glass cloth has a way of getting everywhere, including up the sleeves of jackets, sweaters, and shirts. What's worse, the arms are more delicate than the hands, and seem to suffer more from the itching when glass fibers collect on the skin. This is bad stuff, as the glass is capable of puncturing the skin and leaving microscopic threads that can cause very itchy and possibly permanent dermatitis, and, incidentally, these skin punctures seem to increase your risk of becoming sensitized to epoxy.

Of course, you're working with the boat upside-down. Having donned your gloves, hang your glass cloth over the hull to see how it will lie and how you will need to trim your material to make it lie smoothly. It's worth taking a black marker pen and marking the cloth at the corners of the boat along the edges between the sides and the transoms. Later, when you cut along those lines, you'll be able to tuck the spare material under the flap to produce a neat fold.

Some boatbuilders like to cut the cloth, drape it over the boat, and apply the epoxy onto the dry cloth with a brush. I prefer to go the other way, applying epoxy to the hull first, waiting for it to slowly react until it gets tacky, then laying on the cloth and brushing on more epoxy. I do this partly because the cloth is less likely to slip around when the hull is tacky, and partly because I have less trouble with bubbles under the cloth this way, so that's how I suggest you do it.

Once you've cut out the glass cloth, position it on the dry boat until you are satisfied with how it will lie. Then take it off the boat and mix up a small batch of epoxy. Half a pint is probably more than enough; otherwise you run the danger of the goop heating up and hardening before you can apply it (as I mentioned in Chapter 1). Epoxy mixture is rather like a lump of nuclear material in that it can reach a critical mass: put too much in one container and it's apt to retain the heat being generated by the reaction, which in turn makes it react more quickly, until it's too hot to hold and has started bubbling and emitting evil-smelling toxic smoke.

At this point, you're about to find yourself the proud owner of a singularly useless piece of solid translucent epoxy. Most amateur boatbuilders have got a trophy lump of epoxy lying around the workshop area somewhere—you only have to be called away for a moment for it to happen—and be reassured that there's one in my little garden. Be warned: always use a flat tray for mixing any but the smallest quantities of epoxy.

So, mix up a conservative amount of epoxy, and brush or roll it onto your hull. If the first batch wasn't enough to cover the hull, make up and apply another batch, sticking with small quantities each time. Wait for it to get mostly dry, then lay your glass cloth in place and brush on more epoxy. As it wets through, you'll find that the cloth becomes almost invisible. There will inevitably be bubbles here and there under the cloth, and the tool to eliminate them is the squeegee—the tool with a long rubber strip that window cleaners use to wipe water off a window after it has been washed.

In addition to eliminating bubbles, wiping a squeegee across the surface of the embedded glass cloth will let you collect excess epoxy and spread it to nearby areas that haven't been coated. I can say with certainty that the cost of a cheap squeegee meant for your car windows will be amply repaid by the amount of epoxy it will save, and it'll help to reduce the weight of the boat, too.

If you don't have a squeegee, an effective alternative is one of the small paint rollers with a disposable sponge roller on the end. I've heard that some sponge rollers disintegrate when used with epoxy, however, so test a roller before you use that brand on a large surface; otherwise, you risk covering your boat with nasty little clumps of semi-dissolved foam rubber.

Once everything is neatly pasted down, it's time to tidy up and leave it to harden. Then you can trim the excess glass cloth hanging off of the edges with a handsaw. Don't leave it too long before you do this, as fully hardened epoxy is tough enough to blunt even the best hardened steel saw. It's best to tackle this trimming job within a few days of laying up the glass cloth and epoxy.

Once the epoxy and glass are in place, you must add another thin coat of epoxy before the first one is more than three or four days old in a cool climate, and even sooner if it's warm.

There are a couple of good reasons for this particular time frame. First, a new coat of epoxy will adhere to a previous one best if it can make a chemical bond, which it can do only if the earlier coat has not yet fully reacted. If the first coat has cured completely, the new layer won't adhere unless you sand the first coat. However, if you do that (in order to paint, for example), you'll immediately cut through the fibers of much of the glass cloth, greatly weakening your hull. Adding the second coat won't help at that point. Ideally, you should apply the second coat of epoxy within hours of the first coat.

ADDING A SKEG

Many of the boats in this book need a skeg to help them track through the water in a tolerably straight line between pulls of the paddle or oars. With a stitch-and-glue hull, it makes sense to fillet the skeg into place using the stitch-and-glue method.

The drawing tells the story. Start by drawing a line along the center of the bottom, from the deepest part of the hull all the way back to the transom. Accuracy is important. If the skeg is not truly straight, the boat will have an annoying tendency to bear off in one direction with every pull on the oars.

Most of the plans in this book include material for a skeg. The exact shape is not terribly important, but the curved edge that attaches to the bottom must be fitted to match the curved bottom of the hull. You will use an ancient woodworking technique, spiling, to draw this curve.

First, find a block of wood ¼ or ½ inch high, and a piece of plywood that's roughly the shape you need for the skeg. Have a helper hold the plywood vertically on the centerline of the boat while you hold the block against the bottom. Now, place a pencil firmly on the block with the point against the plywood skeg blank, and drag the block along the curve of the hull's bottom as you draw your line on the plywood, making sure the pencil stays perfectly perpendicular to it. The result should be a line that matches the profile of the curve you are spiling from, but transposed upward by an amount equal to the thickness of your block.

You can now cut out the skeg following this line with a small handsaw or a jigsaw.

If you are working with epoxy, you can attach the skeg with standard fillets on both sides.

Make sure the skeg stands perpendicular from the boat's bottom. This latter point isn't strictly necessary for hydrodynamic reasons, but it looks good, and it's easy to do. The drawing shows a scrap of plywood temporarily tacked with small nails to the transom as a brace against which you can tack the skeg in a perfectly perpendicular orientation until the glue sets. The nails should be easy to remove, and you'll need to cover the supporting piece of scrap plywood with a piece of plastic shopping bag or plastic sheet to prevent it from permanently sticking to the hull.

Small nails or adhesive tape may be used to hold the skeg in place

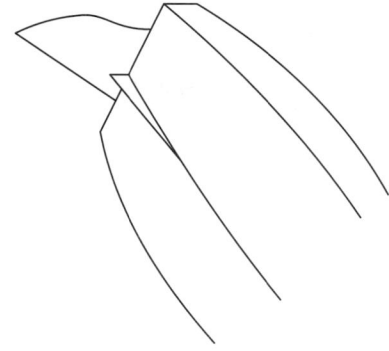

Mark centerline on boat bottom and perpendicular on supporting scrap plywood

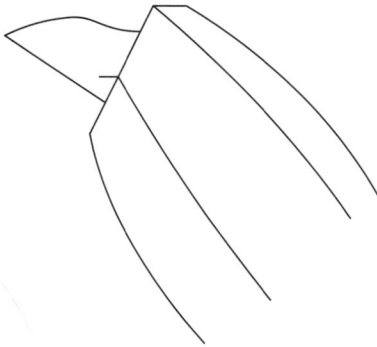

When the glue has set, remove the scrap plywood support—the polythene should make this easy

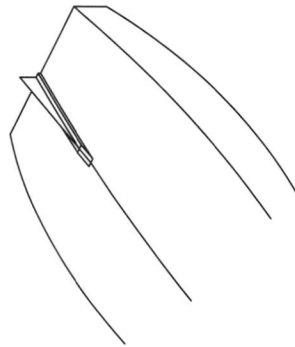

Cover a piece of scrap plywood with polyethylene material and nail to transom

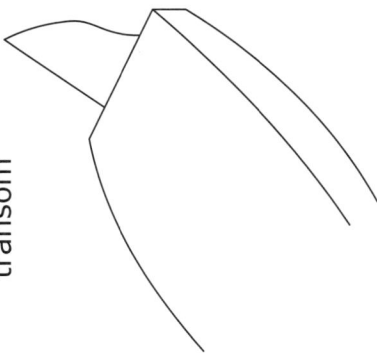

Lengths of 1/2 by 1/4 inch wood are glued and attached to each side and to the bottom

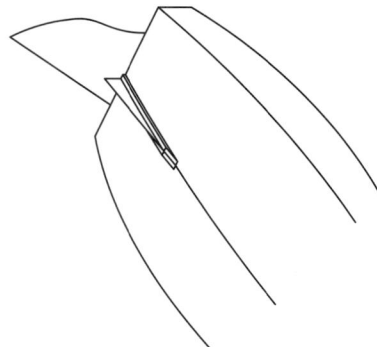

Installing a skeg is similar on flat-bottom and V-bottom hulls. A scrap of plywood tacked to the transom provides a brace to keep the skeg upright. You can either fillet it in place on the bottom, or sandwich it between pieces of $1/2$-by-1-inch (or thereabouts) timber nailed or screwed and glued to the bottom.

Using a scrap of plywood to hold the skeg of a Mouse upright. Notice that small gap between the skeg and the hull left by the builder? It's not a problem. Fill it with thickened epoxy and fillet and tape the joint on both sides, and it'll be strong enough. This Mouse was built by the author.

If you're using polyurethane glue it would be better to support the skeg on both sides using two pieces of framing lumber. Hold the skeg in place along the centerline, and mark the skeg's outline on the bottom of the boat in pencil. Glue a piece of ½-by-1-inch framing lumber along the outline on one side, then sandwich the skeg between it and a second similar piece of lumber. Of course, all the mating surfaces should be slathered with PU glue.

The framing lumber pieces can be weighted down while the glue sets, or you can drive screws from either inside or outside the bottom. Once the adhesive has hardened, shape the forward end of the framing with a Stanley Surform or rasp to make a neat, streamlined job. Be careful not to damage the bottom with the tools.

Whichever method you use, it is well worth applying some epoxy and a strip of glass tape or cloth to the edges of the skeg since this area is subject to wear during launching and beaching.

Anthony Smith's Cruising Mouse skeg installed with fillets.
(Anthony Smith)

INWALES, GUNWALES, BREASTHOOKS, AND QUARTER KNEES

Gunwales are long lengths of lumber fastened along the upper, outer edges of the sides of the boat. Inwales are the same thing fastened to the inside edges. Both stiffen the sides of the boat and provide a strong place to mount oarlocks, leeboards, and other items.

Quarter knees are small triangles of lumber or plywood that reinforce the upper edges of the joint between the sides and the transoms. Breasthooks are the same thing between the two sides of the boat where they come together at the stem, if the boat has a

Aft starboard quarter of small boat

Quarter knee drawn using kitchen jars, cups, etc., and then cut out

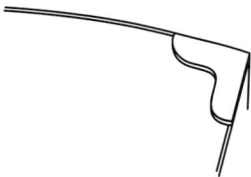

Hold quarter knee in place and draw round

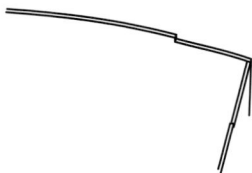

Cut recess for quarter knee

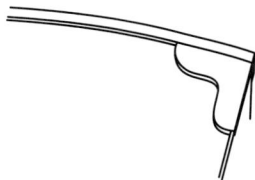

Attach quarter knee using fillet or cleats, then add gunwale

Fitting a quarter knee.

conventional pointy bow. Boats with full decks fore or aft may be built without breasthooks or quarter knees, as they will perform the same service.

Several of the boats in this book have no inwales at all, but Cinderella, Cruising Mouse, Summer Breeze, and Doris do, or can benefit from them. Because being lightweight is one of their key virtues, you should install gapped inwales, which are correspondingly lightweight. The purpose of the gapping blocks is the same as the purpose of the web (the tall, thin vertical part) of a steel girder: they're just spacers, and the real strength is in the long, flat parts—in this case the inwales and the gunwales. If you were to build solid inwales, they would have to be about the same overall thickness as the gapped version for the same strength, but they'd be much heavier.

Make the inwales from some light, straight-grained, knot-free, rot-resistant softwood like spruce or pine. For the inwales on a boat like Cinderella (pages 146–153), for example, you'll need two pieces about 12 feet long by 1¼ to 1½ inch deep, by about ⅝ to ¾ inch thick. You'll need an additional 5 feet or so for the gapping blocks.

You could get all of this out of a few 2 by 4s if you can find them reasonably clear of knots and poor grain. Larger boards, like 2 by 10s, cost a lot more, but they tend to have better wood in them.

Where knot-free material in 12-foot lengths is hard to find or expensive, it can make sense to laminate thin pieces of the best of whatever is available to create the inwales and gapping blocks; it's quick and effective. Say, for example, you need a 12-foot gunwale that's 1½-by-1-inch in cross section, and you find there's nothing available in that length without knots. The answer is to buy or mill 24 feet (or a little more) of 1½-by-½-inch lumber, and glue and clamp it together in two layers. Make sure the glue is well spread so that there are no gaps or "holidays" in the glue line between the two layers. This takes a lot of clamps.

Traditionally, the gunwales should be of hardwood of the same dimensions as the

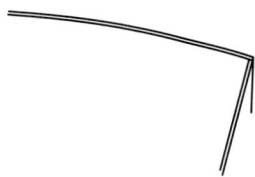

inwales, and I've often used luaun, which is just about the only hardwood available at a good price in ordinary building supply stores where I live. However, with very small boats such as Mouseboats, where appearance doesn't matter very much, I have certainly used softwood, and for a child's boat I have protected the painted softwood with a length of rubbery plastic or even some garden hose. It's all a matter of your priorities. If you want a boat that people admire, buy the hardwood, but if you're building a toy for your kids that they'll grow out of in a year or two, then there's no need to go the expensive route. For good looks, the inwales, blocks, and gunwales should all be the same thickness, and they must not be larger than the dimensions I've mentioned or they will appear too heavy for these small, light boats.

To start off, let's make and attach the gapping blocks. Get out your ruler, square, and pencil, and mark off and cut the gapping material into 2-inch lengths. In the case of a 12-foot boat, you'll need about 24 of them.

Using your tape measure, mark a point halfway along the length of the boat on the top edge of one of the side panels. Mark a similar point on the opposite side. The cautious boatbuilder will at this point get a helper to hold the end of the tape measure at the bow to make sure the marks on each side line up.

Mark a centerline on two of the gapping blocks, and clamp and glue them to the inside of the hull so that the upper edge lines up flush with the top edge of the hull (the "sheerline") and the centerlines line up with the marks on the top edges. You could drill holes from the outside and screw and glue them, but screws would simply add weight and work with little benefit.

If a frame coincides with one of the blocks, cut the block to accommodate it and place the next block along as if the previous one had not been modified.

Once these first two gapping blocks are clamped into place, measure 10 inches along the inside of the hull, and glue and clamp the next block into place, then another, and so forth in both directions, fore and aft, until there are blocks up into the bow. Make doubly certain that these blocks are flush with the edge of the plywood before the glue hardens, and use the tape measure to ensure the blocks are even and symmetrical on both sides. The final blocks at each end will be under the breasthooks and not visible, so their exact position is not that important. In fact, they shouldn't jam up into the stem or against

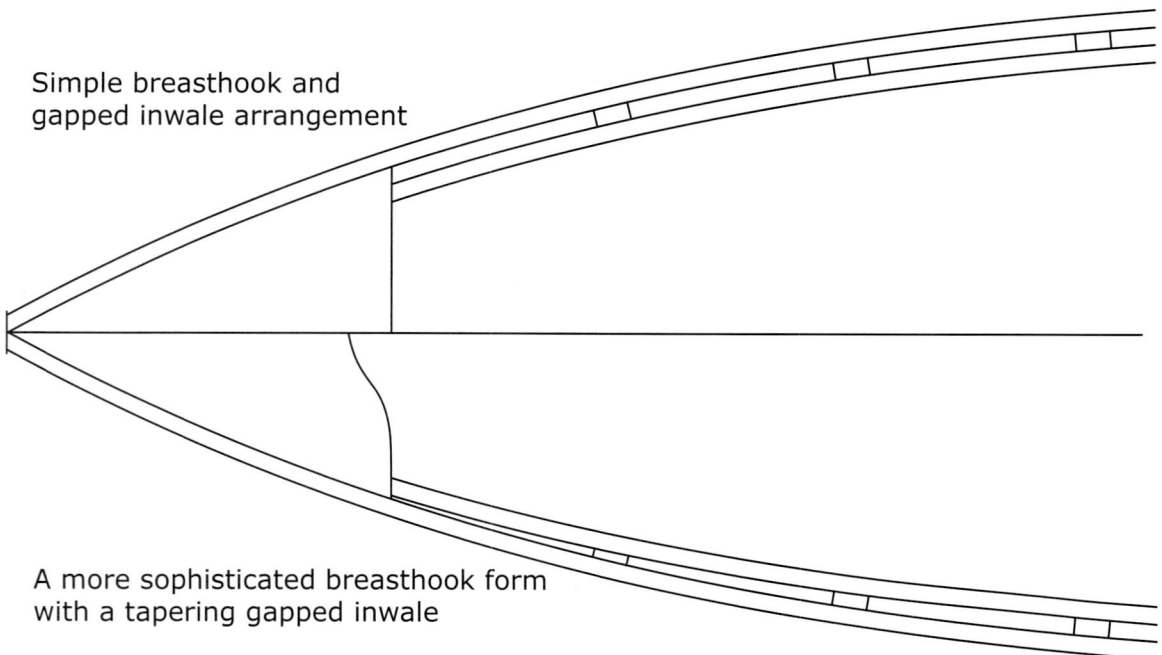

Simple breasthook and gapped inwale arrangement

A more sophisticated breasthook form with a tapering gapped inwale

Two styles of gapped inwales and breasthooks.

each other, so it may be necessary to depart from the 10-inch rule.

Adding the inwales is easy now, even if you have to laminate them. Take the port or starboard inwale, if it's to be solid, or the first layer of it, if you are laminating in place, and clamp one end to one of the gapping blocks in the boat's bow. Make certain that it won't interfere with the inwale that's about to be glued to the other side. Then proceed to glue and clamp to two or three more gapping blocks. Go no further, for you must now start clamping and gluing the inwale on the opposite side. If you do all of one side before beginning the other, you can end up with a lopsided boat. Fasten each side in alternate sections until you reach the opposite end of the boat. You will have to cut some inwale off to make everything fit in the stern the same way it did in the bow.

If you want a gold-plated job, taper the gapping blocks at each end so that both ends of the inwale bear directly on the inside surface of the hull. To do this, leave off the final gapping block at each end, make the one before that about a third of the original thickness, and the next one about two thirds. You'll have to taper the outboard surface of the inwale for about 6 inches at both ends, so that it has a broad enough gluing surface to hold reliably against the hull panel.

The breasthooks are usually made from a single thickness of the hull plywood, or a double thickness in the case of a very light boat made from very thin plywood (⅛ inch). To get the proper shape for a boat with a pointy bow, take your tape measure and measure and mark a point on the upper edge of each side of the hull about 10 to 15 inches from the bow, depending on the size of your boat. Then find a piece of scrap plywood with a good straight edge, and line it up with the ticks you have just made, with the board extending forward over the bow. Reaching underneath the board with a pencil, mark the outline of the shape of the breasthook by tracing against the sides of the hull. You should now have a roughly triangular shape. Cut this out well to the waste side of the line.

Any excess will be cut off prior to adding the gunwales. You might also cut a graceful arc in the aft edge of the breasthook: this is a small touch that will look very elegant.

In a double-ended boat (one with pointy ends fore and aft), repeat the entire process at the stern, but make the aft breasthook slightly shorter for aesthetic reasons—9 to 12 inches should be about right.

Much the same procedure applies to the quarter knees at transoms, both stern and bow. Use the same tracing technique to get the shapes of the edges. The forward ends of the breasthook, however, must be cut square across the width of the inwales. Elegant curves can then be drawn on the hypotenuse of the roughly right-angled piece. A proper boatbuilder would no doubt use a set of geometric drawing instruments for this, but I find I can usually create something perfectly good by tracing pencil lines around objects found in the kitchen and garage. Jar lids and one-quart paint cans usually will form tighter curves, while gallon paint cans, plates, and saucepan lids serve for the larger ones. Since the breasthooks will appear opposite one another, for aesthetic reasons it's worth taking a few moments to ensure that they're symmetrical. You could make a breasthook and use it to trace the shape for the second one, or you could carefully draw one, and then clamp two pieces of stock together and cut both breasthooks at the same time.

The next step is to mark and cut recesses for the breasthooks and quarter knees into the top edges of the side panels and transoms so that they sit flush to the sheerline when viewed from the side. Double-check that the breasthook or quarter knee is the right shape, then use it to trace out where to cut into the sheerline and transom.

Cut this recess very carefully and slowly by hand, using a sharp saw and perhaps a big, sharp chisel. You will be cutting into hull material, gapping blocks, and the ends of the inwales. Check that both ends of the cut are going right as you proceed. Cut on the wastewood side of the line, and take care not to cut deeper into the hull than you intend. If anything, err on the conservative side in cutting the recess. You will be able to use a rasp or Stanley Surform afterward to cut the last little bit down to the line and generally tidy everything up so that the breasthooks will seat properly. Cut and try, cut and try, until they fit perfectly.

Clamps are not likely to be any use for holding the breasthooks and quarter knees in place for gluing: there's simply nothing to clamp against. Instead, use screws or small nails driven through pieces of scrap plywood to hold them down. If you wish, once the glue has set, you can apply fillets to the underside of the breasthooks and quarter knees to marry them firmly to the inwales; if you do, cut the tape a little short to make a neat job that doesn't show.

Finally, glue and clamp the gunwales in place. Clamping should be relatively easy because you're working on a rigid structure. Leave a little extra material at the ends that you can neatly trim when finished.

Once this is done and the glue is set, cut the ends of the gunwales flush with the stem and stern, and then round off the shape to make them look nice. It's pretty easy to do with a Stanley Surform, and it's worth taking a few extra moments to make the two rounded corners symmetrical. We usually cut the gunwales flush at the stem and stern for aesthetic reasons, even though gunwales function as fenders.

CHAPTER 5

FINISHING YOUR BOAT

Even the roughest boatbuilders have to coat their boats with paint or varnish. Apart from looking good, paint and varnish protect your boat from damage by the sun, wind, and water, and help protect you and your passengers from splinters and the rest.

The techniques are all easy; however, it's useful to approach the job in the right way. It's all too easy to make a mess of a painting or varnishing job.

PAINTING

Hopefully, before this point you have decided whether to apply an epoxy coat, epoxy and fiberglass, or neither, as discussed in the

Anthony applies a spit coat—a coat of unthickened epoxy—which provides an excellent base for paint and significantly improves the boat's rot-resistance. A roller is the easiest method of application, but test any new rollers to make sure they're compatible with epoxy—some cheap foam rollers fall to pieces. Note that while the young man is wearing gloves, his arms are bare. Long sleeves are highly advisable when working with epoxy, and goggles aren't such a bad idea either when using a roller. As always in painting, preparation is the key to getting good results. Once your hull is complete, and the epoxy has cured for over a week, it's time to sand. As previously stated, you should *never* sand half-cured epoxy; the dust is quite dangerous because it still contains significant quantities of "live" resin and hardener that have not yet formed a chemical bond. (Anthony Smith)

previous chapter. In the United States, where epoxy is a lot cheaper than in Europe, covering the hull with epoxy and fiberglass cloth is a very popular option. If you choose not to cover in fiberglass, you might consider giving your boat a spit coat of unthickened epoxy; if so, you can then use water-based paint with confidence and the epoxy will seal the plywood against water ingress.

Applying a spit coat is a pretty simple task. Mix small quantities of the epoxy on flat open trays and apply it with a brush or roller. It has pretty good self-leveling properties and is easy to work with, even when using a cheap, disposable brush. Once it has thoroughly hardened, lightly sand the hull with fine sandpaper before moving to the next stage.

Sandpaper can cost lots of money. Try searching the Internet for suppliers catering to builders. You might find some great deals. Here in the United Kingdom, I've found that I can slash the cost of sandpaper by as much as 50 percent if I buy it from the right source, and the same is true for paint.

I never buy the cheapest sandpaper. It costs more in the long run because it wears out well before better brands, and I end up using more of it than necessary. In spite of your desire to get on with the job, to get the most out of sandpaper, don't sand before the previous coat of paint is completely dry, as your sandpaper will clog up and become useless. Sand using a sanding block, which may be bought made from rubber or cork-faced softwood, or simply cut from a hand-sized piece of softwood the length of the paper you are able to buy. Fold the sandpaper around the block so that you can grip it tightly. Don't rush or rub too hard and tear the paper. Regularly knock the sanding block on something handy to get the dust out of it. I only give up on a sheet of sandpaper when areas of it have started to go bald.

Sanding is hard work, and one of the most annoying jobs in boatbuilding. An orbital sander is very helpful, especially if you're *not* working with an epoxy-coated hull or plywood with very thin faces (see discussion below on sanding these types of surfaces). Orbital sanders imported from Asia can be inexpensive (they were on sale at less than $14 at my local hardware store at the time of writing). I've found solid rubber sanding blocks that cost almost as much, so I'd say that the electric orbital sander is now an excellent investment when used with medium or fine sandpaper. I avoid using coarse sandpaper with an orbital sander, though, because it can produce obvious scratches across the grain.

Generally, with rougher sandpapers, always remember to sand with the grain to avoid ugly scratches in the lumber or plywood. If you have used good quality plywood with generous surface layers to build your boat, and you are not covering it with epoxy, sanding should not be too much of a hassle. Start with a medium paper and go on to fine paper in the usual way.

However, not everyone can safely use an orbital sander on a thin boat hull. If your plywood faces are paper-thin, or you have given your boat a spit coat of epoxy, you can damage the face or coating in a heartbeat with an orbital sander. These power tools are aggressive and will quickly sand through the top layer of plywood or remove most of the epoxy coat. Furthermore, if you have invested in covering the hull with fiberglass and epoxy, it's crucial to do no more than lightly sand the very top surface to avoid quickly destroying the glass fibers of the embedded cloth.

So when thin plywood top layers, epoxy spit coats, and epoxy and fiberglass coverings are involved, sanding the hull is going to be more difficult and time-consuming. The upside, of course, is that hulls treated in this fashion are more durable than a plywood hull sealed only with paint. Is it worth the work? That choice is yours, and it should be made at the beginning of the project, not at this point. For many building the small boats in this book, I suggest steering clear of orbital sanders when sanding a coated hull, unless you're a professional and know exactly how to use the tool. Be patient and use the finest grits to get the job done little by little.

If you seek quality results, while you are sanding you'll almost inevitably need to fill some odd dings or gaps, which you might consider doing with filled epoxy, or you may prefer the spirit-based wood fillers sold in hardware stores.

Sand until the hull is as smooth and silky as your patience allows. Once it's done, brush the dust off and thoroughly clean your work area. Even small amounts of dust will create a surprising amount of roughness in your first few coats of paint. Not only will this ultimately show through in your final coat; roughness represents a weakness in the coating that may make it more vulnerable to wear and water ingress.

As I've said before, I think it's forgivable to put an unpainted boat in the water for an hour after completing the hull because you can't bear to wait until the paintwork is done before trying it out.

In fact, it may be a good idea to water-test the boat before painting it because it wets the wood surfaces. Some carpenters say that wood surfaces should always be wetted and allowed to dry before being sanded and painted. Wetting brings up the grain in the same way as the first coat of paint. Once the grain is raised, you can quickly and easily remove the roughness by sanding the bare wood. Do this before applying the first coat of paint; sanding down the raised grain after the wood has been painted is much harder. If you do launch your boat, have a good time, but make sure your boat's good and dry before you start painting!

How much paint will you need? That depends on the size of your boat, of course, but don't imagine even with the small boats included in this book that you will be able to buy just a single can of paint and hope it will do the whole job. You'll need at least two cans, one of topcoat and one of undercoat. The topcoat you'll use gives color and shine to a finish. The undercoat penetrates the surface, covers, and sticks well, but it has fewer pigments and fillers than paints used for topcoats. That's why you need them both to work together in covering the surface of the hull and giving it the finish you want. There are also primers, which are rather like an undercoat, but more so.

A more upscale boatbuilder might use a combination of primer, one or more layers of undercoat, and then a couple layers of topcoat, but a backyard boat might have to get by with a couple of undercoats and maybe one or two topcoats.

As with deciding on the hull coatings, or just going with a good quality marine grade plywood without an epoxy or fiberglass and epoxy coat, you should consider the paint you'll use at the beginning of the project. Are you going to use oil- or water-based paint? You may find it surprising that the decision isn't really a simple quality issue or even just a matter of price; both types of paint can produce acceptable results, and the water-based paint option isn't necessarily the cheaper way to go.

I've changed my own approach over time. I used to buy only oil-based paints, but I now often favor water-based paints because the solvents used in oil-based paints irritate my asthma, the brushes are horrible to clean, and water-based exterior enamel or gloss paints are far better than I ever imagined was possible. Even though the range of water-based paints is much smaller in the United Kingdom than in the United States, I still try to use them whenever it makes sense.

However, oil-based paints come into their own when you're not using epoxy in building your boat. By adding a little mineral spirits (about 10 percent by volume) to the first coat, it's usually possible to make the paint penetrate the surface of the plywood and lumber, which can help to make the oil-based paint seal and adhere well.

Spirit-based paints may be better when only a few coats of paint are to be applied, although I do recognize that this may just be a prejudice born of spending almost my whole life in a country where water-based exterior paints were rarely used on wood until quite recently. If I'm building a quick and cheap boat, and finishing it with just a couple of coats of undercoat and a couple of gloss topcoats, then I'll sometimes go with oil-based paint.

They're also good if you're a perfectionist who wants to make a boat that gleams. If this is your game, take your time: let the paint harden for several days between coats, and rub it down with increasingly fine wetted sheets of wet and dry emery paper as each coat goes on. If you work hard and use quality materials, the results will be stunning. However, if your boatbuilding is more modest, you can apply three coats or so as quickly as the instructions on the can allow. This will get you on the water sooner, but the difference in the finish will show.

All you need to apply these paints is a 2- or 3-inch brush and a can of paint, but the smoothest finishes seem to come from using one of the small paint rollers and a roller tray. You can buy these items in any hardware store for very little money, though replacement rollers aren't usually as cheap as the original one. Regardless of the price, they make an excellent, quick, and smooth job of painting with no runs. It's well worth letting the paint dry well before sanding lightly and applying the next coat. Whichever type of paint you use, once the last coat has been applied, it's wise to let the boat stand for a week or so; both are soft and very easily damaged when fresh.

Water-Based Paint

On the water-based paint side, gloss or enamel water-based exterior paints are a good choice when you're working with epoxy and have decided to cover your hull with either a single thin coat of epoxy or a coat of glass cloth and epoxy. The epoxy surface should be lightly sanded, as discussed earlier in this chapter.

I've had good results using a couple of coats of unfilled epoxy followed by two coats of water-based exterior gloss or enamel paint. (It must be gloss or enamel, by the way. Why? I can't prove it, but I believe they're more durable. I would also be concerned

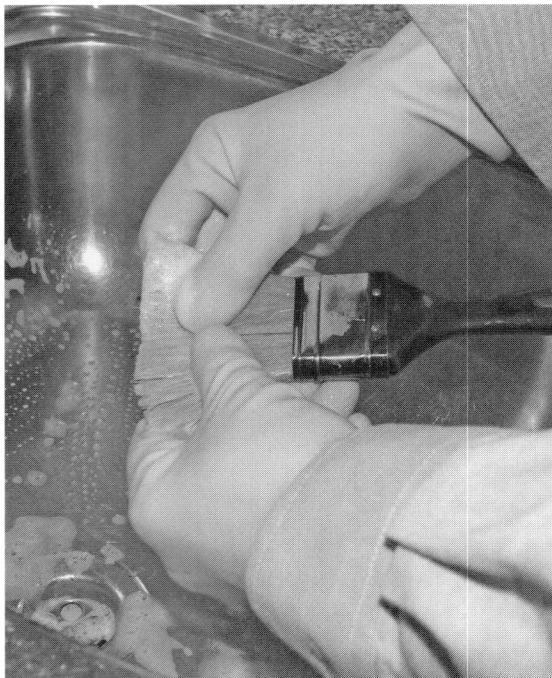

Among the advantages of water-based latex (emulsion) paint is you can easily clean brushes with soap and water in the kitchen sink, with no environmental concerns.

about the possibility of staining with matte water-based paint.)

One of the great joys of working with these paints is that they can easily be washed off in an ordinary sink with the help of a little dish detergent, as long as you rinse well afterward. This is such a major advantage that I can't emphasize it enough: there's no smell, any mess can quickly be cleaned up, and there's no bad feeling about allowing even small amounts of paint solvents into the local sewage system.

Oil-Based Paint

As I've said, this is the stuff to use if you're going very cheap or expensive. A coat of primer, two of undercoat, and two of topcoat are ideal, but if looks and durability aren't terribly important to you, you can certainly get by with less.

As usual, it's worth reading the manufacturer's instructions to find out whether the paint needs to be stirred and how long it should dry between coats, although you may often find you have to wait longer than the manufacturer recommends before you can

fine-sand it ready for the next coat. Brushes and rollers can be used to apply the paint.

Cleaning brushes and rollers when you have been working with these paints is a smelly nightmare involving wiping and squeezing off as much paint as possible with the help of old newspapers, then dousing them in mineral spirits and carefully washing the spirits and remaining paint out with warm water and dish soap.

If you're dreading this job, you'll be pleased to know that I have good news: there is something you can do to keep this brush cleaning to a minimum! By wrapping brushes in plastic bags between coats of oil-based paint you can often avoid cleaning them until the job is finished. Plastic grocery bags and sandwich bags both work fine and usually keep a brush or roller soft for a few days. Don't leave them *too* long, however, or they'll set as rock-hard as if you had left them in the open air.

If you plan to go for a perfect finish, use fine wet-and-dry emery sandpaper that has been sealed so that it can be used wet. Rubbing down with this wet sandpaper can produce spectacular results and it can be even more stunning when used with varnish, so long as you have the patience!

VARNISHING

Varnishing is very much like painting. In fact, varnish is very much like paint without the thickeners and color. However, it's always expensive, where paint can sometimes be found cheap; you have to apply many more layers; and once in place varnish has to be replaced much more often than paint because it is so vulnerable to damage by the sun.

If you're getting the idea that I'm trying to discourage you from using much in the way of varnish, you're quite right. Varnish can look spectacular, but it's a hassle to maintain, and in my considered opinion it should be kept to a minimum unless your boat is a museum piece. A plain, simple boat will generally look just too dull if painted all over, but if you varnish just a few elements, say, the gunwales, knees, and breasthooks, you can make a potentially plain boat look a little special.

People sometimes say a great deal about how to best varnish large areas, but I find there's nothing very special about getting a good finish when varnishing small areas. The key is to sand with very fine sandpaper between coats and to let each coat harden properly (two or three days is a minimum) before sanding each one. All of this takes time, which underlines my basic point about keeping varnish to a minimum, particularly in boats that are meant to be inexpensive.

CHAPTER 6

FITTING OUT FOR ROWING AND SAIL

By this point, the boat you've been building must seem very nearly complete. For the simpler boats in this book that would be true: add a painter and a pair of rowlocks or a doubled plywood pad for an outboard, and you might be ready to go. Sailing boats, however, are inevitably more complicated. But in line with this book's fundamental principle that simple boating provides more fun, this chapter seeks to make rigging as simple as possible, too.

FITTINGS

Almost every boat needs at least a few fittings, and plywood boats are no exception. Even the smallest paddler needs a painter. A rowboat must have oarlocks and an outboard boat needs a place to attach a safety line to prevent the motor from being lost overboard. A sailboat requires a variety of fittings to control the sail, fasten the rudder on the stern, and keep the daggerboard and mast attached to the boat in a capsize.

There's an entire industry dedicated to selling up-market fittings you probably don't need, so please keep in mind that the boats in this book don't require expensive, high-tech fittings designed for the highly competitive racing world. An expensively engineered block intended for the latest Olympic-class wonderboat may be very nice, but it's out of place on a Flying Mouse.

So, before you part with real money on posh boat gear, it's worth shopping around for the less glamorous brands. A standard block with no ball bearings and no ratchet will be a lot cheaper than one that has both, but it will do the job perfectly well if your sail is small and optimal performance doesn't matter. Cheap plastic blocks, fairleads, cleats,

Inexpensive plastic deadeyes can serve a multitude of purposes on small boats, from providing an attachment point for a bow painter, to serving as rudder gudgeons. Make sure you back up the hull panel where these are attached with a reinforcing scrap of plywood—likewise with mounting cleats and such.

An inexpensive non-roller-bearing block suitable for a mainsheet. Blocks like these work fine in small boats—there's no need to buy more expensive models with ratchets or races of bearings.

and such have more than sufficient strength for the stresses that these boats impose. As with other components for your boat, the Internet is a good place to look for fittings. Ask for help around the e-mail forums, too; someone will know just where the best deals are.

It's also worth looking for galvanized iron fittings, which are often sold for farming and gardening by agricultural suppliers and in hardware stores. Although they're made to last in the outdoors, they're much, much cheaper than the stuff the chandlers sell, and in general they'll last as long as you need them to.

Fittings raise a technical issue that you must plan for during hull construction. Because plywood boats are relatively thin-skinned, you must strengthen the areas where you intend to mount the fittings; otherwise there won't be enough material for the screws to grip. As you build your boat, it's well worth taking a moment to consider where to rein-force for the installation of an eye, cleat, or other type of hardware. This includes at the bow to take the painter on just about every boat. In any sailboat, you will have to place reinforcements at the aft corners of the aft deck to take the deadeyes for the traveler (horse), and on the transom for the rudder.

Typically, I deal with this reinforcing issue during construction by gluing a 4-by-4-inch patch of scrap plywood on the side of the plywood opposite where the fitting will be mounted. When the time comes for mounting, I place the fitting where it needs to be, poke a pencil or sharp nail through the mounting holes in the fitting to mark the position of the pilot holes, drill them, and then screw the fitting into place, knowing that I have at least a double-thickness of ply-wood to give my screws the bite they need.

Most fittings can be safely attached with quite short stainless steel screws. It's amazing what these little beauties with their tight screw threads can do. Even a fitting with just two or four ½-inch stainless steel screws driven into a double thickness of ¼-inch plywood makes a robust, reliable point for tying a line or sup-porting a rudder. You need not spend lots of money on silicon bronze screws. These are perfect to use with equally posh bronze fittings on classic and reproduction boats, but are somewhat out of place on our humble craft. However, stay away from sheetrock screws and other screws made from inferior materials, such as brass or zinc-plated steel. These will rust, and they have coarser threads, which don't provide nearly the holding power of stainless screws.

PADDLING AND ROWING GEAR

The oars I make are the same as Jim Michalak's and I hope he won't mind my sharing them with you. (Jim is the designer of Piragua; see pages 143–145.) In any case, he got them from the renowned traditional-style boat designer and builder, R. D. "Pete" Culler. Whatever its provenance, the method produces oars that are sophisticated, well balanced, and narrow-bladed to match the needs and strength of the ordinary Joe who doesn't have the muscles to make much use of a big spoon oar.

The Culler-Michalak method produces a pair of oars at a fraction of the price you will be charged for a single badly balanced and clunky ready-made oar bought from a yacht chandler. So these real paragons among oars are cheap as well as right for the purpose.

The method for making Culler-Michalak oars begins with 1-by-6-inch pine boards free of all but the smallest knots. You'll need two 7-foot lengths of reasonably clear material to make two useful 7-foot oars. They can be shaped using a medium-sized hand plane, but I think it would be challenging to go that way unless you're very practiced and bristling with well-toned muscles. In spite of my usual diatribes against power tools, I think you will need access to an electric hand planer to make your oars. A power planer and a workbench fitted with a means of clamping the lumber will make the job pretty easy as long as you're not a perfectionist.

Three parts are cut from a single board and laminated together to make the oar: a long cen-tral blank incorporating the narrow blade and the main part of the loom, and two face pieces that are then glued and clamped to each side of

As well as demanding muscles, big spoon blades require the rower to feather his or her oars—that is, with each stroke you have to turn your oars forward through 90 degrees so that the blade is horizontal on the recovery stroke to reduce air resistance and minimize the effect of splashing into a wavelet. It's an activity that's probably only popular with Sea Scouts and those who enter rowing races, and to my mind the insistence that rowing with feathering is the only "real" rowing may be one of the key reasons we see so few people rowing around harbors and rivers in the United Kingdom. I see more people sculling over their tender's stern than rowing.

What I have to say on this issue will, therefore, be heresy to most "proper" rowing people, but that's just fine with me. I think an ordinary Joe doesn't want to be bothered with learning to feather his oars on every stroke, and I don't think it's necessary if you use narrow blades. There's lots of traditional justification for this. If you make your blades narrow, feathering becomes much less important, or even useless. Consider the traditional oars you see used with Irish Curraghs from the Dingle peninsula. Developed for use on the often rough Atlantic coast of Ireland, which features great trains of long Atlantic rollers, these oars are no wider at the blade than at the loom (oar shaft) and they're mounted on a solid metal tholepin in such a way that they can't be feathered at all.

the loom to make up the necessary thickness. Mark and cut out the shapes as shown in the drawing, and glue and clamp the whole lot together as accurately as you can. Since fully hardened epoxy can blunt steel tools, I'd use polyurethane glue for this job.

Once the glue has hardened, the shaping can begin. The loom can remain square in section, but the handle must be rounded to around 1¼ inches in diameter, and the shaft beyond the loom, from the point where it sits in the oarlock down to the narrowest point, needs to be rounded. First plane it to 8 equal sides, then to 16, and then to 32, at which point you should be able to easily sand it round. Take care in using the plane, however, since you don't want to weaken the oar in the area where the blade meets the shaft.

As shown in the drawing, the blade of the oar needs to be chamfered slightly from the center, with the blade ending about ½ inch thick along the sides and at the end, and rising to the full thickness of the lumber at the center. It's essential to mark out the centerlines along the edges of the blade and across the tip, and then to mark two parallel lines ¼ inch on either side of these. The parallel lines are the marks you will trim to. In section, the blade should be a kind of elongated diamond.

Once shaped roughly with the power plane, each oar blade will need to be shaped more carefully with a Stanley Surform and finally with an orbital sander starting with the coarse grit and working gradually to a finely sanded surface that's fit for varnishing. Don't be tempted to varnish the handle, however. For the sake of your hands, it is best left as bare, smoothly sanded wood.

After the oar has been varnished, whip it with cord where it sits in the oarlock to prevent wear. The whipping can be plain cheap white polyester cord bought from any hardware store.

John Welsford (see page 217) once told me that when rowing over distances he uses a length of fine cord to prevent his oars from sliding away. Bore a small hole near the top of the blade (but still a good way from the narrowest part of the shaft) and tie a length of

Michalak oars, 7 feet

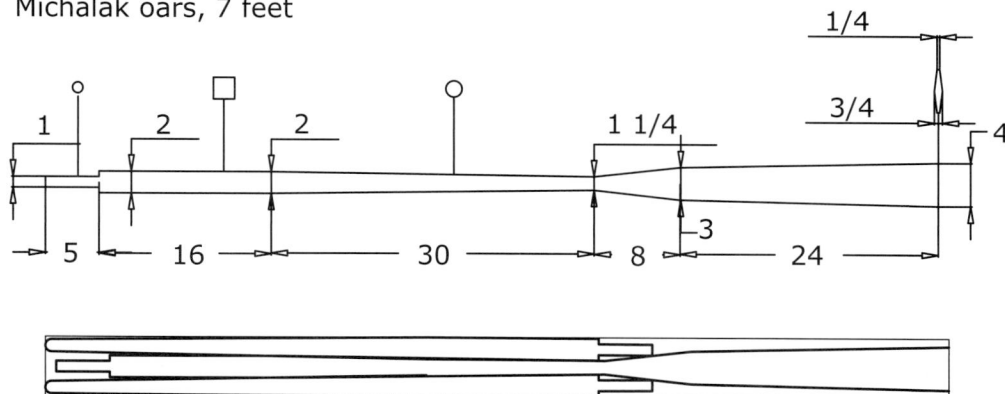

Jim Michalak's oar design, derived from Pete Culler's, is simple, inexpensive, and elegant. The oar is cut from 1-by-6-inch lumber, with cheek pieces laminated to the sides of the oar shaft (the "loom") to build it out to the proper thickness. Note that the top end of the loom is left square—the extra weight helps balance the oar on the oarlock.

A pair of laminated oars under construction, showing the taper of the blades. The handles or grips should be round when finished. (Ben Crawshaw)

cord from this hole to the oarlock. This will keep the oars from falling in the water, and when exhausted you can lean on them to rest with confidence.

Oarlocks need to be placed at just the right distance from each of the thwarts you're planning to row from. A good traditional rule of thumb is to make the fore-and-aft distance between the aft edge of the thwart and the center of the U of the oarlock the same as the length of the rower's forearm. If you're adding an extra rowing position, you will, of course, need to add a second set of oarlock sockets using the same method.

The oarlock itself needs to be well supported. Make up two blocks, each laminated from two 4-by-8-inch pieces of ¼-inch plywood (more, if your inwales are deep), then glue and screw them to the hull sides (driving the screws from the outside of the hull) so they are flush with the sheerline. Center the fittings in the middle of each block. The oarlocks can be standard models. I use the kind where the casting mounts with screws on the top surface of the block. Others prefer the kind where the pin of the oarlock fits through a hole bored in the block itself. I'm pretty sure the latter are quieter than the surface-mounted type.

I've never used plastic oarlocks, so I can neither recommend nor warn against them. I've had a pair waiting to be used in my junk box for years, but they look so shoddy that I have never installed them. However, they're cheap and widely available, and I'd suggest that they would work best on the lighter boats in this book, such as the Rowing Mouse, as users of these light rowboats are probably the least likely to stress plastic oarlocks to the point of breaking.

"Stretchers" are footrests, and they make rowing considerably more efficient by helping fix your body in place so that you can use the powerful muscles in your legs without sliding around on the seat. The stretcher itself is just a bar that stretches across the hull, made from doubled plywood or any other piece of suitable lumber strong enough to push against. To allow adjustment, I make them up so that the ends of the bar sit in brackets made from double-thick plywood cut into a "comb" shape then glued and screwed onto the sides of the boat where they meet the chines. In use, you'll need to experiment with the position of the stretcher to find the most comfortable position. Because of the way the boat's sides curve, you will need to make stretchers of different lengths for each rowing position and for different rowers.

I've drawn up a plan for a single-blade paddle that can be made from a 48-inch piece of softwood, 6-by-1-inch (true dimensions), but you could use a standard lumberyard 1-by-6-inch (actually ¾-by-5½-inch). Smooth the resulting slight irregularities accordingly, as there is nothing sacred about these dimensions. As always, small, tight knots are often

acceptable, but larger knots must be avoided.

To make this paddle, draw a centerline along one face of the wood, and from the centerline square off the whole board in 1-inch squares. Once this is done you will be able to copy the profile from the drawing fairly easily.

Cut out the profile of the paddle and then on pieces of the scrap mark and cut out two pieces measuring 29½ by 1¼ inches. Taper one end of these fillets or cheek pieces to an eighth of an inch deep or less, and glue and clamp them to either side of the handle as shown, starting about 6 inches from the end of the handle, and with the tapered end toward the blade.

Then start shaping: this is the kind of job for which long winter evenings are made, and the objective is to create something that feels good in the hands and will be comfortable without weakening the handle and its shaft. The edges of the blade should be about half an inch thick or a little less, and nicely rounded. The shaft should be a nice oval along most of its length until it starts to widen toward the handle end. (The long axis of the oval should be at 90 degrees to the blade.) The grip should be well contoured and rounded to make it comfortable to hold and to push against with the palm of your hand. Some people might use a power planer for this job, and others might go for a Stanley Surform. But if you can get hold of a spokeshave and a stone to sharpen its blade, I

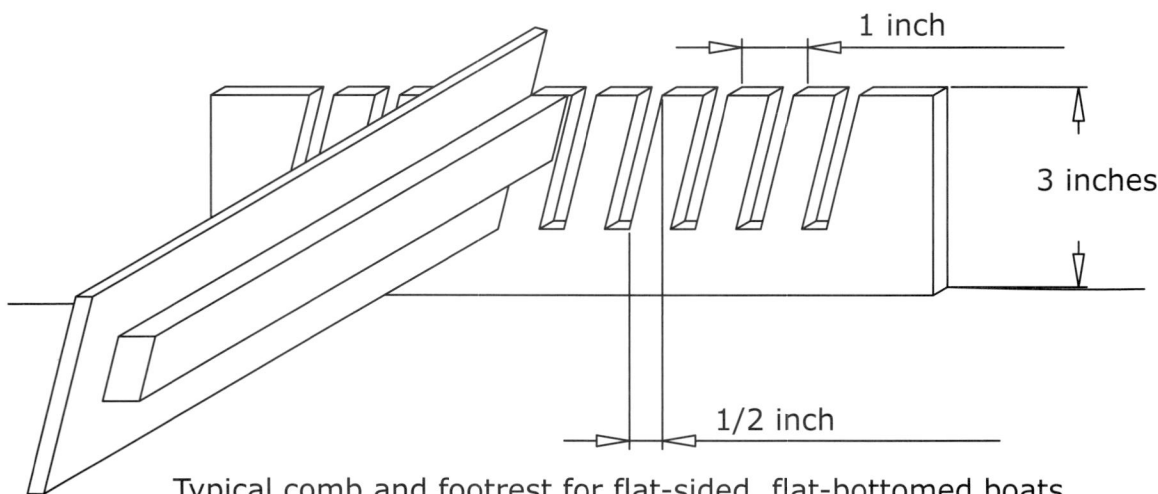

Typical comb and footrest for flat-sided, flat-bottomed boats.
This bracket and stretcher arrangement will work on many of the boats in this book.

Paddle half-profile: trace onto 6-by-1-by-48 inch pine or ash, or substitute softwood reasonably clear of knots.

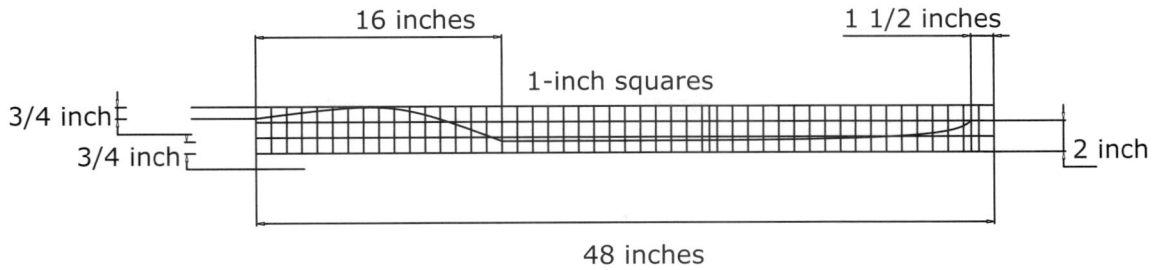

16 inches

1 1/2 inches

1-inch squares

3/4 inch

3/4 inch

2 inch

48 inches

When the paddle profile has been cut out, mark out and cut out a rectangular section measuring 29 1/2 by 1 1/4 inches, and cut this in two to create two pieces of wood measuring 29 1/2 by 1 1/4 inches by half the thickness of the wood.

Fillet 1

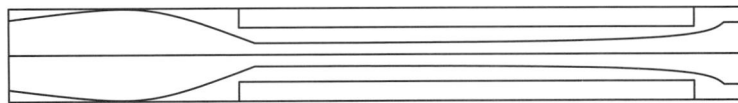

Fillet 2

Clamp and glue the fillets on each side of the handle.

Shape using plane, rasp, or Stanley Surform to create a round shaft and hand-friendly smooth surfaces that are comfortable to hold.

1/2 inch

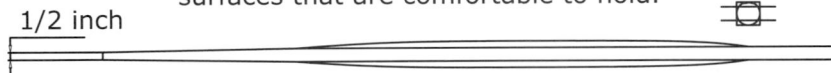

Sand and varnish the finished paddle.

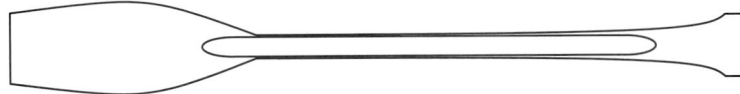

Single-blade paddle plans.

Handle: 60-by-1/4-inch pine, possibly a closet pole or dowel. Angle between blades is to taste but may be between 50 and 60 degrees.

Blade from 1/4-inch plywood

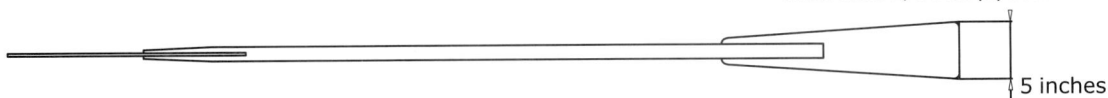

5 inches

Pole material may be chamfered slightly at ends to improve appearance. Blade may be glued and clamped into place, or it may be glued and screwed.

Double-blade paddle plans.

swear doing this job in the old-fashioned way will be satisfying and relaxing.

Double-blade paddles suitable for the small Mouseboats can be made simply and effectively using 1¼-inch softwood dowels or the better kind of closet pole bought from a hardware store or lumberyard. Make sure that the pole is solid and not jointed, as jointed ones might be manufactured with non-waterproof glue.

Mark and cut on 8-inch slot in each end of the pole. Add an angle of twist between them if you're an experienced paddler and know that you prefer a feathered paddle. Cutting the slot requires good clamping and careful work with the handsaw to ensure you follow the lines you marked. Cut out plywood blades and glue and screw them into place.

SAILING GEAR

It's in the nature of things that sailing rigs add complexity to a boat, but the rigs you'll find in this book are about as simple as they can be.

Rigs consist of two basic parts: boards and sails. Daggerboards, centerboards, and leeboards mounted on the side of the boat, jut under the water to help the boat track straight rather than slide sideways. The rudder, which can be considered a board, also juts under the water and contributes to keeping the boat from sliding sideways. When the rudder is amidships (centered), it presents little resistance to the water, but when it is turned even a little, more of its surface area is presented and the resistance increases, turning the boat in one direction or the other. Sails, of course, catch the wind and propel the boat: what happens is that the force on the sail and the force on the board combine to propel the boat forward, much as you might make a fruit pit fly by squeezing it between two fingers.

There's no need to know exactly how sailing works from an aerodynamic standpoint. But you do need to know that the designs for the sailboat rigs in this book must be followed exactly in terms of positioning the boards and sails, or the magic won't work. If you make arbitrary changes, the boat will not be in balance, and it won't sail properly.

Leeboards, Daggerboards, and Rudders

Let's get down to the details, starting with the underwater part of the rig. Depending on the design, leeboards in small boats can be as simple as a shaped piece of plywood (see Summer Breeze on page 187) fastened to the port side of the boat with a large stainless steel bolt, the biggest washers you can find with holes to exactly match the bolts, and two nuts (the second is a locknut). It's nice to have a large rubber washer on the outside to prevent leaks and distribute the stresses. Also, to further distribute the stresses the bolt should pass through a robust reinforcement on the inside of the boat made from a piece of scrap plywood.

Because of the orientation of screw threads, a single bolt-on leeboard should be on the port side of the boat rather than on the starboard side. This way, as the leeboard lifts it with tend to tighten the nut and bolt, as opposed to loosening it.

Where you have an external chine along the outside bottom edge of a boat, the leeboard can bear against it. With a stitch-and-glue boat, add two bearing strakes ½ inch thick to distribute the stresses and to prevent the board from scratching the boat's side. One of these strakes should be along the chine itself, and the other just below the gunwale. These can each be a couple of feet long and will look better and

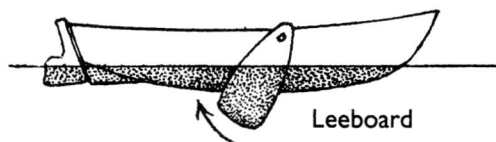

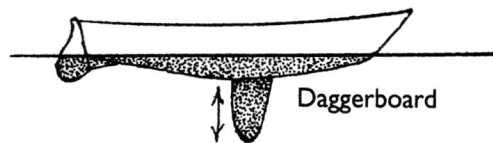

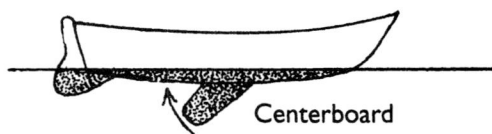

Leeboards, daggerboards, and centerboards all act like a wing in the water, preventing the sailboat from slipping sideways. (Kelly Mulford)

distribute the stresses better if they are tapered at the ends.

Leeboards have only become common in small homemade boats in the last couple of decades. Before that daggerboards pretty well ruled the roost, and in fact they are still very popular today, even though they take up space inside the boat and are more complicated to make because they require a slot in the bottom of the hull. The worst feature of a daggerboard is that when you hit something underwater it won't kick up, as will a leeboard. Instead, the daggerboard will slam against the slot in a grounding, which can break the slot and hole the boat. It's worth avoiding these situations when sailing, but in my experience in the kind of very small boats we're discussing, the speeds involved are rarely enough to cause serious damage in a grounding situation. Nevertheless, the case or slot that the daggerboard slides up and down in should be strong and solidly made, and so far as is reasonable it should be protected from rot.

On the plus side, daggerboards can be shaped to a greater extent than leeboards to improve their performance, and the designer is not forced to place a daggerboard at the widest point of the beam, which is often the case with leeboards. Because the designer has more freedom in determining where a daggerboard can be placed to balance the sail plan, he or she also has more freedom in terms of where the rig goes—the implication of this is that boats with daggerboards frequently look nicer than those designed to have leeboards. They also appeal to the conservative in us because they're the solution we feel is expected and traditional.

To build a daggerboard case (trunk), begin by marking and cutting out two pieces of the same plywood you used to make the hull. Make these slightly deeper than specified in the plans. This is partly in accordance with the "waste-wood side of the line" approach to cuts that should underlie all your woodworking, and partly to leave some jiggle room when it's time to trim the bottom of the case to match the profile of the bottom of the boat.

After the case is assembled, you won't be able to do anything to protect the inside surfaces from rot, so you must protect them before you put the pieces together. The best approach is to cover them with epoxy and fiberglass cloth and paint. First, cover the whole interior surface of the case side pieces with fibreglass and epoxy, including the areas that will be glued to the vertical end logs of the case. Then glue one side of the case to the end logs. Next cover the end logs with epoxy and cloth. Thoroughly paint all of the inside of the case, except for the areas that need to be glued to the remaining side of the case.

If you're not using epoxy, put masking tape over the gluing surfaces and paint the exposed surfaces with as many coats of primer, undercoat, and gloss paint as your patience allows. Five or six properly applied coats will not be too much. Finally, remove the tape, lightly sand the gluing surfaces, then glue and screw the end logs of the case in place.

How the upper end of the case is engineered varies with the design. It might be butted up against a frame to emerge at the deck, or it might end in a thwart. At the lower end of boatbuilding, cases don't vary much. Two logs are glued and screwed or clamped into place, and these and the case they are attached to are then fitted by either spiling or cut-and-try until the profile matches the shape of the inside of the bottom of the boat. Bed the foot of the case in a mash of polyurethane or epoxy and cloth (the cloth gives the joint a bit of useful wiggle room), and clamp and screw it firmly into place, making sure that it is accurately lined up with the centerline of the boat. This may seem a pretty rough-and-ready procedure, but it should be done as carefully as possible because this part of the boat is subject to more stresses than almost any other part. Any rot in this area can quickly be fatal to your boat. If there's any place to over-engineer a boat, this is it.

It's only later, when the epoxy has hardened, that I return to cut the slot through the bottom of the boat from the outside. I begin with a small hole drilled in the middle of where the slot should be. Once I have broken through to the open area inside the dagger-board case, I slowly enlarge the hole with a rasp or a small saw until I've opened the whole slot. This must be carefully done to avoid damaging the inside of the case, and once it's done, the cutaway area must be sealed with more paint or epoxy.

The board can be a very simple proposition, and is even simpler than the leeboard discussed earlier. In fact, the daggerboard is often just a flat plywood board with rounded-off edges and a handle on top to prevent it from sliding out of the boat through the slot.

The very best daggerboards have a carefully calculated airfoil shape that has the effect of increasing its effectiveness when the boat is travelling through the water at low speeds, but anyone who has bought this book is unlikely to be in the market for such high technology. If your designer has specified a simple, flat board, I'd say you should go with what's drawn, as most beginning to intermediate non-racing sailors are unlikely to notice any difference in the boat's performance.

However, if your designer has proposed a plywood daggerboard that's a minimum of ¾ inch thick, here is a good compromise that lies between the simple and less efficient and the complex but highly efficient.

Draw a line about a third of the way back from the leading edge: this will be the thickest part of the board. Use a Stanley Surform to create a nicely rounded leading edge that widens sweetly up to the line I've just talked about. Next, draw a centerline along the trailing edge, and then draw two lines parallel to it about ⅛ inch on either side. Then use whichever tool you favor to cut away the excess until you have a flat, smooth surface on both sides of the board from the line denoting the thickest part of the board back to the ¼-inch-wide flat spot on its aft edge.

Once a handle has been attached and the board has been sanded and painted, this board will give you most of the advantages of a high-tech board without your having had to work too hard.

One last thing on daggerboards: once on the water they will tend to float upward in their slot. Some people weight a daggerboard by building in a few ounces of lead or fishing weights. Working with lead, however, is difficult and dangerous, and I prefer to use a piece of bungee cord (elastic shock cord) rigged from some convenient part of the boat and stretched over the top of the daggerboard. In Doris the Dory (see page 197), for example, I use a simple loop of shock cord tied around the mast and stretched over top of the daggerboard and onto its aft edge. Simple as it is, it gives no problems.

Rudders are often a more complicated piece of work, and each boat that requires a rudder in this book has its own design. David Beede's design for Summer Breeze (see page 185) is different from the one I drew for the Flying Mouse (see page 160) and looks rather better in return for a little more work. If you prefer this approach of rounding the corners of the rudder stock, use it on any of the designs, but please make sure you retain the underwater profile of the blade specified in the design.

The drawings tell the story of the components that have to be made up and then glued and screwed together. A simple stainless steel bolt, two washers, and two nuts (one is a locknut) provide the rudder with something to pivot on, but should not be done up so tightly that the rudder won't freely lift. You want the rudder to kick up if you hit an underwater obstruction, such as when beaching.

Boatbuilders sometimes scrimp on rudder hardware and make cheap alternatives to rudder pintles and gudgeons, including gadgets based on seat belt strapping or various bits of plumbing hardware. But because rudders are among the most important aspects of a boat's performance, I'd suggest that a good strong, built-for-the-job set of pintles and gudgeons and a purpose-made clip to keep the rudder in place are good choices.

When installing rudder fittings, ensure the fittings are in line with each other, or they will work themselves loose. To avoid splits, do not position fittings such that the screws are close to the ends or edges of the wood or plywood material. Also, you must

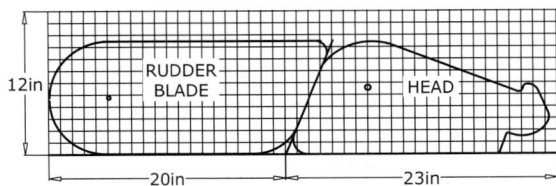

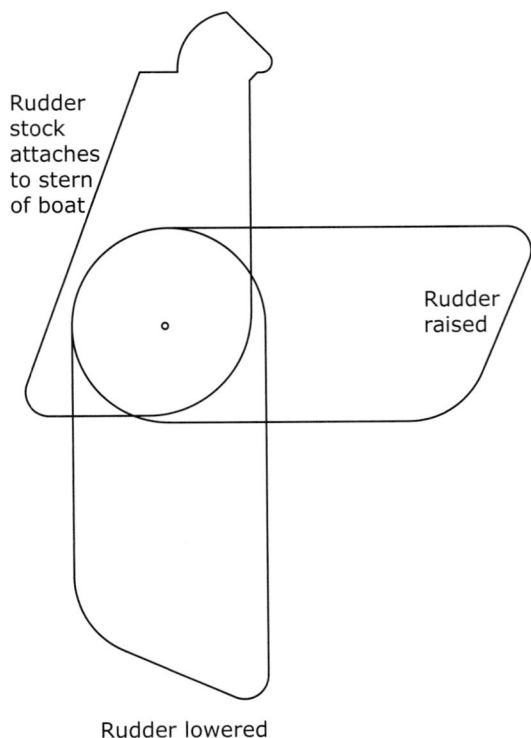

The cutting diagram (top) shows how to get the maximum amount of rudder and head stock from a limited amount of lumber. The assembly sketch (bottom) shows the kick-up principle. The slope of the angled edge of the head stock is adjusted to suit the angle of the boat's transom.

good when cut down to size). An even better option is to cut a large piece of lumber in two, reverse one of the pieces, and laminate them together. That way, any tendency to warp will be counteracted by the two pieces trying to move in opposite directions.

For a larger boat, I'd take one of these approaches, but for the small boats included in this book, starting with a chunk of wood of this size would be overkill. With smaller boats, I've had some good results building up masts for small boats from several thin boards cut from the same tree and sold as a bundle. Warping can be nearly eliminated by aligning the boards so that the curve of the grain in each piece of wood opposes the grain in the others and then laminating them together. A mast for a very small boat made up of two pieces of 1-by-2-inch lumber glued together is likely to remain straight, and a mast made up from more pieces will be better still.

The lower end of the mast should be square in section, but from above the mast's supports—the mast partners—it should be planed on its corners to make it first 8-sided, then 16-sided, and perhaps even 32-sided. Then sand it with progressively fine sandpaper to make it smooth and round.

Take care not to plane too much wood away, or the mast will become overly thin and flexible. Try to sand along the length of the grain rather than across it to avoid unsightly scratches that always show under varnish.

Almost all small sailboats need at least one more spar (usually a boom) to support the sails. The spar is best made up in the same way as the mast, by laminating pieces of 1-by-1-inch or 1-by-2-inch material. The edges of the spar should be rounded to reduce the chance of injury if it hits you.

mount the gudgeons (the eye-shaped part) on the transom so that the lower pintle (the pin that fits into the eye) engages its eye ½ inch or so before the upper pintle engages. This is essential when hanging the rudder on the stern. If you locate the gudgeons so that the pintles engage simultaneously, you'll find it almost impossible to install the rudder when afloat even in calm water.

Spars

I make masts from spruce when I can get it and other softwood when I can't. It is possible to use a single particularly straight and straight-grained piece of lumber (2 by 12s can be particularly

Sails

Sails can get complicated and expensive, if you want the very best. But luckily for us the difference between a well-made and perfectly-shaped sail and one that is less than ideal is reduced given the small sails you'll need for the boats included in this book. Sails that are adequate for your boat can be made the old-fashioned way with darts (i.e., folds at the corners) and curved edges to create the crucial rounded shapes rather than broad-seamed from strips of sailcloth cut in subtle arcs, and laboriously and expensively sewn together. What's more, because the stresses involved are much smaller, it's quite possible to make effective sails that will last a season or two from a good quality plastic tarp—so long as you don't ask too much of them.

The sails for my designs in this book can be made from either tarp or sailcloth.

However, tarp sails and those made from sailcloth have to be made in slightly different ways because of the properties of the materials and the way in which they are bonded: adhesive cloth-backed tape for tarps and stitching for sailcloth.

Tarp sails are best made from the heavyweight kind of plastic tarp material that builders and roofers use. The cheap blue kind sold in hardware stores has no UV protection and is made from material so thin that it's only fit for prototype sails expected to last no more than a few outings.

Having obtained the material, I mark out the shape provided by the designer, including the darts, 3-inch-deep hems, and curved sail edges using a felt pen, a ruler, and a tape measure. For the curved sections I use a flexible batten held in place with a few heavy objects, such as books, furniture, or bricks. Next, I use the cloth-backed tape to stick a piece of rope into the seam between the hem and the sail edge. With this boltrope firmly in place, I use double-sided carpet tape to stick down the hems, and I then use a single-sided cloth-backed tape to cover the seam where the hem meets the sail.

I use double-sided cloth tape inside the darts, and single-sided tape to cover them on the outside. I then go on to use the same materials to paste reinforcing patches of the sail material into the corners, and if I'm in the mood, I cut another long patch to tape onto the sail about a quarter of the way up from the boom to allow me to add a line of reefing points, though since most of my tarp sails are

prototypes made from the cheap blue stuff I rarely bother with this.

Finally, I use a grommet kit sold in chandleries and sewing shops to put grommets where they are needed: in each corner of the sail, at 1-foot intervals along the luff, wherever else lacing is required, and along lines of reefing points.

Where the grommets are close to the edge of the sail and therefore the boltropes, I use whipping cord to whip the grommet to the edge of the sail and around the boltrope to ensure that any stresses are communicated to the boltrope rather than the sail material.

Sails made from sailcloth can be made in very nearly the same way with darts, reinforcing patches, and the rest, but in the case of "proper" sails the boltrope goes on the outside of the hems, not the inside, and has to be stitched into place by hand. It's a nice, peaceful job for a long winter evening, and it will produce a very attractive and traditional sail.

The seams of a sail made from sailcloth must be stitched with a strong cast iron or industrial grade sewing machine set to zigzag stitching. I use double-sided adhesive tape

A grommet (cringle) set includes a hole punch, the two-part grommets themselves, and a tool for clenching them together. This set has been in my toolbox for some time.

(packing tape) to stick down the hems and seams before stitching with the machine. There's quite an art to the task of rolling up a sail in such a way that you can get the seam to a sewing machine's needle, and I'd strongly suggest that if you do make your own sails from real cloth it would be well worth enlisting some help, particularly from someone who knows a little about setting and using a sewing machine.

The grommets can be added using the same grommet kit I mentioned earlier, and, again, where grommets are near the edge of the sail they should be firmly whipped to the boltrope.

Rigging

Rigging presents a range of interesting challenges, including buying and installing small fittings, sealing or serving the ends of lines so they don't unravel, and tying knots.

On the fitting front, because these small boats with their old-fashioned rigs involve only relatively small forces, they don't require more than a few inexpensive pieces of hardware. All complex technology is eliminated. For example, a simple deadeye or a small hole drilled through the mast will support the mainsail, so long as the line is looped around the mast a couple of times. Winding the line around the mast ensures that the forces exerted by the sail are spread around the mast and not concentrated on half of it—distributing the stresses in this way makes the whole rig much more robust.

The looping trick is also easy to do, as you'll see in the three accompanying photos. The mast is about 2 inches wide at the top, so I drilled a hole 2 to 2½ inches from the top of the mast just wide enough to accept a doubled piece of polyester or Spectra line, say 18 inches long. I tied a bowline in the end, folded the line in half, and then pushed the folded end through the hole, turning it over and looping it over the mast two or three times such that the resulting tail is over (not under) the looped-around part. Finally, I used the loop created by the bowline to tie on the sail. With the small sails you'll find in this book,

The lashing at the top of the mast goes around the mast a couple of times to prevent stresses that could cause the mast to crack. One end has a bowline tied in it.

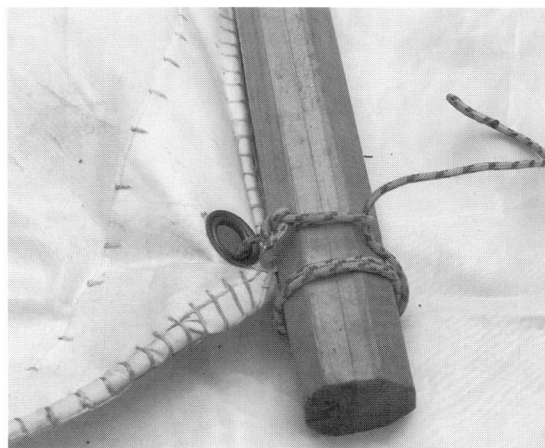

Tying the temporary bight used to attach the head of the sail to the mast.

The same temporary bight used at the tack (bottom of the sail nearest the mast) to tighten the luff of the sail.

the same approach works well with sprits and yards.

I also try to use the same kind of stress-reducing method whenever I'm working with deadeyes. Because the forces on the mast end of a sprit pull forward, I'll put a deadeye on the aft side of the mast, loop a line a couple of times through the deadeye and around the mast, and tie it to a stout ring or the becket of a small block. The snotter line, which tightens the forward end of the sprit in a spritsail so that it holds the sail out to make a good shape, falls through the block and is then tied off to a cleat on the sprit itself.

This way, all the forces on the deadeye tend to pull the fittings directly into the mast and not out of it—at most there might be some sideways forces, but that's all. The method seems to work, as I can't remember a fitting dragging out of any of my small-boat spars.

With the low stresses involved in such small boats, this kind of solution can be sufficient, so long as the ring is sufficiently substantial. In place of a small block in these situations, I've even used a beefy metal or plastic ring or deadeye bought from a chandlery, tied at one side and with the running line fed through it and lashed to the sail or spar.

On the subject of blocks, if you're using a sail with a gaff or yard at the top, you'll want to be able to lower your sail using a halyard. Resist the temptation to buy an expensive block. A cheap galvanized block, the kind people use when rigging a clothesline, will work fine for years.

Where I do use moderately good quality blocks is on mainsheets. Sheets need to be thick and soft so that they're comfortable to hold for long periods, and because a little thickness in a line tends to discourage it from tangling. These thick lines can be used with deadeyes instead of blocks, but in my experience they don't really run smoothly through anything less than a block. Cheap galvanized blocks sold for use in gardens and on farms are useful for halyards, snotters, and other applications on board a small boat, but they're heavy and hard and can give you a good clout if they hit you, which could happen if you use one on your mainsheet. In contrast, a lightweight plastic block is a gentle thing that will usually leave you with only a little light bruising if you're careless enough to get hit by your boom.

As always on a boat, when driving screws to fit blocks, always drill a hole of the correct size so that you don't start any splits in which your screws might eventually work loose.

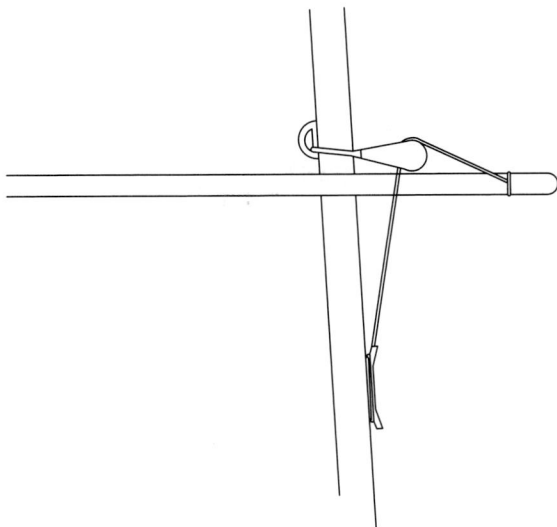

With a little care, it's possible to make all the stresses in a sailing rig bear on the whole spar (avoiding stresses that might crack a spar or pull out a fitting) or pull sideways. Here a block is suspended from a deadeye behind the mast, which means that it is attached by a line that extends around the mast. A line looped around the sprit and lashed to a hole in the sprit falls through the block and finally to a cleat, where the stresses are sideways.

Finally, you may be wondering about reefing. How is it possible that the sail in the pictures is tied to the mast and not able to be lowered? Can this be safe?

The explanation is that the sail is a spritsail that uses a sprit to hold up its peak—that is, the highest part of the sail. When the time comes to reef, what happens is that the sprit is taken in, which effectively reduces the sail to a small triangle generally about half the size of the original sail. This is called scandalizing, and in small boats it can generally be done very quickly—more quickly than any other kind of reefing. I think that this ability to put in a deep reef very quickly is one of the advantages of the spritsail in a small boat, and it's one of the reasons I'm a fan of this type of sail.

When the hull and rig are ready, it's time to launch the little boat!

You'll no doubt spend the first three hours or so adjusting this or even that string to get a good shape out of the sail, or even moving this or that fairlead or chainplate. That's only to be expected and is part of the fun of these projects. Once the tweaking has been completed to your satisfaction, there is one last very important step: checking the sailing characteristics of the boat. You do this by

Preventing Rot

Once all the effort of construction is over and you've had your first exhilarating sail in your newly built boat, it's time to look at protecting it from rot. Given that wood materials are not what they once were (see Chapter 2), there are some precautions you can take that will make a huge difference to your boat's useful life. One is to make sure your boat is protected from the elements. Apart from running it full speed onto rocks, perhaps the fastest way to destroy a home-built wooden boat is to allow it to fill with rainwater then let it stand for a few months.

At a minimum, store your boat upside down on blocks so that it won't draw moisture from the ground, preferably with a tarp covering the hull to keep sunlight off the paintwork, and especially any varnish.

Storing the boat indoors is even better and may be easier than you may think, particularly with small boats. For example, the Mirror dinghy I built with my dad all those years ago lived slung upside down from the garage roof; a system of string and blocks allowed it to be lowered or put away in a moment. Similarly, people who build the shorter Mouseboats of 8 feet or less often keep them propped up against a wall in a garage, shed, or utility room.

Another important precaution is to fit not just one but two drain plugs in every buoyancy tank and to remember to open them whenever the boat is put away—fitting two instead of one seems to increase the flow of fresh air, greatly reducing the likelihood of fungal growth and rot. I've also used rotating plastic hatch lids sold in chandleries once or twice, and my impression is that they're even better than the plugs. However, you have to be careful to ensure they're properly sealed, as a buoyancy tank that fills when the boat is swamped is a lot less useful than it should be.

If you take these precautions, even a quickly built homemade boat made from cheap materials can last many years. (But if you want to go a step further, there is a wealth of experience to show that encapsulating vulnerable lumber with glass and epoxy is a very effective way of keeping rot to a minimum.)

One final point is that many designers show limber holes in these plans. These are small gaps between frames and the hull designed to allow water to flow through rather than stand at a point where it might start rot. Limber holes are helpful in a boat that's going to be kept on the water in a position where spray or rain will collect in it. If, however, the boat you're building is going to be stored as I've described, I don't think limber holes will be particularly useful.

observing the position of the tiller while the boat is sailing to windward to see if you have a "lee" or a "weather" helm. Lee helm is bad, and a slight weather helm is good.

While you're under sail and sitting comfortably in the middle of the boat (*not* at the stern), see if the tiller has to be held to leeward (away from the wind) to sail in a straight line to windward. If you have to hold the tiller to leeward in anything but the lightest breeze, you have a "lee" helm that has to be corrected. Boats with lee helm turn away from the wind and may sail off if you fall overboard, leaving you stranded in the water. This can be potentially lethal, especially if you are in cold water without a life preserver. Depending on the design of the boat, you'll need to either move the mast aft or find a way to lean it back a little to counteract lee helm, and in some cases this may mean you have to fiddle with the placement of any standing rigging as well.

Ideally, you want a boat that requires you to pull the tiller slightly toward the wind when sailing to windward; this kind of balance in a boat is called "weather helm." A boat that is set up with a slight weather helm will be optimized for sailing upwind. There is a safety issue here as well, for a sailboat that needs the tiller pulled a little to windward when sailing upwind will always turn into the wind and sit still if you let go of the lines and tiller, which

is useful in difficult circumstances or if you happen to fall overboard and need to climb back in the boat.

So, as a final double-check, go sailing and see what the boat does when you let everything go. If the boat turns slowly upwind and sits still, the mast is in the right spot, but if it turns slowly off the wind and tries to sail away, you will need to move the mast aft a few inches. Hopefully, you'll find that your boat has a slight weather helm and you can get on with the fun!

Drain plugs are essential in small boats with buoyancy tanks. Remember to tighten them before launching, and, almost as important, undo them and drain any water out of your tanks when you've finished. Each tank should have two drains to promote air circulation, and it's a moot point whether the drains should be at the bottom of the tanks to drain into the bilges, or near the deck in a boat that is normally stored upside down.

CHAPTER 7

MAKING MODELS

Making a model is not absolutely essential in building a small boat, but most people find it's a very useful step in the process and I recommend it. It helps the builder understand how the boat works, and in particular establishes confidence in the new boatbuilder that the combination of curved panels will together create a boat of the right shape and really will make a rigid and strong structure. I've discovered that many people find it difficult to believe that floppy plywood can so easily be converted to a stiff small boat that feels safe while in use.

Making a model also helps in deciding which boat to build. If you're choosing among three or four designs, it's particularly satisfying to be able to make models of each one, and to be able to think about how its three-dimensional form will work as it sits on and moves through water, and, of course, how it will look from every angle. What's more, making the model will reveal any possible problems or mistakes in the original plan, and these can sometimes be found even in the work of the best-known designers. Phil Bolger once wrote that if he didn't make mistakes from time to time the gods would be jealous, and that might go for all of us who draw plans for small boats.

Building a model can also help when the boat bug has bitten hard and a budding builder desperately wants to make a pretty and specialized design, such as a dory, that is unsuited to the kind of boating the builder has in mind. When this happens making a model can sometimes cool the ardor.

You can make a model of one of the designs in this book using various materials. The high-tech, high-cost way begins with a trip to your local model shop to buy plastic glue and plastic modeling sheet, or even the lightweight plywood used for making model airplanes. That's a great way to go, but there's no doubt that the easiest and quickest route to success is to use materials you'll find around the house, such as the cardboard from a box of breakfast cereal and a roll of tape. I've often done this sort of thing, and you can even put the resulting model in water for a few minutes before it falls to pieces.

Let's use the accompanying simplified plans for the Mouse to build a model. Trace or copy the plans on page 79 using a photocopier, or even use your scanner to copy them. Print at 100 percent, or whatever size will fill your sheet of model-making material.

Once the printing job has been done, I often paste the printed paper directly onto the cardboard using either PVA glue (polyvinyl acetate), a not-very-sticky stationery adhesive (for example, Elmer's Glue-All in the United States, or Pritt glue sticks and roller adhesives in the United Kingdom), or flour glue, depending on what I have in the house. Sometimes I simply tape the marked sheet to the cardboard, which makes accurately cutting out the panels a little more difficult, but it does have the special advantage that there's no difficulty if I wish to remove the paper and glue from the panels afterward.

Once the printout is safely attached, I cut out the shapes using kitchen scissors, remove the remnants of the plans from the cardboard, and assemble the structure. If you don't remove the remnants of the plans, it's often good to assemble the parts such that they're all either on the inside or outside of the model.

Try to assemble the model as accurately as possible. On the scale of these little models a small inaccuracy can equate to an inch or even several inches on the full-scale boat,

Model Mouse

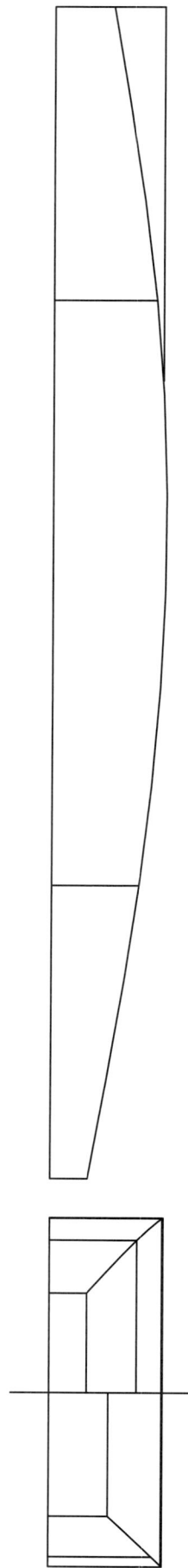

Simplified Mouse plans for model making. Lines drawing.

Foredeck

Aft frame

Forward frame

Bow

Stern

Aft deck

Side panel

Bottom

Side panel

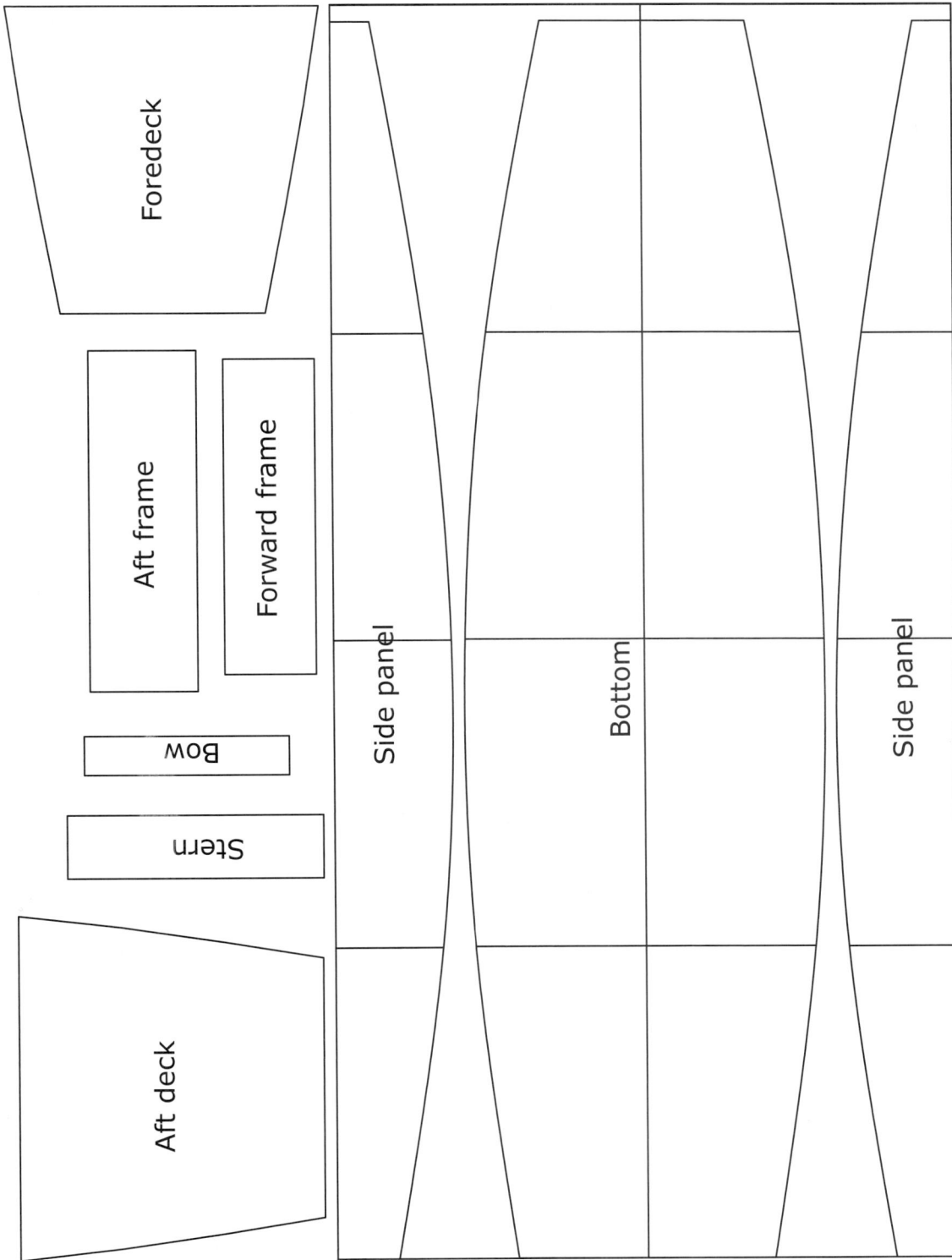

Simplified Mouse plans for model making. Panels drawing.

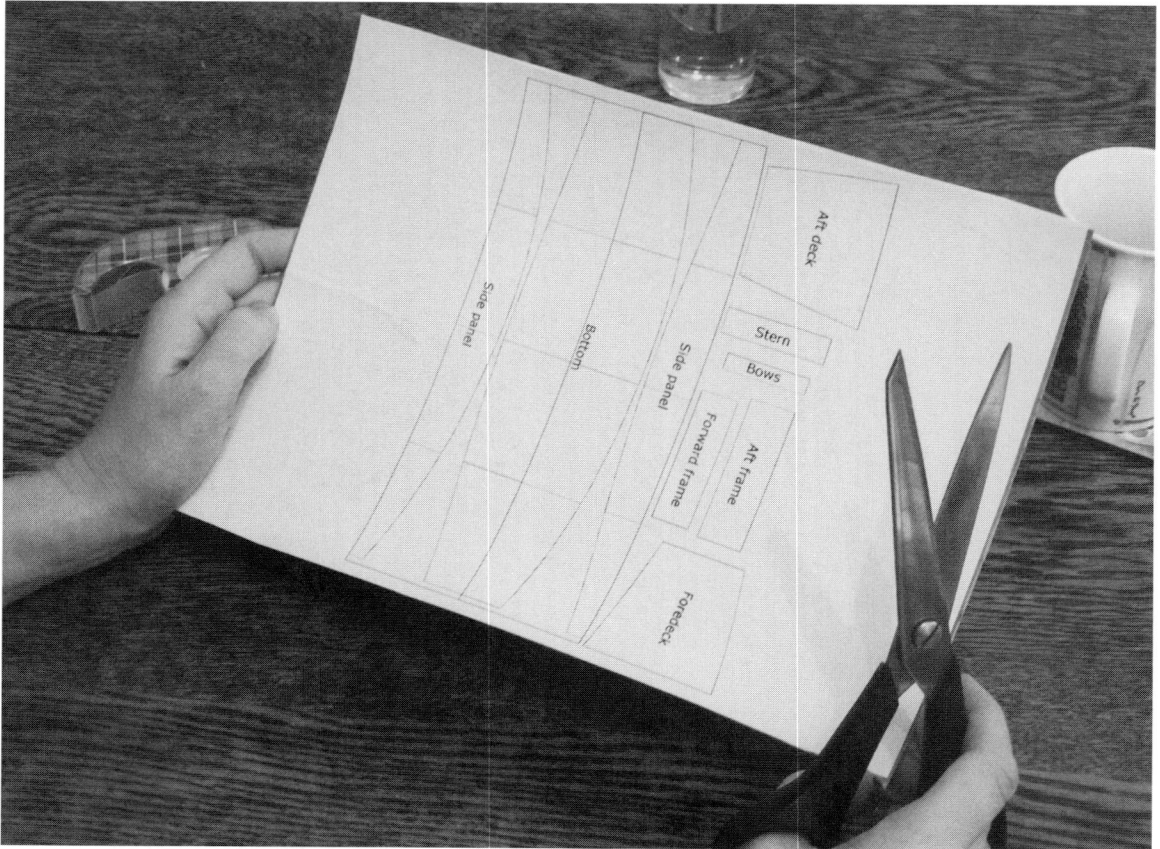

Photocopy any of the plans in this book at full size, or reduce or magnify as you wish.

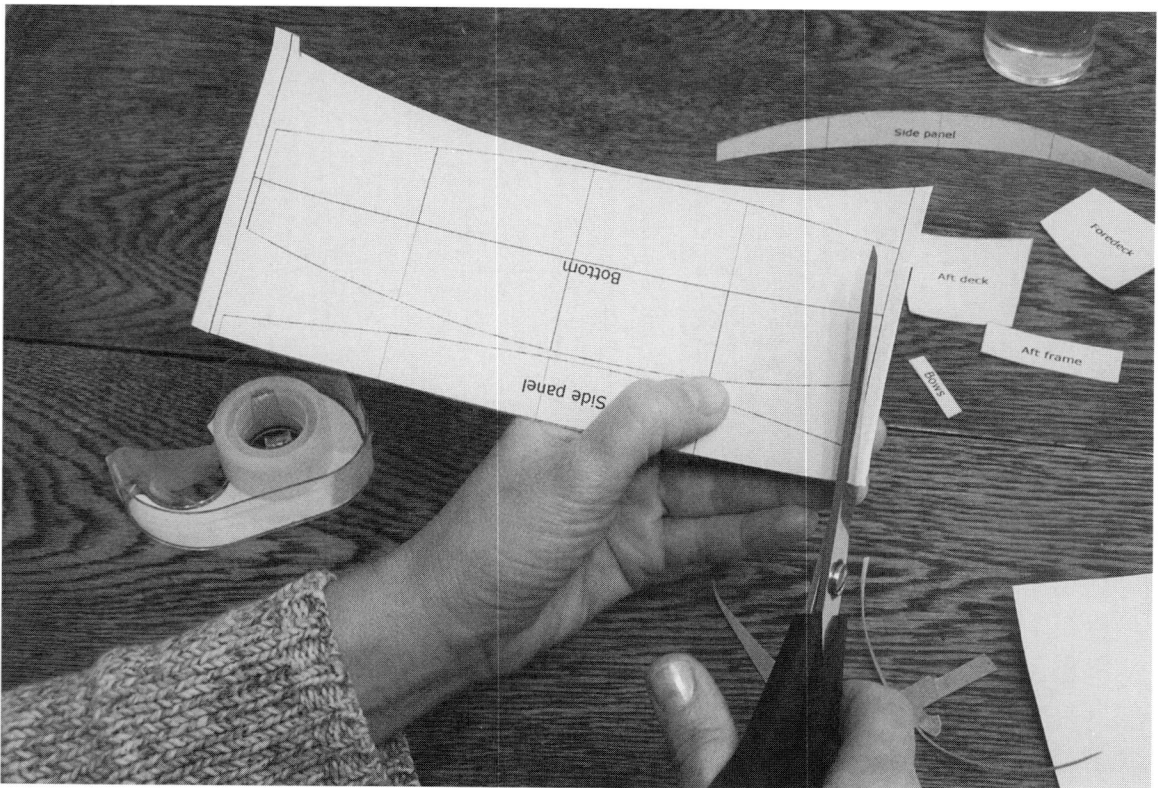

Cut out the components of your boat. If you picture yourself very small, you can pretend you're sawing plywood.

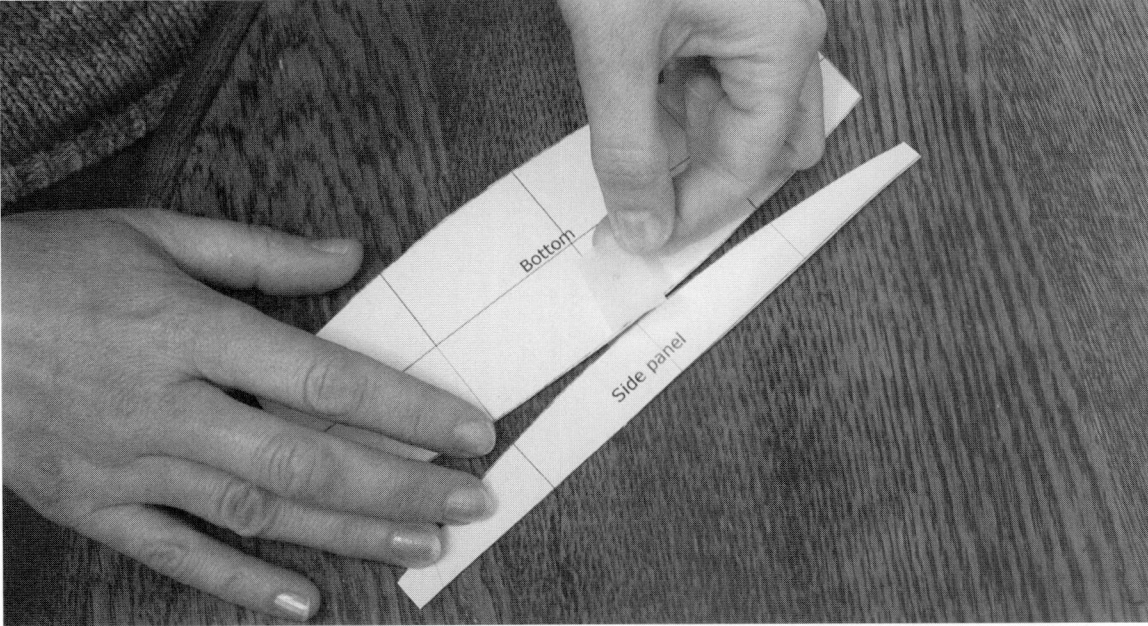

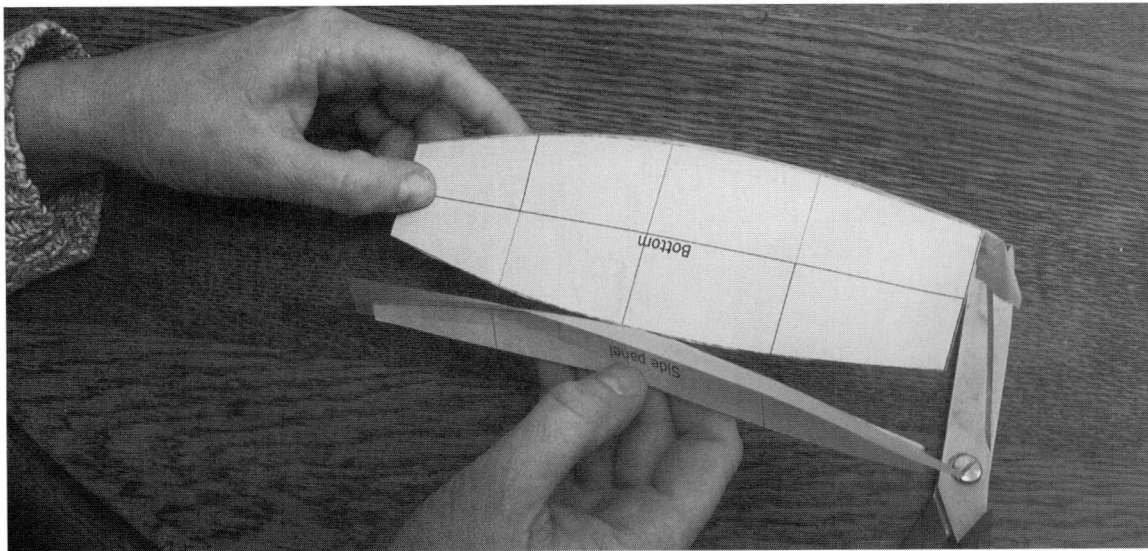

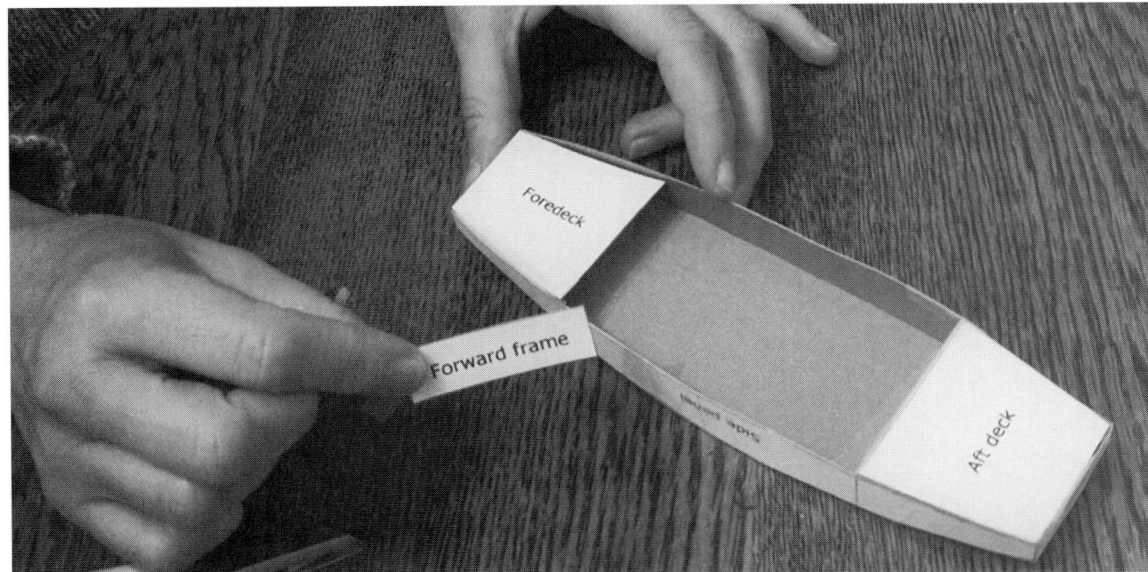

"Build" your boat with clear tape. The exercise will give you a very clear idea of how the curves go together to make a boat.

When you're done examining the finished boat for construction possibilities, give it to a child to play with. If you glue it together and paint it, or make it from waterproof materials, it can be used as a bath toy.

and may make a significant difference to the overall appearance, by, for example, flattening the bottom profile.

If you lose track of which piece is which, take another look at the original in the drawing you have been working from. Generally it's not too difficult to make sure that you are attaching the edges together correctly since the dimensions have to match. For example, the width of the bow transom matches the bow end of the bottom panel. With a V-bottom boat, it's often good to tape the two halves of the bottom together first.

Once it's assembled, spend some time checking that everything's square and admire your handiwork. If you would like to go a step further, you may run a bead of hard plastic glue along each of the interior seams like a miniature fillet, let the glue set, and then remove the tape for a really good-looking model.

(Anthony Smith)

The Boats

MINIMOUSE AND MICROMOUSE

TWO TINY, FLAT-BOTTOMED DOUBLE-PADDLE PRAMS

	Minimouse	Micromouse
Length Over All (LOA):	7'10"	6'6"
Length Waterline (LWL):	7'1/2"	5'10"
Beam:	27³/4"	29¹/2"
Weight:	35–40 lb.	30–35 lb.
Displacement at Design Waterline:	200 lb.	160 lb.
Crew:	1 adult	1 child
Propulsion:	Double-paddle	Double-paddle
Construction-Methods:	PU stitch-and-glue	PU stitch-and-glue
	Epoxy stitch-and-glue	Epoxy stitch-and-glue
	Simplified chine log	Simplified chine log

The Minimouse has its origins in David Colpitts' discovery that at least some of the kids he worked with had a few problems with the original V-bottomed Mouse (see the Introduction, page 103). He requested an even simpler flat-bottomed version, and I was glad to oblige. The resulting boat couldn't be much smaller or simpler while remaining suitable for most adults weighing up to 180 or 200 pounds. Moreover, Minimouse paddles pretty well and provides a stable platform that's just the thing for kids and adults enjoying their first boating experiences. It can be carried by an adult with one hand, yet has all the rigidity you could want. There's a large volume of built-in buoyancy to keep you afloat in the event of a capsize—so much, in fact, that even when it's full of water, it's not difficult to tip some water out, climb back in, and bail from inside.

It's probably the boat I should have designed in the first instance, but I don't regret designing the V-bottom Mouse for a moment because it remains the most popular member of the Mouse family. It is certainly the better boat because it is faster and more suited to coping with waves.

The Micromouse is similar but smaller, making it more suitable for children and smaller adults, and building it requires a proportionately smaller quantity of materials.

Constructed cheaply from inexpensive (but carefully chosen) ¹/₅-inch or ¹/₄-inch (5 or 6 mm) water- and boil-proof (WBP) or marine plywood, these boats can be used for fishing, exploring, or simply playing. But perhaps an even stronger justification is that they are about the simplest, useful small boats imaginable. They require the least investments in time, skills, tools, and materials, and they can be built so quickly that the disruption around the house can be kept to an absolute minimum.

They are both perfect for a first experiment in plywood boatbuilding, which is why I've placed them here, as the first designs in this book.

Once your Mouseboat has been built, you can find a thousand uses for it, and that's part of the fun. It's a boat you can carry under your arm, yet can still be used to paddle or row; it will comfortably allow you to lie in the sun or creep up a tiny stream, and will also let you stand up long enough to climb into another boat or up on to a dock.

Both Minimouse and Micromouse can be built three different ways:

■ Very cheaply and quickly using polyurethane stitch-and-glue

Minimouse
7'10"

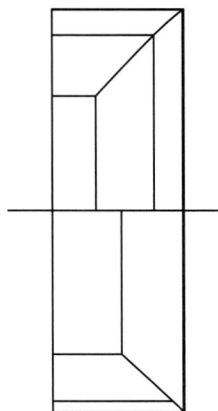

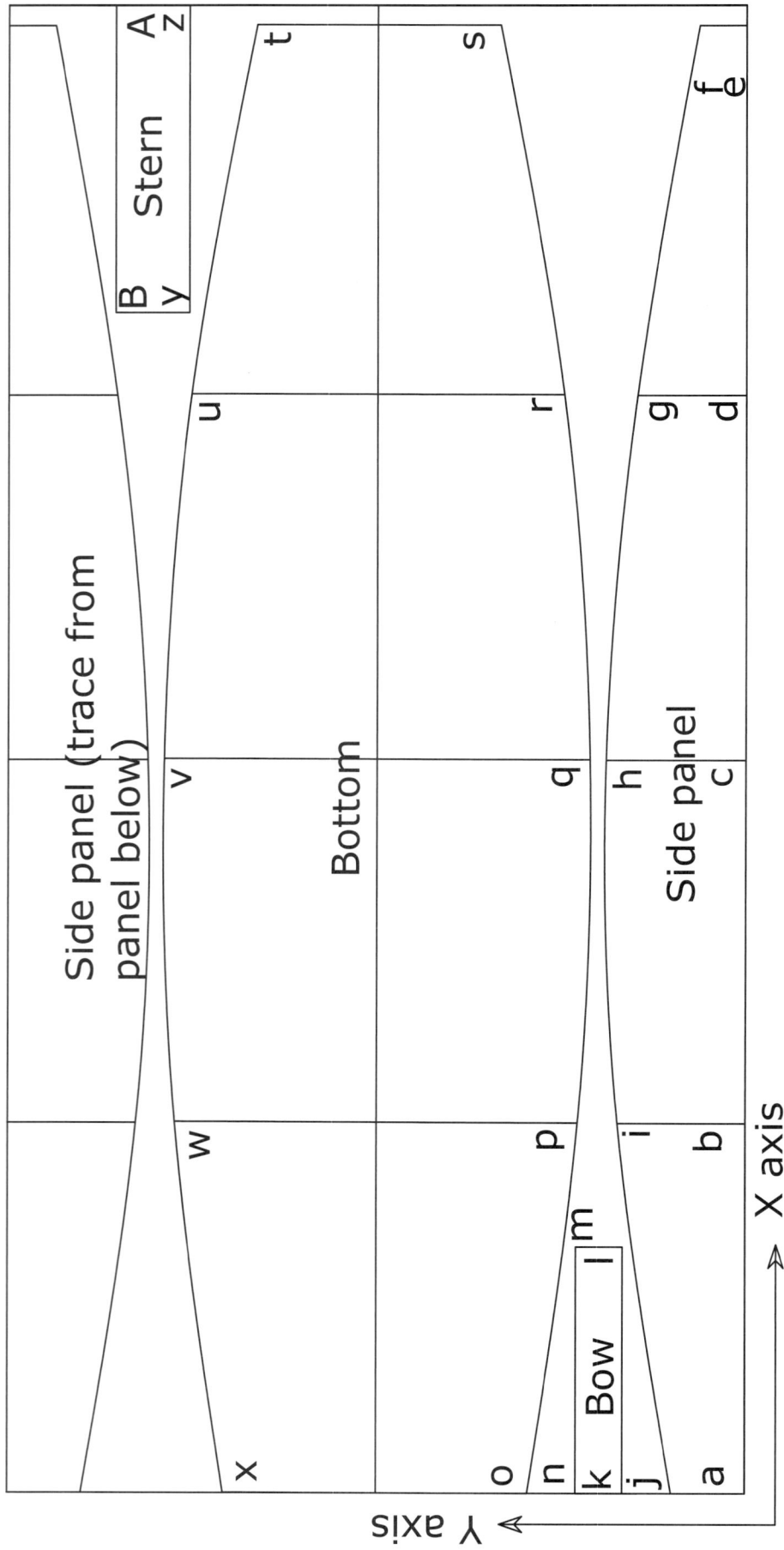

Stern

A
z

B
y

t

s

f
e

Side panel (trace from
panel below)

u

r

g
d

Bottom

v

q

h
c

Side panel

w

p

i
b

m
l

Bow

o
n

k
j

x

a

X axis

Y axis

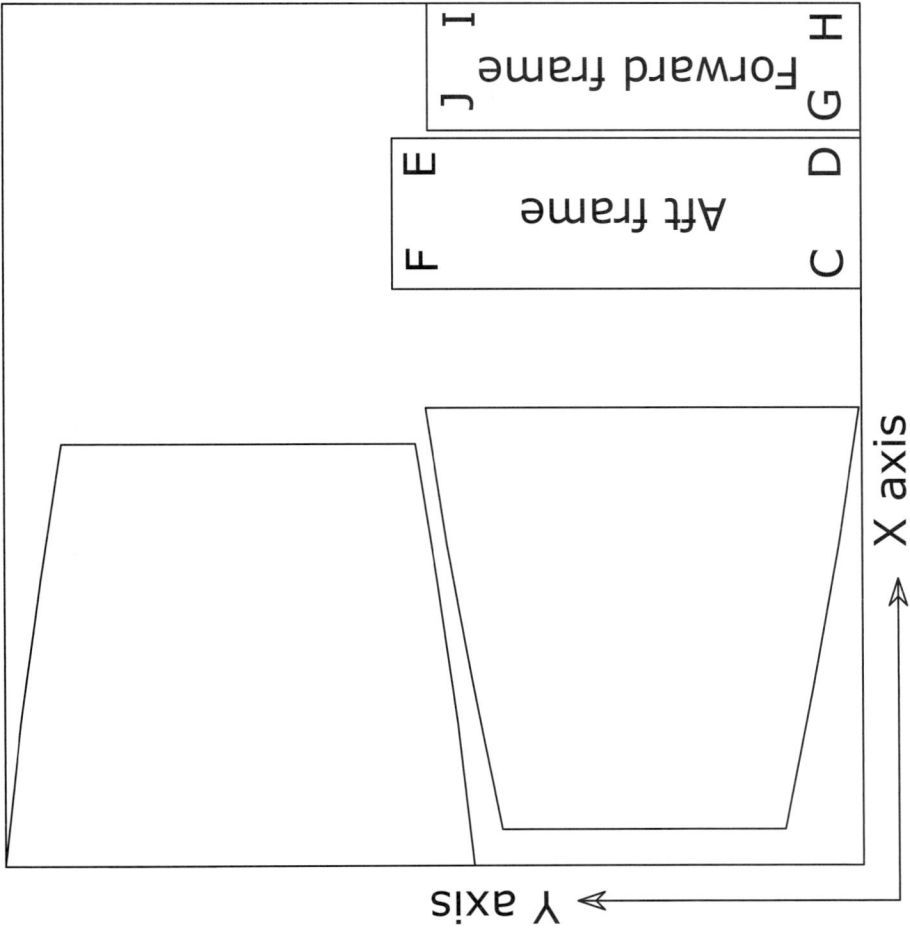

X axis

Y axis

Forward frame

I J

H G D C

Aft frame

E F

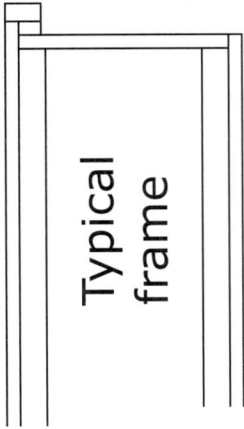

Typical frame

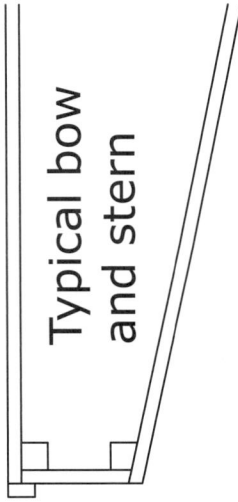

Typical bow and stern

- Very cheaply and somewhat less quickly by the simplified chine log method

- More expensively by epoxy stitch-and-glue

All of these options will produce a boat that, with occasional painting, could last for many years, but the second two are likely to last longer than the first. As described in Chapter 1, the polyurethane stitch-and-glue method is the one that will get you afloat in days or even hours. If you're just beginning to get the boatbuilding bug, this is likely to be the way for you. The simplified chine log method will appeal to those who enjoy a little light carpentry and are perhaps less keen on slathering glue, while the epoxy stitch-and-glue method is better for those who enjoy gluing more than carpentry and have decided to go for better materials in general.

Minimouse Coordinates (in inches)

	x	y
Side panel		
a	0	0
b	23³/₄	0
c	47¹/₄	0
d	70⁷/₈	0
e	94³/₄	0
f	94³/₄	3
g	70⁷/₈	7¹/₈
h	47¹/₄	9¹/₈
i	23³/₄	8³/₈
j	0	4³/₄
Bow transom		
k	0	8
l	15⁷/₈	8
m	15⁷/₈	11
n	0	11
Bottom		
(Only mark out if building by stitch-and-glue)		
o	0	14¹/₈
p	23³/₄	10⁷/₈
q	47¹/₄	10¹/₈
r	70⁷/₈	11⁷/₈
s	94³/₄	16
t	94³/₄	31⁷/₈
u	70⁷/₈	36¹/₈
v	47³/₈	37⁷/₈
w	23³/₄	37¹/₈
x	0	34
Stern transom		
y	76¹/₄	36¹/₄
z	96	36¹/₄
A	96	41
B	76¹/₄	41
Aft frame		
C	32¹/₈	0
D	40¹/₂	0
E	40¹/₂	26¹/₄
F	32¹/₈	26¹/₄
Forward frame		
G	40⁷/₈	0
H	48	0
I	48	24¹/₄
J	40⁷/₈	24¹/₄

Minimouse Coordinates (in millimeters)

	x	y
Side panel		
a	0	0
b	581	0
c	1158	0
d	1735	0
e	2320	0
f	2320	75
g	1735	174
h	1158	223
i	581	204
j	0	118
Bow transom		
k	0	196
l	388	196
m	388	270
n	0	270
Bottom		
(Only mark out if building by stitch-and-glue)		
o	0	347
p	582	268
q	1159	249
r	1737	290
s	2321	392
t	2322	780
u	1737	884
v	1159	928
w	582	910
x	0	833
Stern transom		
y	1867	888
z	2352	888
A	2352	1006
B	1867	1006
Aft frame		
C	817	0
D	1029	0
E	1029	666
F	817	666
Forward frame		
G	1039	0
H	1219	0
I	1219	616
J	1039	616

Note: Small discrepancies may exist between the millimeter and inch tables. See page ii.

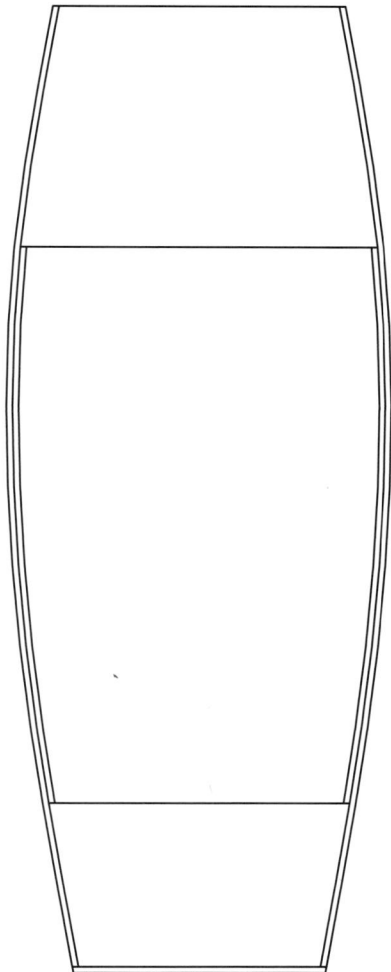

Micromouse
6'6"

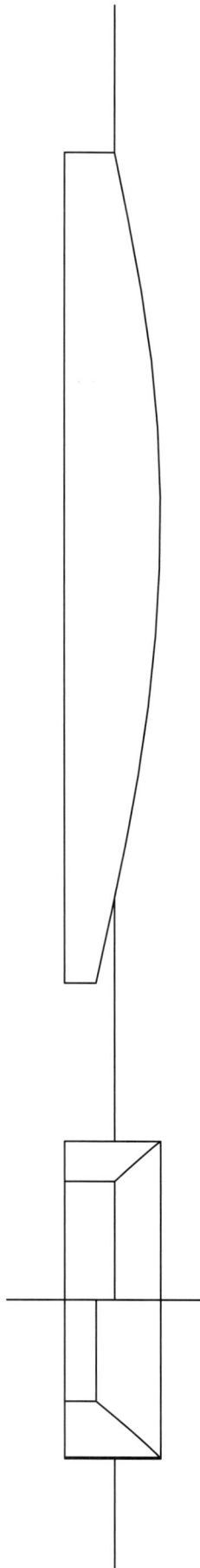

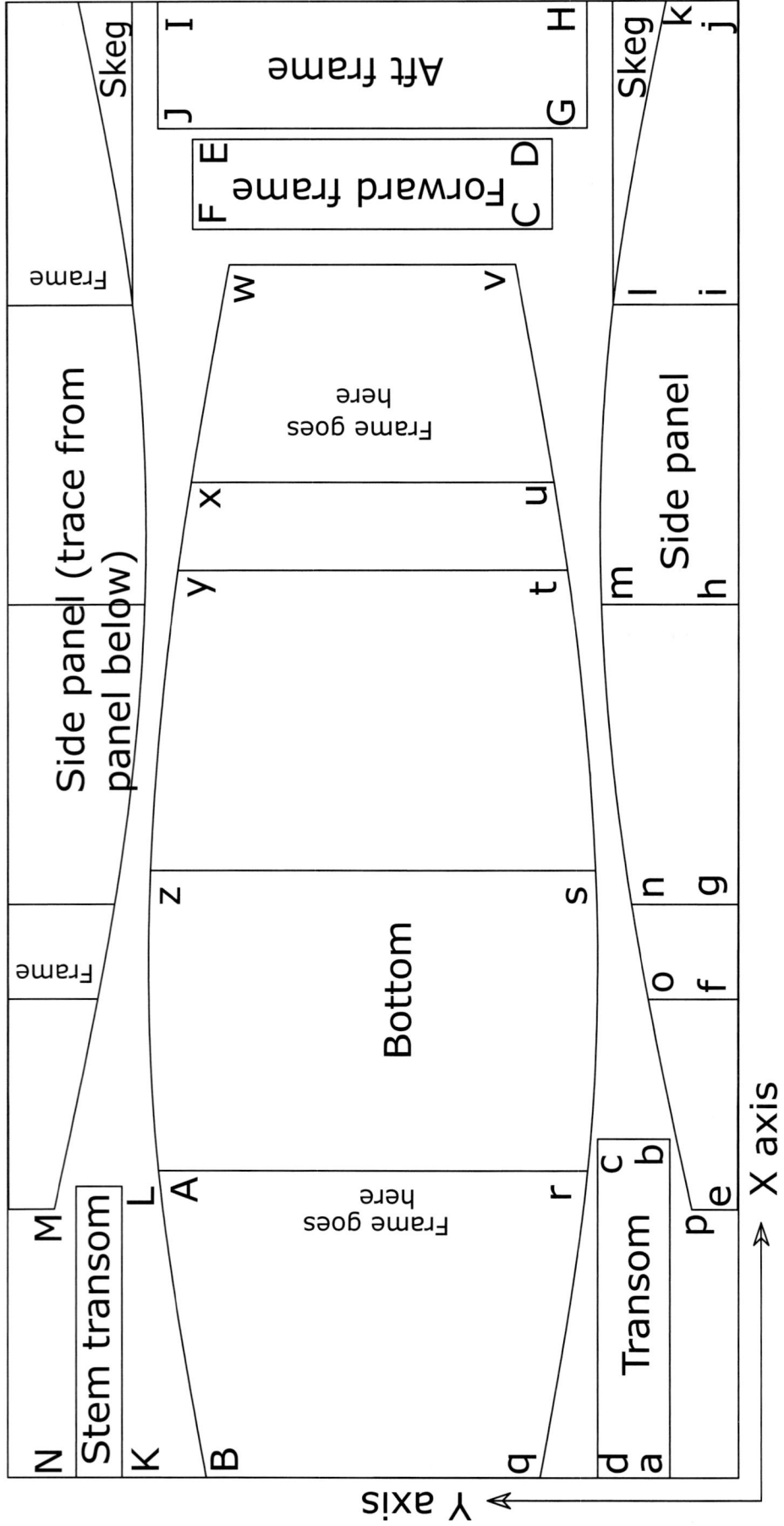

Skeg

I

Aft frame

J

H

G

F E

Forward frame

C D

Skeg

k

j

Frame

w

v

l

i

Side panel (trace from panel below)

Frame goes here

x

u

Side panel

y

t

m

h

z

s

n

g

Bottom

o

f

Frame

M

L A

Stem transom

N

K

B

Frame goes here

r

c

b

Transom

q

d

a

p

e

X axis

Y axis

Approximate sizes and shapes of fore and aft decks; these should be traced onto the material using the hull as a template.

X axis

Y axis

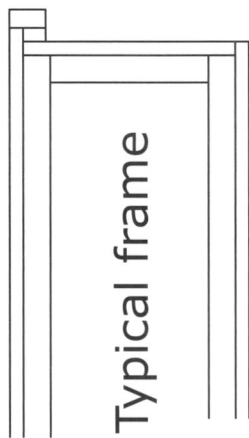

Typical frame

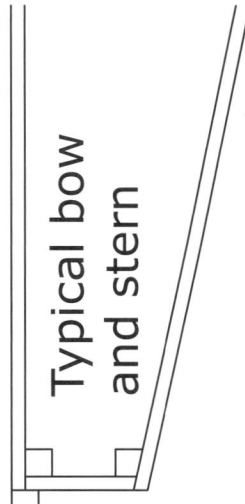

Typical bow and stern

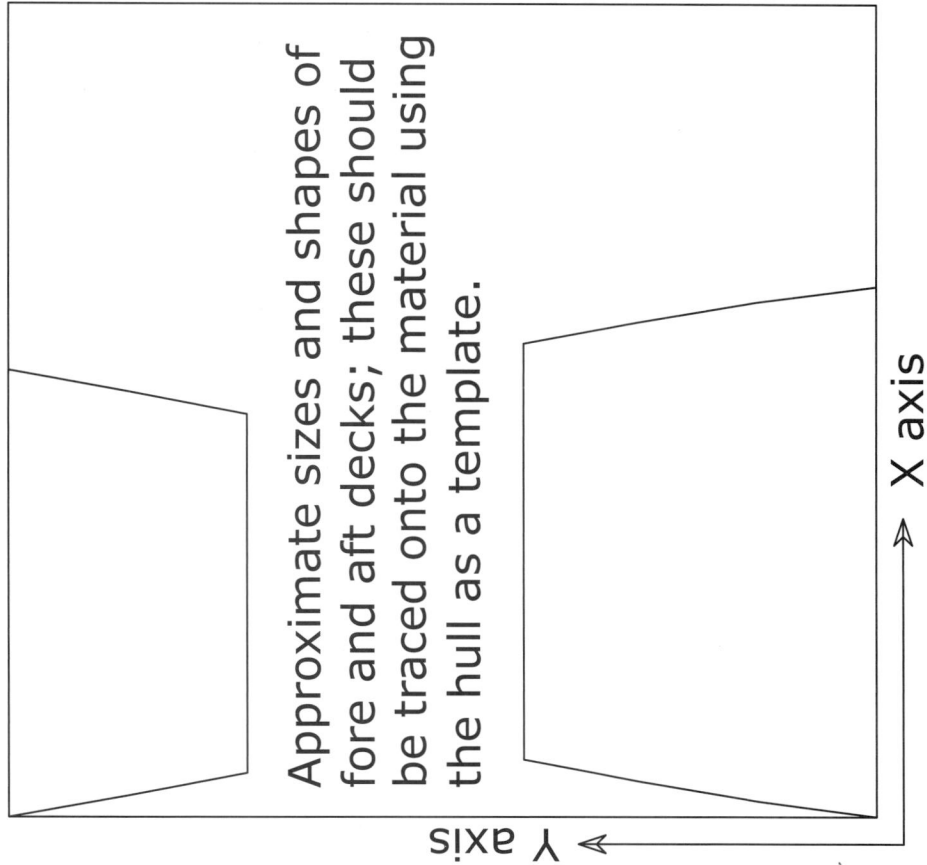

HULL

The details on all three building methods are found in Chapters 3 and 4. I won't repeat them here, but I will point out some aspects of how they apply to these two particular boat designs and emphasize important points since this is likely your first project.

If you go for either of the stitch-and-glue methods, you'll find that stitches aren't in fact necessary for Minimouse and Micromouse. The curves are so gentle that the ¼-inch plywood panels bend together very easily, and they can be held together temporarily with duct tape on the outside while you stick them together permanently with epoxy or polyurethane on the inside.

If you choose the simplified chine log method, first mark out and cut the sides, transoms, and bulkheads (labeled "frames" on the plans), but not the bottom. Make up the structure of the boat complete with the external chines. If you haven't already done so, glue wooden cleats along the lower edges of the insides of the bow and stern transoms, and along the lower edges of each of the bulkheads. These should be on the forward side of the forward bulkhead, and on the aft side of the aft bulkhead. To put it another way, they should both end up on the insides of the two buoyancy tanks, as this will make sure they stand a little "proud" and can be beveled to mate with the bottom and create a good gluing surface.

With the hull upside down, and without putting any glue on the work, lay the plywood bottom on the hull and test the angles of the gluing surfaces by rolling the plywood from fore to aft (or vice versa) and watching how it lies against the bottoms of the transoms and bulkheads. Trim the bottoms of the cleats with a Stanley Surform or rasp to match the angle of the sides. This should be done carefully, as polyurethane glue won't retain its strength if the gap is more than about ⅛ inch. The result of your labors should be a flat gluing surface all around the sides of the boat and across the transverse members.

The transoms are subtly different. As I explained earlier, when making them up the cleats should be placed so they stand "proud" of the bottoms and sides of the transoms since you'll need some stock for beveling to make the transom cleats mate with the boat sides and bottom.

Once the external chines are in place, mark out the bottom using the actual boat as a template. It helps to have a family member or a friend hold the plywood in place, but heavy weights will also work. First, though, check that the hull is square by measuring the distance between each set of opposite corners, with the boat upside down. If it isn't quite true, joggle the corners a little so that the diagonals are equal.

With the plywood held firmly in place (having some helpers here is a big plus), draw around the external chines on the underside of the plywood. Once marked, it should be cut out gently with a sharp saw to avoid creating jagged splinters around the edge. Also, remember the woodworker's mantra: measure twice and cut once, and always saw on the waste-wood side of the line. In this case, I'd cut as much as a quarter of an inch outside the lines to allow a margin for error!

Gluing the bottom of the boat is a job that's much easier if you have access to a good collection of clamps, but clamps are expensive, even at the prices charged by shops that sell tools imported from India or Korea. Luckily there are some alternatives. One is the PVC pipe clamps described on page 19 in Chapter 2. Another is to screw the plywood into place, using small stainless steel screws and working progressively from one end of the boat to the other, forcing the bottom to the curve in small increments. This is highly effective, but it's time-consuming. As always in boatbuilding, each hole should be carefully drilled to make a hole about the size of the screw's central spindle and about ¼ inch shorter than the screw.

Finally, if you get into trouble with putting on the bottom and find you have gaps that are larger than they should be, you can usually safely fill any gaps with filled epoxy.

Micromouse Coordinates (in inches)

	x	y
Stern transom		
a	0	4$\frac{1}{2}$
b	21$\frac{7}{8}$	4$\frac{1}{2}$
c	21$\frac{7}{8}$	9$\frac{1}{4}$
d	0	9$\frac{1}{4}$
Side panel		
e	17$\frac{3}{8}$	0
f	30$\frac{7}{8}$	0
g	37$\frac{1}{8}$	0
h	56$\frac{3}{4}$	0
i	76$\frac{1}{4}$	0
j	96	0
k	96	4$\frac{3}{4}$
l	76$\frac{1}{4}$	8$\frac{1}{4}$
m	56$\frac{3}{4}$	9
n	37$\frac{1}{8}$	7
o	30$\frac{7}{8}$	5$\frac{7}{8}$
p	17$\frac{3}{8}$	3
Bottom		
(only mark out if building stitch-and-glue)		
q	0	13
r	19$\frac{7}{8}$	9$\frac{7}{8}$
s	39$\frac{3}{8}$	9$\frac{3}{8}$
t	59	11$\frac{1}{4}$
u	64$\frac{3}{4}$	12$\frac{1}{8}$
v	78$\frac{7}{8}$	14$\frac{5}{8}$
w	78$\frac{7}{8}$	33$\frac{3}{8}$
x	64$\frac{3}{4}$	35$\frac{7}{8}$
y	59	36$\frac{3}{4}$
z	39$\frac{3}{8}$	38$\frac{5}{8}$
A	19$\frac{7}{8}$	38$\frac{1}{8}$
B	0	35
Forward frame		
C	81$\frac{1}{8}$	12$\frac{1}{4}$
D	87	12$\frac{1}{4}$
E	87	35$\frac{3}{4}$
F	81$\frac{1}{8}$	35$\frac{3}{4}$
Aft frame		
G	87$\frac{3}{4}$	9$\frac{7}{8}$
H	96	9$\frac{7}{8}$
I	96	38$\frac{1}{8}$
J	87$\frac{3}{4}$	38$\frac{1}{8}$
Stem transom		
K	0	40$\frac{1}{2}$
L	18$\frac{3}{4}$	40$\frac{1}{2}$
M	18$\frac{3}{4}$	43$\frac{1}{2}$
N	0	43$\frac{1}{2}$

Micromouse Coordinates (in millimeters)

	x	y
Stern transom		
a	0	110
b	537	110
c	537	227
d	0	227
Side panel		
e	425	0
f	758	0
g	910	0
h	1390	0
i	1868	0
j	2352	0
k	2352	115
l	1868	201
m	1390	220
n	910	172
o	758	145
p	425	75
Bottom		
(only mark out if building stitch-and-glue)		
q	0	319
r	486	243
s	964	230
t	1445	275
u	1586	296
v	1932	358
w	1932	819
x	1586	880
y	1445	902
z	964	946
A	486	933
B	0	857
Forward frame		
C	1989	299
D	2133	299
E	2133	877
F	1989	877
Aft frame		
G	2150	243
H	2352	243
I	2352	933
J	2150	933
Stem transom		
K	0	992
L	461	992
M	461	1066
N	0	1066

Note: Small discrepancies may exist between the millimeter and inch tables. See page ii.

DECKS

Without decks and seats to prevent twisting, a hull can often seem very floppy; however, once these are in place, the small plywood boats included in this book will become rigid. The Minimouse and Micromouse are no exception. As always with Mouseboat building, the principle is to make everything as easy as possible. In the case of the decks, install them in almost exactly the same way that you installed the bottom. You'll find that the decks are even easier to deal with.

Add strips of wood around the tops of the sides on the outside of the boat to create a good flat gluing surface to which you can attach the deck. I use knot-free $\frac{3}{4}$-inch- or 1-by-$\frac{1}{2}$-inch pine material and it has always proved adequate for the purpose. I call this lumber the "inner gunwale" because another layer of material will usually be applied over it later to create a visible outer gunwale. Note: don't confuse this inner gunwale, which is installed on the outside of the hull, with an inwale, which is installed on the inside of the hull.

Let's deal with the aft deck first. Take the factory-cut edge of the plywood and hold it against the forward edge of the aft bulkhead so that the material covers the aft buoyancy tank and all of the gluing surfaces on the inner gunwales and the tops of the cleats on the bulkhead and transom. With the deck material held firmly in place, you should have no difficulty using a pencil to draw around the outside of the inner gunwales and transom onto the underside of the deck material.

The forward deck should be tackled in exactly the same way, the only differences being that it is slightly smaller and needs a reinforcing patch of scrap plywood on the underside where the painter fitting will be attached. A piece of plywood of perhaps 4 inches by 6 inches should do it. Glue it along the deck's centerline far enough back from the forward edge so that it won't interfere with the bow transom. Make a note of where it is so you can be sure of placing the fitting and drilling its screw holes in the right place. A galvanized eye intended for farm or garden use will do fine here.

Although clamps or screws are needed to fix the bottom while the glue sets, in the case of these straight and level decks it's possible to use clothespins (clothes pegs) because the stresses are so slight. There is, however, a knack to making them work. Don't clamp them nose-on to the side of the boat: they tend to pop off that way. Instead, put them on at an angle that's almost sideways-on (but try to avoid getting them snagged up with glue so that they have to be chipped off later). Also, be prepared to use *lots* of them. You may find as many as 30 to 40 are needed. It would be wise to avoid using the clothespins you normally use for the laundry, but, luckily, clothespins are cheap. Buy a packet for your boatbuilding projects.

Alternatively, I'd suggest using a few small stainless steel screws to anchor the deck to the bulkheads and transoms. Drive them in at 6-inch or 8-inch intervals.

Yet another approach is to use heavy objects from around the house to hold the deck down until the glue sets. Bags of gravel and sand are good for an even distribution of weight, and the plastic bags the stuff comes in don't stick to the glue.

Once the glue holding the deck to the gunwales, sides, bulkheads, and transoms has set, it's a good time to attach a decorative second (outer) gunwale to finish the job. This is my small concession to the urge to gold-plate Ultrasimple boats, but it's just about the only one I'd recommend to everybody as they look good and they do help to seal the decks from any rot-inducing dampness that might creep in from the sides.

The outer gunwales should be the height of the inner gunwales plus the thickness of the plywood, so that the plywood is flush with the top edge of the outer gunwale. They can be glued and fastened with small stainless steel screws. Even easier, you could drive small galvanized steel or bronze finishing nails through the outer gunwale into the inner gunwale, as the structure is now rigid enough to allow a little light hammering.

You may wish to tidy the gunwales a little. Many people, I know, find the "stepped effect," where the outer gunwale is a small fraction of an inch higher around the cockpit than the inner gunwale, less than attractive. One solution is to add a fillet of epoxy along the length of the "step" on the inside of the outer gunwale. Another approach that I like better is to fit a light inwale running from the forward to aft bulkheads along the inner side of the hull (on the inside of the cockpit). This will leave a small "trench" between the various wales, which can be easily filled using epoxy or a proprietary filler.

SKEG

A big skeg is worthwhile to keep such short boats tracking more or less in a straight line. For maximum effectiveness and strength, its bottom should run horizontally from near the deepest part of the boat to the stern. The "waste" plywood left over from cutting out the main hull pieces will yield some good skeg stock that nearly fits the bottom with hardly any shaping. With just a little work with a saw, Stanley Surform, or rasp, you will have it fitting neatly along the centerline of the aft part of your hull.

For more on fitting skegs and all other hull construction procedures, see Chapter 4.

SEATS

By now your little boat will be looking a lot like those you see in the pictures and on the Web. You should be pleased with it, with yourself, and with your helpers. Probably for the first time in your life you've built a real boat, and I'm pretty confident in saying that for many of you it won't be your last.

People generally use one of two kinds of seats in their Mouseboats: the simple flat seat or the cut-down resin lawn chair (plastic yard seat).

The yard seat is probably the easiest and may even be the cheapest. All you have to do is buy the seat for a few dollars or pounds, cut the legs until they fit the bottom profile of the boat, and fasten and glue some plywood pads under the legs to reinforce the bottom. The seat bottom should be 2 or 3 inches below the sheerline. The height is important: make the seat too low and a double-paddle will clash with the Mouseboat's sides; make it too high and it will adversely affect the stability of the boat.

The seat should be located just aft of the middle of the cockpit. A crew of the maximum

design weight should submerge the boat so that the bottom edge of the aft transom just kisses the surface of the water when the boat is lying level on the water both fore-and-aft and side-to-side. Take care to avoid placing the seat too far aft; you'll induce transom drag, especially if you're an adult of average size. Your Mouseboat will go faster and paddle more easily if you aren't dragging the transom. Get someone to take some photographs while you're sitting in the boat to confirm it's in proper trim. From the photos in this book, you'll see that some boatbuilders are getting this right and some are not. I'll leave you to judge which ones are a bit off in terms of trim!

The built-in seat is only slightly more difficult. You'll need a piece of plywood at least ¼ inch thick, long enough to reach across the boat, and about 18 inches wide, and some more 1-by-1-inch lumber for framing. In making the seat for my daughter's original Mouseboat, I laid the plywood on top of the cockpit of the boat so that it lined up with the aft bulkhead, then reached underneath with a pencil and traced the inside shapes of the sides onto the plywood. This works because the Mouse has vertical sides, and nothing could be quicker or simpler.

I then screwed and glued two ½-inch framing lumber straps to the inside of the sides, and another on the front face of the aft bulkhead, so that when the seat was in place its top was 4 inches below the sheerline—that is, the top of the framing lumber strap was 4¼ inches down from the upper edges. Four strong stainless steel screws on each side and

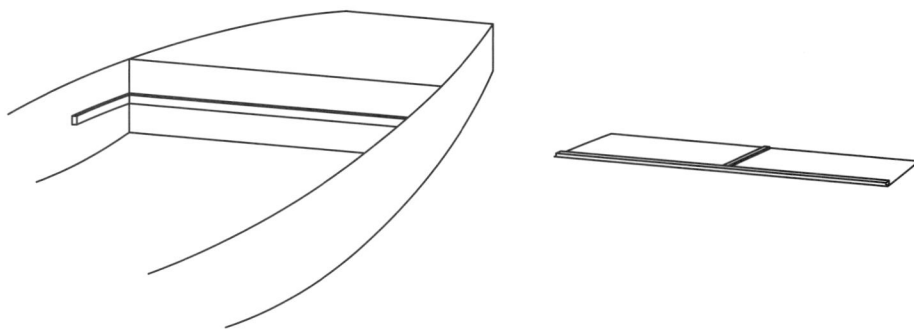

When using thin plywood for a built-in seat in a Micromouse, Minimouse, or any similar boat in this book, supports and bracing can be done as shown. If thicker lumber is used for the seat, you may be able to dispense with some of the bracing and rely entirely upon cleats securely glued to the side panels.

half a dozen along the aft support would be about right. I've found it will support my weight of 200 pounds reasonably well.

Making the seat was a breeze. I cut out the seat using an ordinary jacksaw and then screwed and glued a length of framing lumber along the underside of the forward edge, with another halfway between the forward edge and the aft-most edge. When everything was ready, I glued and again screwed the seat directly onto the framing lumber straps, and once the glue set it proved perfectly solid for my family's purposes.

Les Brown's Micromouse *Bumblebee* carries its builder's 160-pound frame quite comfortably. Les modified the plans to achieve the gracefully arched decks, which provide stowage space for a collapsible cart that he uses to get his boat from his home to the water nearby. To accomplish this, he arched the tops of the transoms and used arch-topped open frames instead of square, solid bulkheads. (Les Brown)

LILYPAD

A STONE-SIMPLE PUNT

Length Over All (LOA): 12'11"
Length Waterline (LWL): 11'3½"
Beam: 34"
Weight: 100 lb.
Displacement at Design Waterline: 650 lb.
Crew: 2 adults and one child or equivalent (550 lb. max.)
Propulsion: Pole, oars, or battery power (batteries will reduce max. crew weight)
Construction Method: Simplified chine log

While we're on the subject of flat-bottomed boats, punts are a type that can often be built extremely quickly, particularly if they have straight sides.

They are also very versatile. They can be poled (stand on the stern and push) or quanted (stand near the bow, plant your quant on the river bottom, then walk to the stern, twist out the quant, and walk back to the bow). With a few modifications they can be rowed or sculled. You could also power Lilypad with an electric trolling motor with the batteries near the bow, or even a very small outboard of 2 hp or less installed in a small well. In the past, people even used to sail punts, although that's a very rare sight now and I don't recommend it for a punt of this size.

Punts are also obviously good for fishing, watching nature, or just messing about. You could even erect a simple tarp tent and go camping. If they're made heavily, punts become quite stable and are great little boats for maintaining ponds and small lakes. If you've ever tried to clear a small lake of weeds and lily tubers without getting into the water yourself, you'll know how useful that stability can be.

Lilypad is straight-sided and very basic, and uses transoms fore and aft to give the bottom a good load-carrying shape within a

short length. There are almost no angles to cut or trim, and because this boat is built using the simplified chine log method, all the joints go together with screws or nails and some glue—polyurethane or epoxy mixed with a little filler.

The deck forward has three purposes. It enables you to launch the boat bow-first at a sharp angle from a steep riverbank without flooding; it lends the structure useful rigidity; and it'll be a good place to sit as an alternative to sprawling on an inflatable cushion or beanbag in the forward cockpit area. There's no deck aft as it won't be needed in poling; because the bottom rises gently to a transom, not steeply to the deck as in other punts, this area will be flat enough to stand on if covered with a nonskid paint.

One good carpenter or two half-good ones could literally bang and glue this boat together in an afternoon. It takes three sheets of ⅜-inch plywood, and because this is an easy boat to build I think I'd consider using inexpensive softwood plywood with the intention of making this a quick and basic build. Using ½-inch plywood would make it stronger and more stable, but I think it would be difficult for you to bend it to the curve of the bottom. On the other hand, using ¼-inch plywood would be rather light and I'd avoid it unless you only intend to sit,

Lilypad
12'11"

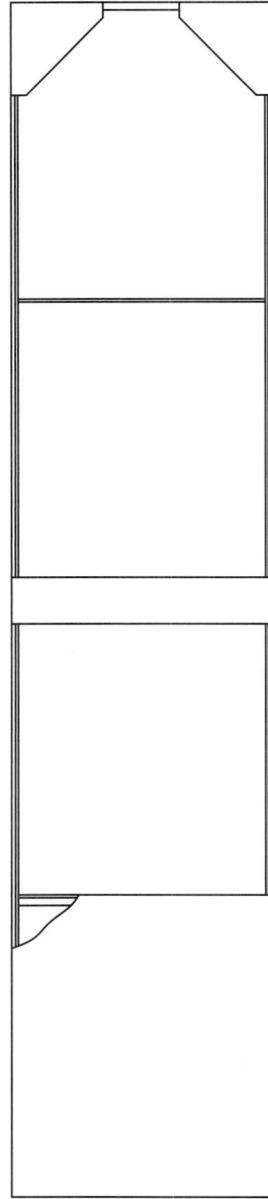

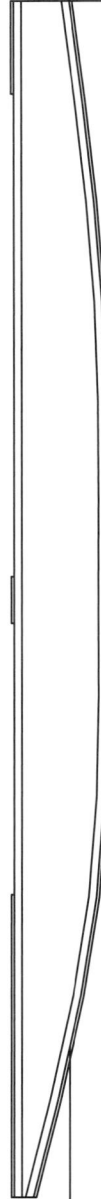

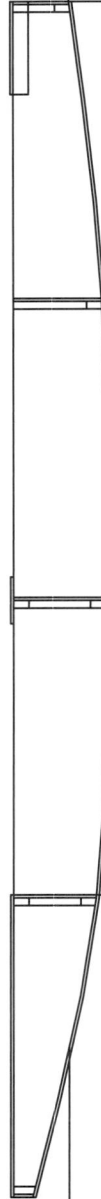

650 lb.

Aft corner braces

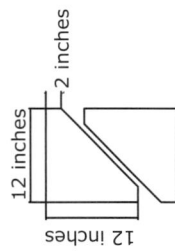

2 inches

12 inches

12 inches

Frame 3

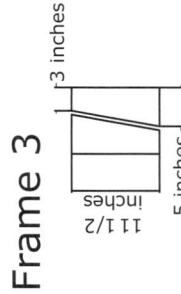

3 inches

5 inches

11 1/2 inches

Bulkhead 3

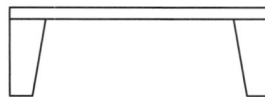

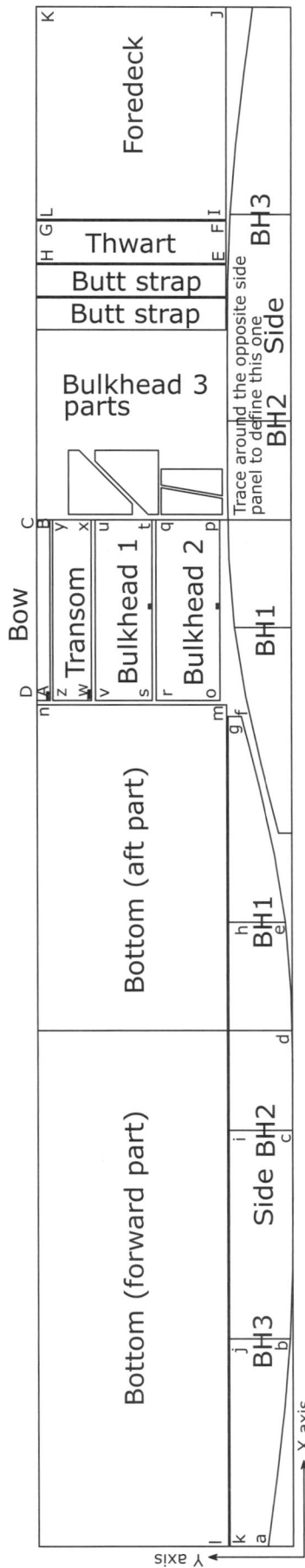

Foredeck

K · J

GL · FI

H Thwart E

Butt strap

Butt strap

Bulkhead 3 parts

BH3

Trace around the opposite side panel to define this one

Side

BH2

C B A D

Bow

z Transom w

y x
u t
v s
q r
o p

n m

Bulkhead 1

Bulkhead 2

BH1

g f

Bottom (aft part)

h BH1 e

d

Bottom (forward part)

i Side BH2 c

j BH3 b

a k l

X axis

Y axis

Lilypad Coordinates (in inches)

	x	y
Side panel		
a	0	4³/₄
b	38³/₄	¹/₂
c	77¹/₂	0
d	96	¹/₈
e	116¹/₄	1¹/₄
f	155	9¹/₂
g	155	12
h	116¹/₄	12
i	77¹/₂	12
j	38³/₄	12
k	0	12
Bottom		
l	0	12¹/₄
m	157¹/₈	12¹/₄
n	157¹/₈	48
Bulkhead 2		
o	158	13¹/₂
p	192	13¹/₂
q	192	25¹/₂
r	158	25¹/₂
Bulkhead 1		
s	158	26¹/₄
t	192	26¹/₄
u	192	37
v	158	37
Stern transom		
w	158	37³/₄
x	192	37³/₄
y	192	45
z	158	45
Bow transom		
A	158	45¹/₂
B	192	45¹/₂
C	192	48
D	158	48
Thwart		
E	240	12¹/₄
F	248	12¹/₄
G	248	48
H	240	48
Foredeck		
I	248¹/₄	12¹/₄
J	288	12¹/₄
K	288	48
L	248¹/₄	48

Lilypad Coordinates (in millimeters)

	x	y
Side		
a	0	116
b	949	12
c	1899	0
d	2352	2
e	2848	31
f	3798	233
g	3798	294
h	2848	294
i	1899	294
j	949	294
k	0	294
Bottom		
l	0	300
m	3851	300
n	3851	1176
Bulkhead 2		
o	3871	331
p	4704	331
q	4704	625
r	3871	625
Bulkhead 1		
s	3871	643
t	4704	643
u	4704	907
v	3871	907
Stern transom		
w	3871	925
x	4704	925
y	4704	1103
z	3871	1103
Bow transom		
A	3871	1115
B	4704	1115
C	4704	1176
D	3871	1176
Thwart		
E	5880	300
F	6076	300
G	6076	1176
H	5880	1176
Foredeck		
I	6082	300
J	7056	300
K	7056	1176
L	6082	1176

Note: Small discrepancies may exist between the millimeter and inch tables. See page ii.

rather than stand, in the punt. As always, it's wise to begin by building a model.

See Chapter 3 for details on working with plywood. Start marking out the plywood by squaring it off. Cut out the sides, frames, stem, transom, and butt blocks as shown, and don't forget to label each component. From one of the long pieces of butt block, cut two pieces to join the sides, and butt-join them together.

You may clamp them together for gluing, or use weights, or small nails, or even screw them together using stainless steel screws. In doing this, remember that the block goes on the *inside* of each side. As you assemble them it may help to lay them out so that you can clearly see that they are mirror images of each other. Similarly, use the long butt block to joint the two sections of the bottom.

Cutting out the long plywood panels with a circular saw. It's possible to cut fair curves with a circular saw by setting the blade just deep enough to break through the bottom of the plywood. Note the long pieces of lumber below the plywood, which are used to support it in the middle of the span between the sawhorses. (Maxwell Holtzman)

Use a waterproof polyurethane (PU) glue for this job. The fast-setting type can be convenient, but if this is your first boatbuilding project you may find the 5-minute PU variety leaves almost no margin for error!

Make up the stem, transom, and frames as drawn. Each one should have 1-by-2-inch or 1-by-1½-inch cleats all around, made up so that the lumber is proud along one long edge—this will later be trimmed to fit the fore-and-aft curve of the bottom.

Now it's time to go to three dimensions, which is easiest to do if you work with the boat upside down. Lilypad is arguably the simplest boat to build in this book because its sides are straight and the frames are left square, so they don't have to be trimmed to match. You could simply clamp the whole lot together and glue it right away, perhaps adding screws once the glue has set, but it's just as good or better to clamp, drill for screws, apply glue, and then drive the screws. Flip the hull over to screw and glue the gunwales, foredeck, thwart, and aft

corner pieces into place, and then flip it back to add the chine logs.

At the end of the clamping, gluing, and screwing process, you should have a reasonably square "ladder" almost ready to accept the bottom. Only one more major task remains before the bottom can be added: trimming the frames and transoms to match the fore-and-aft curve of the bottom.

With the boat upside down once again, begin by lining up a straightedge across the bottom, and trim the frames with a Stanley Surform or a rasp until they just kiss the straightedge. Then screw and glue the bottom into place, driving the screws at regular intervals through the plywood bottom into the frames and the chine logs. If you've got enough clamps and heavy weights, you can get by with fewer screws, and may be able to manage without any at all.

Finally, it's essential to add a skeg using the same methods described earlier in the book. Its exact shape is not terribly important, but a fin

Hull assembly in process. Bow and stern transoms have been glued and ring-nailed to the side panels. The #2 bulkhead is clamped in place; ring nails will be added after the glue sets. A temporary strut has been clamped from the hull to a nearby solid object to counteract a bend in the warped plywood side panels while the glue dries. Note that the bow transom has been left "proud"; it will be planed down to match the curve of the bottom later. (Robert Holtzman)

like those already discussed for the Minimouse and Micromouse would be fine. It should be made to be a good, big size without extending below the deepest part of the hull, and can be made from plywood left over from cutting out the rest of the boat's panels.

Finally, you need to make a pole. I'd suggest starting with a 12- to 14-foot length of 4-by-4-inch or 4½-by-4½-inch softwood. Spruce would be good because it's relatively light, and laminating it together from several thinner pieces of wood would reduce warping. Plane the corners of the wood to make an 8-sided pole, and then again to make 16 sides. This is a job for a small power planer, and it would be wise to use polyurethane glue rather than epoxy to avoid blunting your tool.

Alternatively, you may find you can buy a perfectly acceptable peeled rustic pole of the right size from a nearby fencing supplier.

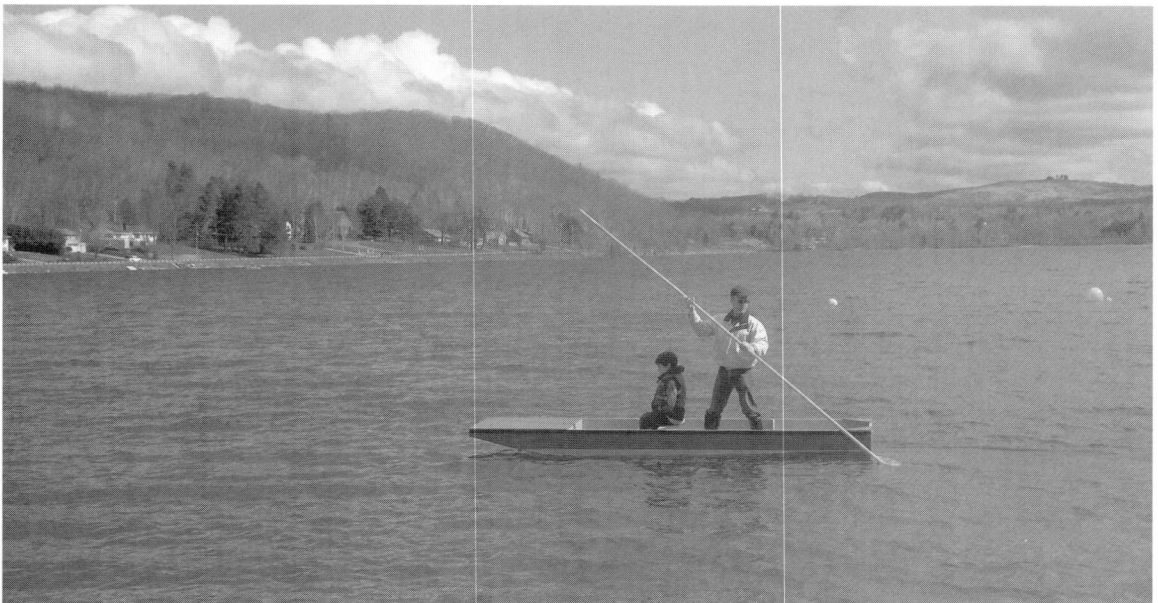

Lilypad goes together easily, carries a substantial load, and is surprisingly maneuverable. (Cate Monroe)

MOUSE AND ROWING MOUSE

A V-BOTTOMED MESSABOUT PRAM
FOR DOUBLE PADDLE OR OARS

Length Over All (LOA): 7'10"
Length Waterline (LWL): 6'9"
Beam: 29¹/₂"
Weight: 35–40 lb.
Displacement at Design Waterline: 250 lb.
Crew: 1 adult
Propulsion: Double-paddle (Mouse), oars (Rowing Mouse)
Construction Methods: PU stitch-and-glue
                                       Epoxy stitch-and-glue

This section presents the original V-bottomed Mouseboat in both its paddling and rowing forms. The paddling version is very popular and has been built many times by Mouseboats enthusiasts.

Although the flat-bottomed Mouseboats discussed earlier (see pages 84–96) work well, they're not quite as fast, well mannered, or strong-bottomed as the V-bottomed Mouseboats. What's more, you get these advantages without having to do a lot of extra work because the V-bottoms are only slightly more complicated to build using the stitch-and-glue method, and there's only one additional seam in the design. Many of them have been successfully built by first-time boatbuilders.

Their only disadvantage is that they are quite difficult to build with chine logs; the best building method for these V-bottomed boats is definitely stitch-and-glue, although some have been built with chines.

There is not a lot to be said about building the standard V-bottomed Mouse that hasn't already been said about the flat-bottomed version. As with the flat-bottomed Mouseboat built with the stitch-and-glue method, the job starts with squaring off the plywood material, plotting the coordinates, and marking out the panels (remembering to mark the insides of the sides where the frames will go). Cut out the panels and use cloth-backed tape to assemble the structure, along with some small nails driven through bits of scrap plywood to hold the frames in place. Then it's a polyurethane or epoxy and fiberglass job just like the stitch-and-glue flat-bottomed Mouseboats. The frame-tops, gunwales, decks, and painter eye work in exactly the same way.

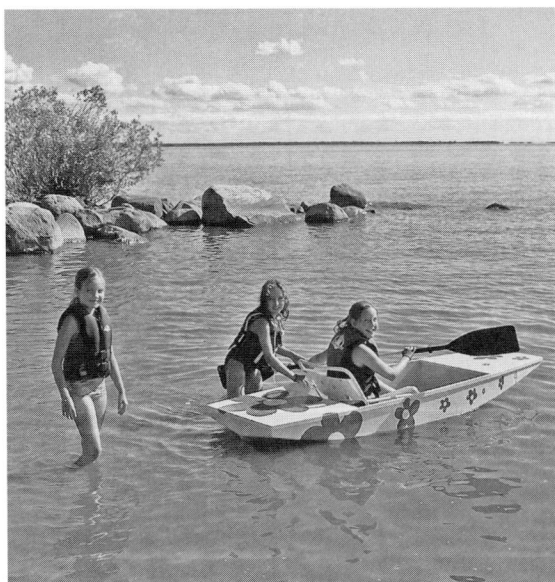

Ultrasimple boats like the Mouse need not be taken too seriously. Butch Kuhn built this wonderful flowery version, and it's clear that the love kids it. (Butch Kuhn)

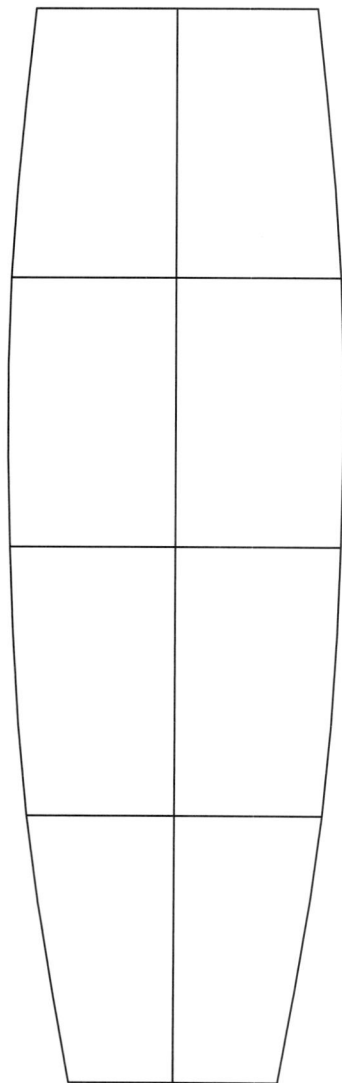

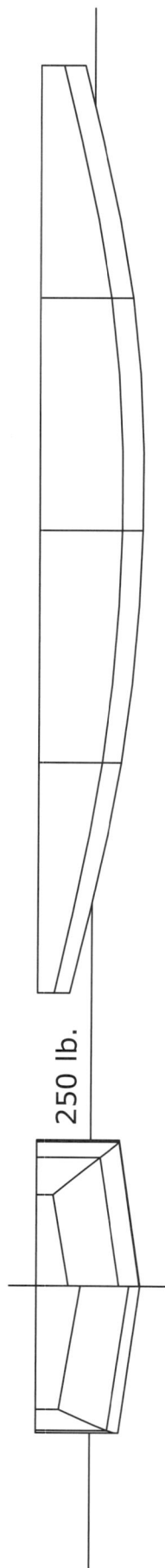

Mouse
7'10"

250 lb.

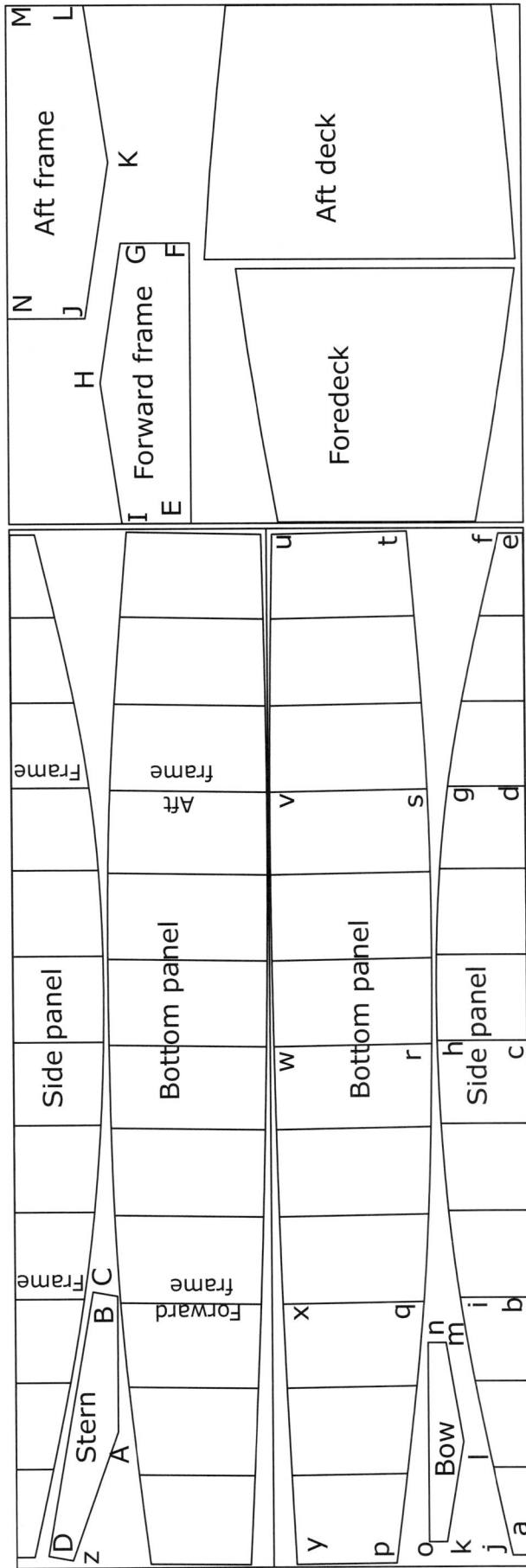

M
L
N
J
Aft frame
K
H
G
F
I
Forward frame
E
Aft deck
Foredeck

Side panel
Frame
Frame
Bottom panel
Aft frame
Bottom panel
w
v
u
t
r
s
g
d
f
e
h
Side panel
c

Stern
Frame
C
B
Forward frame
Forward frame
x
q
n
m
i
b
Bow
l
Frame
A
D
z
y
p
o
k
j
a

105

Mouse Coordinates (in inches)

	x	y
Side panel		
a	1	0
b	$24^7/_8$	0
c	$48^3/_8$	0
d	$71^7/_8$	0
e	$95^1/_2$	0
f	$95^1/_2$	$2^3/_8$
g	$71^7/_8$	$7^1/_4$
h	$48^3/_8$	$8^3/_8$
i	$24^7/_8$	$6^1/_2$
j	1	$1^5/_8$
Bow transom		
k	$2^1/_4$	$7^3/_4$
l	$11^1/_2$	$6^1/_4$
m	$20^5/_8$	$7^3/_4$
n	$20^5/_8$	$9^1/_2$
o	$2^1/_4$	$9^1/_2$
Bottom panel		
p	$^3/_8$	$12^1/_2$
q	$24^1/_2$	$9^7/_8$
r	$48^1/_8$	9
s	$71^5/_8$	$9^1/_4$
t	$95^3/_4$	$10^7/_8$
u	$95^1/_2$	$23^1/_2$
v	$71^3/_8$	$23^7/_8$
w	$47^7/_8$	$23^5/_8$
x	$24^1/_4$	23
y	$^1/_4$	$21^7/_8$
Stern transom		
z	$^3/_4$	$42^3/_4$
A	$12^1/_2$	$38^1/_2$
B	$25^1/_8$	$38^1/_2$
C	$25^1/_2$	$40^3/_4$
D	$1^1/_8$	45
Forward frame		
E	0	31
F	26	31
G	26	$37^1/_2$
H	13	$39^3/_8$
I	0	$37^1/_2$
Aft frame		
J	19	$40^3/_4$
K	$33^1/_2$	$38^1/_2$
L	48	$40^3/_4$
M	48	48
N	19	48

Mouse Coordinates (in millimeters)

	x	y
Side panel		
a	25	0
b	609	0
c	1186	0
d	1762	0
e	2340	0
f	2340	58
g	1762	178
h	1186	206
i	609	158
j	25	41
Bow transom		
k	56	191
l	282	153
m	507	191
n	507	232
o	56	232
Bottom panel		
p	9	307
q	600	241
r	1179	219
s	1756	225
t	2345	268
u	2339	575
v	1750	584
w	1173	579
x	595	563
y	6	535
Stern transom		
z	17	1047
A	307	942
B	615	942
C	625	997
D	28	1103
Forward frame		
E	0	759
F	637	759
G	637	917
H	319	965
I	0	917
Aft frame		
J	465	999
K	821	944
L	1176	999
M	1176	1176
N	465	1176

Note: Small discrepancies may exist between the millimeter and inch tables. See page ii.

A MOUSE FOR ROWING

As I've said, the Mouse also comes in a rowing version, and it's the perfect small boat for messing around. True, it won't go fast, but then again at its proper cruising speed it's a very light rower and plenty of fun!

If a young boy or girl wanted to row their own boat, a Rowing Mouse could well be the right one to choose, although very young kids generally seem to do better with paddles.

Rowing Mouse

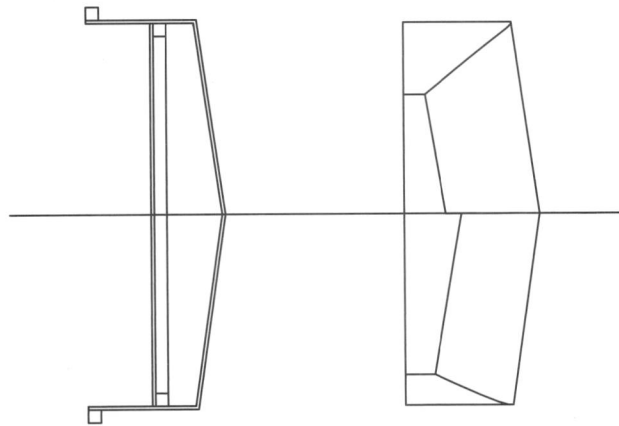

7'10"

Length of rower's forearm from elbow to wrist

5 inches

8 inches

Ideal position of seat may vary depending on rower

Forward frame

Aft frame

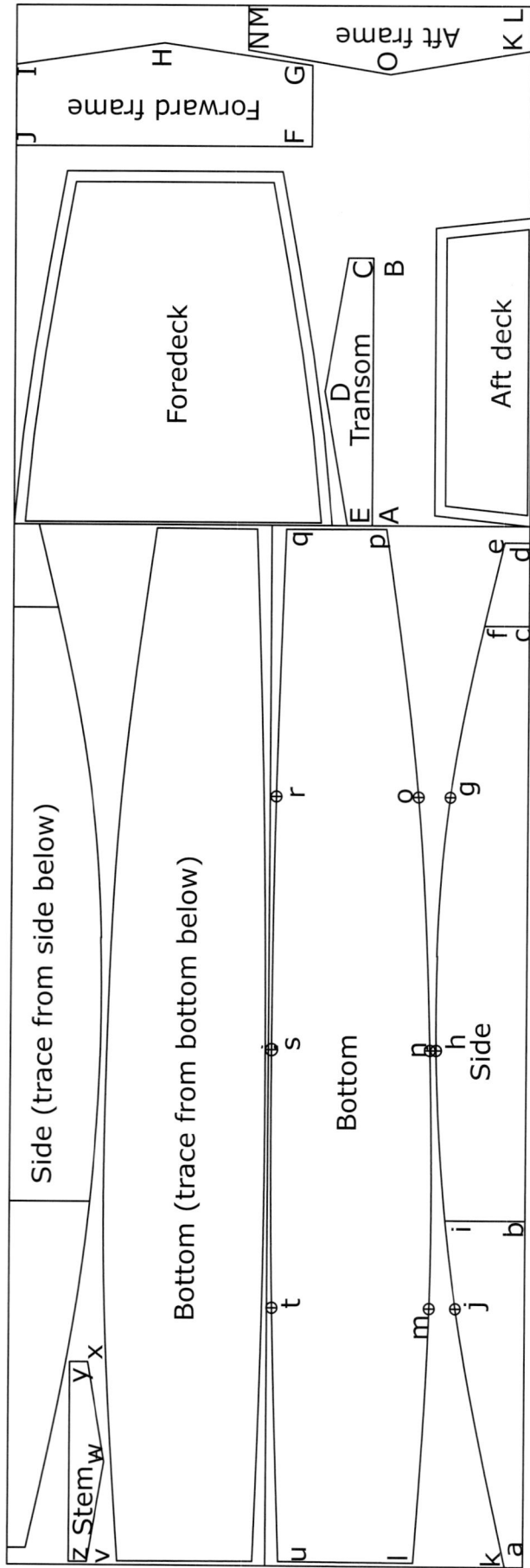

Forward frame lines up with i–b,
aft frame with f–c

Foredeck

Transom

Aft deck

Side (trace from side below)

Bottom (trace from bottom below)

Bottom

Side

Stem

Rowing Mouse Coordinates (in inches)

	x	y
Side panel		
a	0	0
b	31 3/4	0
c	86 3/4	0
d	94 1/2	0
e	94 1/2	2 3/8
f	86 3/4	4 1/8
g	70 7/8	7 1/4
h	47 3/8	8 3/8
i	31 3/4	7 1/2
j	23 3/4	6 1/2
k	0	1 3/4
Bottom panel		
l	3/8	10 1/4
m	23 3/4	8 7/8
n	47 3/8	9
o	70 7/8	10 1/4
p	95 5/8	13 3/8
q	95 5/8	22 5/8
r	70 7/8	23 3/8
s	47 3/8	23 3/4
t	23 3/4	23 1/2
u	3/8	22 3/4
Stem transom		
v	1/4	40 5/8
w	9 1/2	39
x	18 3/4	40 5/8
y	18 3/4	42 1/4
z	1/4	42 1/4
Stern transom		
A	0	14 5/8
B	24 3/4	14 5/8
C	24 3/4	17
D	12 3/8	19 1/8
E	0	17
Forward frame		
F	35	20 1/2
G	42 1/2	20 1/2
H	44 1/2	34 1/4
I	42 1/2	48
J	35	48
Aft frame		
K	43 7/8	0
L	48	0
M	48	26 3/8
N	43 7/8	26 3/8
O	41 5/8	13 1/4

Rowing Mouse Coordinates (in millimeters)

	x	y
Side panel		
a	0	0
b	778	0
c	2126	0
d	2314	0
e	2315	57
f	2126	102
g	1736	177
h	1161	206
i	779	183
j	582	158
k	0	41
Bottom panel		
l	8	251
m	582	218
n	1161	220
o	1736	250
p	2344	327
q	2343	555
r	1736	574
s	1161	581
t	582	577
u	10	559
Stem transom		
v	7	995
w	232	957
x	458	995
y	458	1036
z	7	1036
Stern transom		
A	2352	360
B	2959	360
C	2959	417
D	2655	469
E	2352	417
Forward frame		
F	3210	501
G	3392	501
H	3442	839
I	3392	1176
J	3210	1176
Aft frame		
K	3426	0
L	3528	0
M	3528	647
N	3426	647
O	3372	324

Note: Small discrepancies may exist between the millimeter and inch tables. See page ii.

Mouse is the perfect boat for enjoying solitude or observing nature up close on the smallest, quietest waters. This is the author's partner paddling *Lion*, a Mouse the author built for his daughter. "It's fabulous," she said, "and no one's ever going to get me out of here. I so wish someone had built me one when I was a child."

CRUISING MOUSE

A TWO-PERSON ROWING OR PADDLING PRAM WITH CARRYING CAPACITY

Length Over All (LOA): 11'10"
Length Waterline (LWL): 11'2½"
Beam: 29½"
Weight: 70–85 lb.
Displacement at Design Waterline: 450 lb.
Crew: 2 adults or equivalent
Propulsion: Double-paddle, oars
Construction Methods: PU stitch-and-glue
                                        Epoxy stitch-and-glue

For some reason the 12-foot Cruising Mouse hasn't been as popular as the smaller members of the Mouse family. One explanation may be that, having built one of the smaller Mouseboats, many builders find their families wanting one of each, rather than sharing a larger boat, and so go on to build more small Mouseboats as we did in our family.

Another explanation could be that having experimented with one small Mouseboat, many find the confidence to go on to build something more complex and demanding than a larger Mouseboat. Frankly, Cruising Mouse won't go anywhere the little boats can't take you. On the other hand, it doesn't demand much more investment in time, materials, or skill than the smaller boats, but its performance is better under paddles or oars, and it has a significantly greater carrying capacity: about 450 pounds, or the weight of two good-sized men, three or more teenagers, or a couple and some basic camping gear. In fact, I think Cruising Mouse is the most versatile hull of the entire Mouseboat family.

Cruising Mouse also makes an ideal tender (dinghy) for those who prefer paddling to rowing. Its length and hard chine give it the kind of stability and carrying capacity required for a tender without overly compromising its paddling qualities; its scow bow should avoid the side-to-side shearing that makes towing sharp-bowed small boats a nightmare; and its narrow beam will help to keep drag low. Furthermore, its buoyant ends should help keep it drier than some boats with a similarly low freeboard.

If fitted with a rowing thwart and oarlocks, the Cruising Mouse should make a light and quick rower with a considerably better turn of speed than the 8-foot Rowing Mouse. Its performance would certainly suit someone who likes a brisk row for exercise each evening.

The basic Cruising Mouse is built the same way as the standard V-bottomed Mouse with the small but important difference that, like the Lilypad punt, it requires panels of 12 feet or so in length to be created from two 8-foot sheets of plywood.

At almost 12 feet, Cruising Mouse would be just a little heavy for most people to carry under their arm, but there's nothing to stop you from building one of these boats with a divide at the center to produce a version that breaks down into two parts that quickly clip together for going on the water. The idea would be to construct two center bulkheads and to stitch-and-glue them into the hull

Cruising Mouse

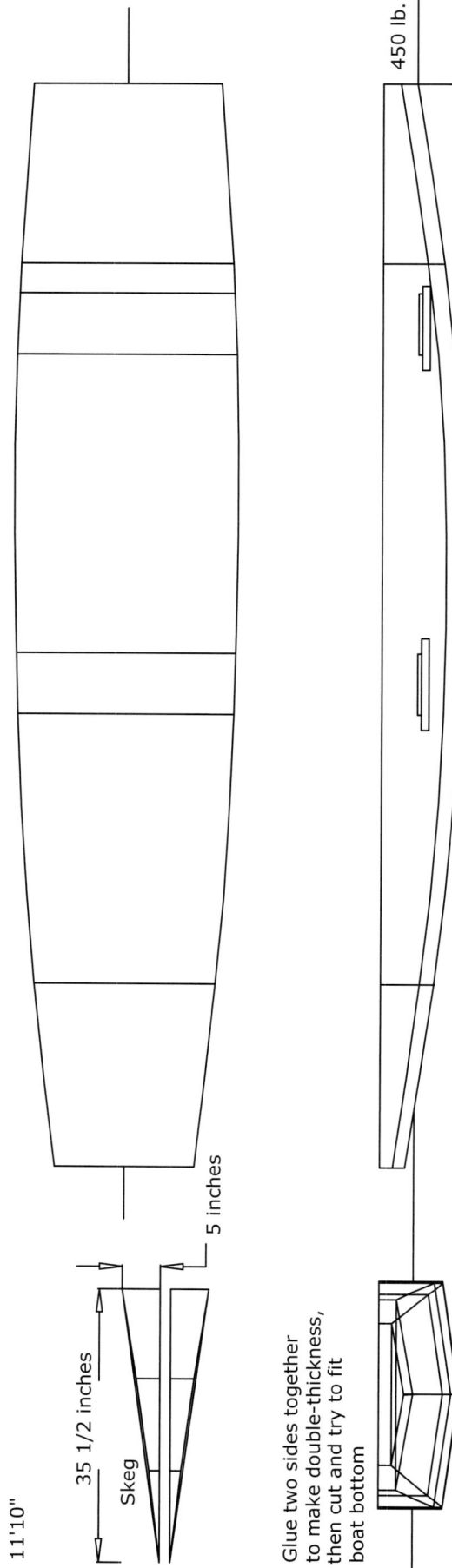

11'10"

35 1/2 inches

Skeg

5 inches

450 lb.

Glue two sides together
to make double-thickness,
then cut and try to fit
boat bottom

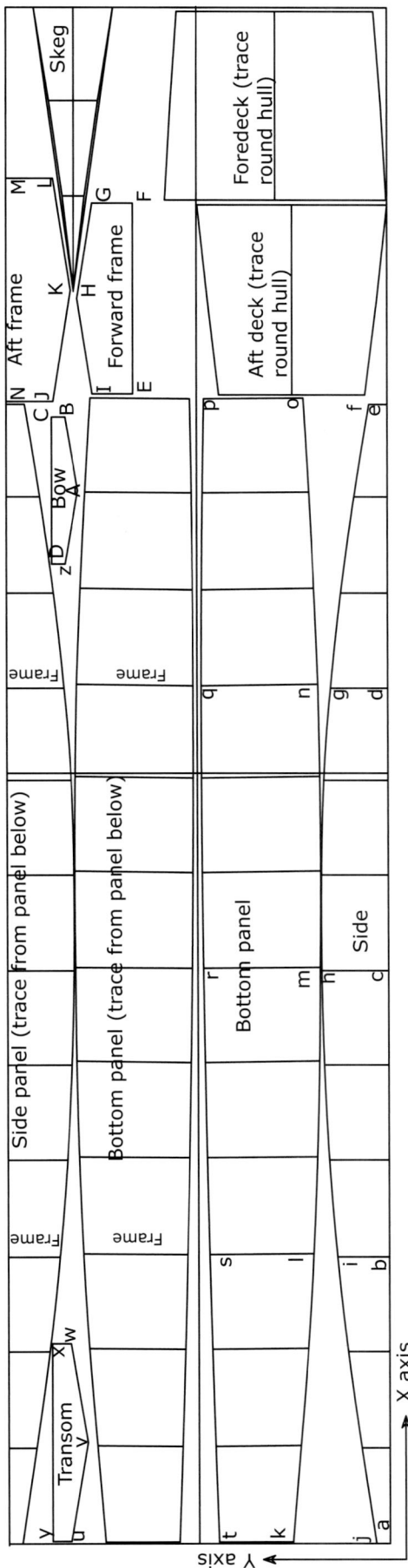

Cruising Mouse Coordinates (in inches)

	x	y
Side panel		
a	0	0
b	$35^3/_4$	0
c	$71^1/_4$	0
d	$106^3/_4$	0
e	$142^1/_4$	0
f	$142^1/_4$	$2^1/_4$
g	$106^3/_4$	$7^1/_4$
h	$71^1/_4$	$8^3/_8$
i	$35^3/_4$	$6^1/_2$
j	0	$1^3/_4$
Bottom panel		
k	$^1/_4$	$12^1/_4$
l	$36^1/_8$	$9^1/_2$
m	$71^3/_4$	$8^5/_8$
n	$107^1/_4$	$8^3/_4$
o	$143^1/_8$	$10^5/_8$
p	143	$23^1/_8$
q	$107^1/_8$	$23^1/_2$
r	$71^5/_8$	$23^1/_4$
s	36	$22^5/_8$
t	$^1/_4$	$21^1/_2$
Stern transom		
u	$^1/_4$	$40^1/_8$
v	$12^3/_4$	38
w	25	$40^1/_8$
x	25	$42^1/_2$
y	$^1/_4$	$42^1/_2$
Bow transom		
z	$122^1/_4$	$40^5/_8$
A	$131^1/_2$	39
B	$140^3/_4$	$40^5/_8$
C	$140^3/_4$	$42^1/_4$
D	$122^1/_4$	$42^1/_4$
Forward frame		
E	$143^5/_8$	32
F	$167^1/_2$	32
G	$167^1/_2$	$37^1/_8$
H	$155^5/_8$	39
I	$143^5/_8$	$37^1/_8$
Aft frame		
J	$142^5/_8$	$42^1/_8$
K	$156^5/_8$	$39^7/_8$
L	$170^5/_8$	$42^1/_8$
M	$170^5/_8$	48
N	$142^5/_8$	48

Cruising Mouse Coordinates (in millimeters)

	x	y
Side panel		
a	0	0
b	875	0
c	1747	0
d	2617	0
e	3486	0
f	3486	56
g	2617	176
h	1747	206
i	875	159
j	0	42
Bottom panel		
k	6	299
l	884	233
m	1757	211
n	2629	216
o	3506	260
p	3504	566
q	2624	576
r	1754	571
s	882	554
t	6	527
Stern transom		
u	6	982
v	311	931
w	613	982
x	613	1041
y	6	1041
Bow transom		
z	2996	995
A	3222	956
B	3447	996
C	3447	1036
D	2996	1036
Forward frame		
E	3519	785
F	4104	785
G	4104	910
H	3813	956
I	3519	910
Aft frame		
J	3495	1032
K	3837	977
L	4180	1032
M	4180	1176
N	3495	1176

Note: Small discrepancies may exist between the millimeter and inch tables. See page ii.

about an eighth of an inch apart at the center of the boat. Glue large plywood patches (at least 3 inches by 3 inches) at each of the exposed corners of the two center bulkheads, and after the epoxy has set, bore holes right through the patches and both bulkheads. Carefully saw the boat in half between the bulkheads, add epoxy and glass tape, tidy up, paint the whole boat, and finally bolt the two ends together with good-sized bolts, wing nuts, and large steel washers to help distribute the stresses. Place rubber washers between the steel washers and the plywood to keep the water out.

The nearly-finished Cruising Mouse built by Anthony Smith in Chapters 3 and 4. So many of the general construction details are clearly visible without a coat of paint. Note the butt-strapped sides and bottom; the "seat bearers" or cleats supporting the thwarts; the neatly-filleted-and-taped interior seams; and the treatment of the decks and gunwales. (Anthony Smith)

Anthony Smith (bow) and his father paddling on an English canal on launching day. Seems they didn't have two double-bladed paddles at hand, so they split one. Not very efficient or comfortable, but even so, look at that bow wave! (Anthony Smith)

POORBOY

A SMALL OUTBOARD SKIFF

Length Over All (LOA): 10' or 11'6"
Length Waterline (LWL): 7'6" or 9'1"
Beam: 41¼"
Weight: 100–120 lb.
Displacement at Design Waterline: 420 lb.
Crew: 2 adults
Propulsion: Small outboard
Construction Method: Simplified chine log

Steve Lewis is someone who, like me, finds an exciting challenge in creating designs that make boating and boatbuilding accessible to very nearly everyone—even to busy parents of families with huge weekly bills and moderate incomes. Since this description applies to both of us, it makes sense that we should have the greatest sympathy for this group of boat users!

However, where I like to travel in near silence and don't like the sound and smell of an engine, Steve is different. He doesn't mind engines and has drawn and built several boats for home building designed for use with an outboard. One of his designs is PoorBoy, and it is a masterpiece of simplicity.

Steve says PoorBoy will handle up to a 10 hp engine, which can be run at full throttle when traveling in a straight line, but must be throttled back before turning. Going too fast into a turn might capsize the boat. Like many other flat-bottomed boats, PoorBoy can trip on its chine and roll over in a sharp turn at high speeds.

However, I'd go further and plead with you not to use anything larger than a 10 hp engine. Youngsters who might be tempted to push this little boat too far should be given a smaller engine for their own safety.

PoorBoy can be built in either 10-foot or 11-foot 6-inch lengths. Don't let the absence of dimensions on the drawings throw you: all the dimensions you need, and there aren't many, are made clear in the text below. Likewise, the design is so simple that you don't need a nesting diagram either (that's the plan that shows how to lay the various pieces out on the plywood). That, too, is explained below.

In the 10-foot version, PoorBoy takes two sheets of plywood (one ¼ inch, the other ⅜ inch); and half a sheet of ¾ inch plywood for the transom, transom doubler (the engine mounting pad), knees, and breasthook. You will also need 18 feet of 1-by-12-inch lumber for seats, some scraps for the central mold, and various lengths of material for the stem, inwales, chine logs, cleats, and so on, along with a bunch of ⅝-inch and ¾-inch screws. The usual choice of adhesives and sealants can be used.

The 11-foot 6-inch version takes a little more material (roughly another half sheet of ⅜-inch plywood and another half sheet of ¼-inch plywood). The extra expense for the materials is well worth it because the larger version of PoorBoy is a much better boat. The disadvantage of going to the larger size is that the boat will be more difficult to car-top.

Both versions have a beam of 49 inches at the sheer and 43 inches at the chine, with sides and transom flared at 13 degrees. The side panels are 16-inch-wide plywood panels joined with a butt block toward the aft end of the boat.

Stern transom

Side

Bottom

Side

PoorBoy Skiff

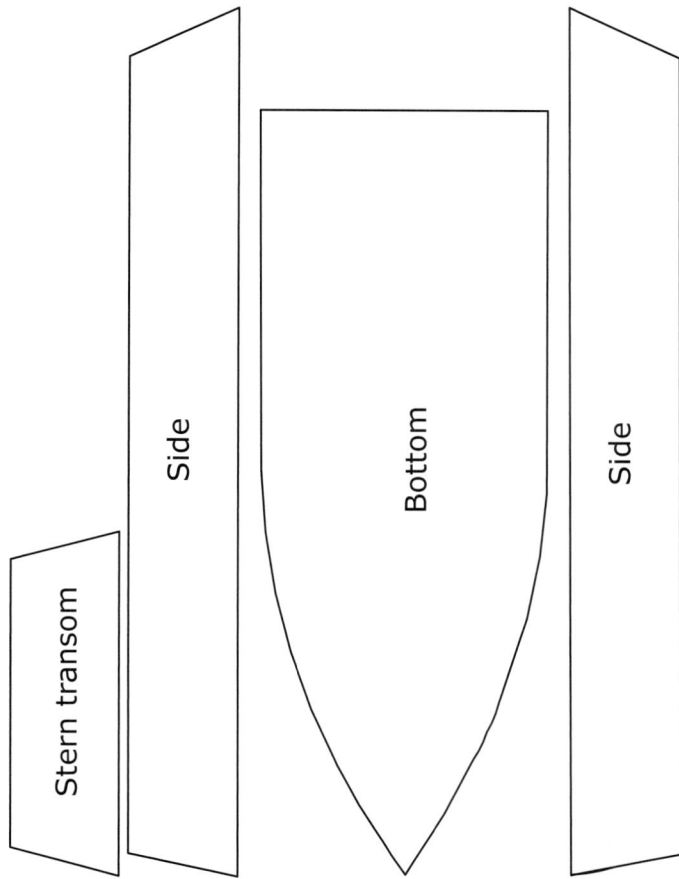

Center mold detail

22.5 degrees

15 degrees

13 degrees

35 1/2 inches

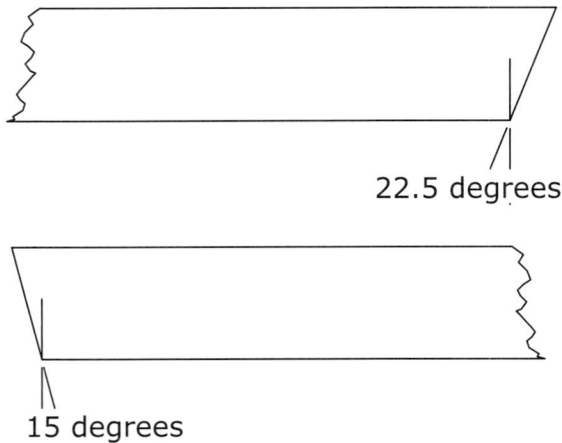

Cut the ends of the side panels to the angles shown. The top image is the bow end; the bottom is the stern.

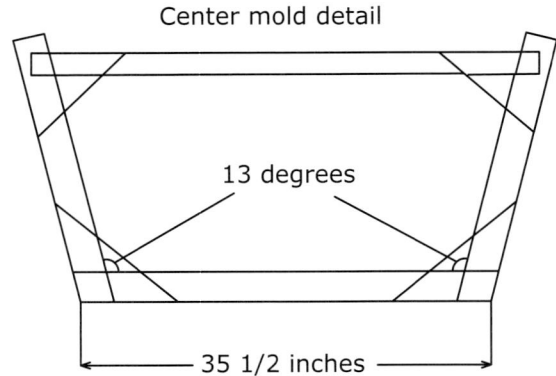

A single center mold ensures that PoorBoy's sides go on at the proper angle of flare. Use a protractor and care to get the angles correct.

I'll let Steve explain his build:

To start building, cut your side panels and butt them together using 3-inch-wide butt blocks or fiberglass butts. Cut the panels out of the plywood, ripping it into three even strips of just a cut under 16 inches wide.

Use two of the full-length panels for the front of the boat, cut the third in half to make the aft half of each completed side panel, and then cut a 3-inch strip from each for use as butt blocks. If you're using glass tape and epoxy on each side to make your butt joints, you can omit cutting the strip off and make a boat that is 3 inches longer.

Line up the panels and glue and screw the butt blocks into place, making sure that they are both on the inward side of each panel. Flip them over and tape the outside of the seam.

Cut out your transom to 17 inches high, beveling the bottom to 13 degrees, cutting from the inside of the panel to the outside—in other words, the outside of the panel will be shorter height-wise than the inside.

The stem piece is cut to a 36-degree half-angle—that is, 36 degrees per side—and the bottom is beveled by 22½ degrees to take the angle of the bottom as it rises to the stem. (Author's note: You'll need a bevel gauge to cut the stem angles.) The angles don't have to be 100 percent accurate, but should be fairly close. Leave the stem long so you can trim it later.

Make the mold next, using scrap—like a piece of old closet paneling or bits of discarded pallet or packaging. The inside of the bottom is 35½ inches across, with sides angled outward at 13 degrees. If your mold is being made up as a frame as shown, cut side pieces and attach them to the bottom at this angle, then cut a top piece to create a trapezoid shape (see picture). You'll need to cut some large plywood triangles to brace

the corners of the mold to keep it square when you bend the panels around it.

Using an adjustable power saw, cut two strips of lumber 1 by 1½ inches wide to 13 degrees across the narrow side to make chine logs to attach the bottom, and glue and screw them on the bottom outside edges of the side panels.

Now you can assemble the boat. Take your side panels and attach them to the mold 4 feet forward of the bottom corner of the transom edge (5 feet 6 inches, if building the larger version). Cut, and then glue and screw pieces of 1-by-2-inch lumber to the bottom and side edges of the transom on its inside surface and install the transom, gluing and screwing from the outside through the plywood and into the lumber. With a helper to assist you, use a Spanish windlass (see page 39, Chapter 4) to bend the side panels inward and upward to the bow, and glue and screw the panels and stem piece together. Be sure the bottom of the stem piece is even with the bottom of the panels.

Flip the boat over and square it up by measuring from the bow back along each edge 2 feet, then mark a point. Measure from the opposite corner of the transom to each point; these should be even. Measure from the opposite corners of the transom to the corners of the mold; these should also be even. Finally, measure from the opposite corner of the mold to the points marked off, and these too should be even. Since the structure will not be rigid at this stage, it should be possible to twist it gently to even up these measurements.

When everything is square, lay the bottom sheet of plywood onto the boat (with the edge at the transom and one side even with a side panel) and trace around the outside of the boat. This will leave a gap in the bottom at the bow,

Cut the material to lie along the sheerline at 15 degrees, but cut the underside of the rail at 30 degrees to allow rainwater to run off when the boat is stored upside down.

Rubrail detail.

This should be 10–12 inches wide and made from $3/4$ inches plywood or solid wood, with cleats at each end attaching the seat to the boat sides, and a center brace along the underside.

Thwart seat detaill.

but there's no need to worry because you will fill it with what's left over when you cut out the main bottom panel. Check again to be sure that the boat is square, then cut outside of the line you traced so you can be sure you will be able to trim it with a rasp or sandpaper to fit. Glue and screw the panel onto the bottom.

Take a piece of the material that you have cut off and butt the factory edge up to the edge of the hole, trace the outline, and cut it out. Using another piece of plywood scrap, make a butt block to fit inside the bow area, then glue and screw everything together. If you plan to fiberglass the bottom, you won't need to tape this joint, but if you're not planning to fiberglass the bottom, it would be better to tape it to properly seal and strengthen the structure.

Bevel cut your inner wales to 13 degrees and your outer wales to 30 degrees (see drawing), and glue and screw the outside one (from the inside).

To make a breasthook, lay a piece of $3/4$-inch plywood scrap on the bow of the boat and trace the inside of the bow onto it. Mark it, cut it out, and bevel it so that it will fit flush with the top edges of the side panels from the bow aft for a distance of about 12 inches, then screw and glue it into place. While you're at it, I recommend you cut off the stem about 4 to 6 inches above the tip of the point of the bow, which will give you some material you can use to attach a painter.

At the aft end of the boat, cut and bevel some knees to brace the corners. These can be plain triangles of 12 inches by 8 inches, or you may choose to be a little artistic with them and give them a little curve.

Cut a doubler for the center of the transom (this is where the motor will be mounted) and glue and screw it into place. Bevel the top of the transom and doubler so that it is level with the sides.

The transom is designed to take a short shaft outboard without modification. You will not need a center brace since the boat should not have more than a 10 hp motor on it and the transom is easily strong enough for an engine of this power and weight. (See the earlier comments about slowing down in turns.) The chines and transom corners should be taped with glass and epoxy for durability and leak protection and the bottom can be glassed for additional abrasion protection.

Fit, glue, and clamp your inwale after the rest is in place for a professional-looking finish. Notch the wale to fit under the breasthook and knees. The seats can be simple planks on cleats attached to the sides, fully enclosed chambers, or anything in between. The front of the rear seat should be 30 inches from the rear transom to allow you to comfortably use a tiller motor. Measure 12 inches up from the floor of the boat and mark on each side the correct distance from the transom.

This will be the top of the seat, so measure and mark a line $3/4$ inch farther down to give you the point where the tops of the cleats supporting the seats will go. The cleats should be $1\frac{1}{2}$ inches deep and $3/4$ inch wide, but doubled once in place to $1\frac{1}{2}$ inches. Another doubled $1\frac{1}{2}$-by-$1\frac{1}{2}$-inch strip will run under the center of the seat athwartship to help support its middle.

Glue your cleats together and screw them to the sides, level with the bottom. Before going any further, look at the end-grain of the seat plank. The growth lines should cup down; otherwise the plank will cup upward and hold water. Once you're sure which side should be uppermost, put the seat on the cleats and drill angled holes through the plank and into the cleats. Glue and screw the plank to the seats, and then fill the holes. The front seat should be a little forward of where the mold is, so you will have to cut the top of the front seat to fit the curve of the boat's sides in this area. The cleats should also be beveled to fit the sides.

Alternatively, you may prefer to use a large cooler rather than a permanent thwart! In this case, simply glue and screw crosswise cleats on the floor to keep the cooler from sliding around. Drinks and ice are good ballast to keep the front end down, or you can use the cooler as a live well.

Finishing requires nothing more than sanding the rough spots and painting. Use a couple coats of primer and a couple coats of exterior oil or latex paint. If you use latex, you will need to let it cure properly (this takes three to four weeks); oil-based paint cures in about a week. Whatever paint you use, let it cure out of the sun where air can circulate.

It's hard to imagine a simpler little outboard skiff, and that's what makes the PoorBoy so perfect. (Steve Lewis)

DOGSBODY

A GARVEY

Length Over All (LOA): 11'6"
Length Waterline (LWL): 9'9½"
Beam: 61"
Weight: 150–170 lb.
Displacement at Design Waterline: 650 lb.
Crew: 2 adults
Propulsion: Small outboard
Construction Methods: PU stitch-and-glue
Epoxy stitch-and-glue

Small outboard boats are hugely popular for fishing and just messing around on the water, and they are interesting and fun to build.

Dogsbody is a good example of a utilitarian small boat that includes buoyancy tanks under side benches fore and aft, and storage space in side benches amidships. The rectangular area on the foredeck is intended as a hatch for the fuel tank. If this boat were holed almost anywhere, it wouldn't sink, which must be regarded as an important safety feature.

Dogsbody has been built successfully several times, and overall reports from the field are very good. The first boat was powered by a 25 hp outboard, which I thought was rather large and heavy for a boat I'd conceived as being about right for a 15 hp engine. Still, the owner assured me that Dogsbody would turn securely and safely even at full throttle, and was quick to boot. He did say that in rougher water it was useful to ask his partner to move forward to keep the bow down, but this probably has less to do with the boat and more to do with that heavy outboard. If you build your own Dogsbody, do bear in mind that equipping it with an outboard that's too big for the design will have an impact on the boat's performance in certain conditions.

At the opposite end of the engine size scale, Ken Newton fitted his Dogsbody with a much smaller 4 hp engine. He too is extremely happy with his boat, and reports that the little engine makes the boat skim very happily. He also loves the way he can step from one side of the boat to the other in complete confidence that his boat won't heel excessively, even when he and his wife are sitting on the same side bench.

Build the boat with ⅜-inch plywood to make it rugged and able to withstand hard use and less than vigilant maintenance. You may want to use a thinner plywood, but if you do it will be important to keep the weight and power of the outboard down!

The build is a little different than what you'll see elsewhere in this book. It's based on an "egg crate" construction technique that begins with creating a form that ultimately becomes part of the structure of the boat.

There are some great aspects to this building method. Because the form has a flat upper surface, by constructing the boat upside-down you can easily ensure that there's no twist in the shape. It also has the advantage that it allows larger boats to be built with the stitch-and-glue method. That's not so important with this boat, but it would be in a boat approaching 16 feet or so.

After marking out, the first step is to cut out the inner sides of the buoyancy tanks and assemble them along with the frames. Then the flat bottom section, which plays an important role in keeping the boat straight; the transom; and the bow are put in place using

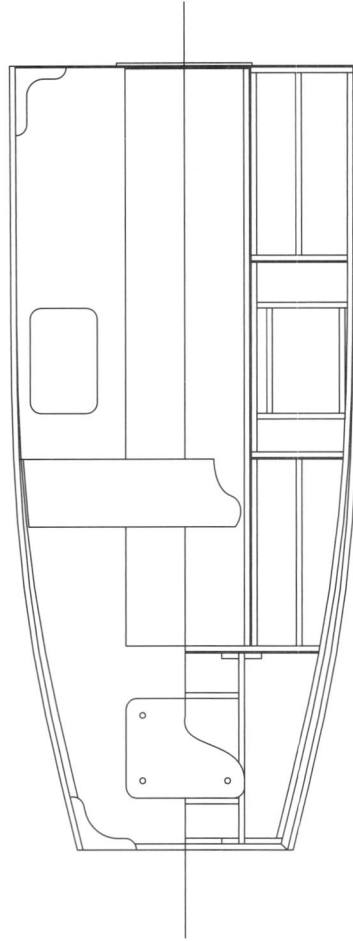

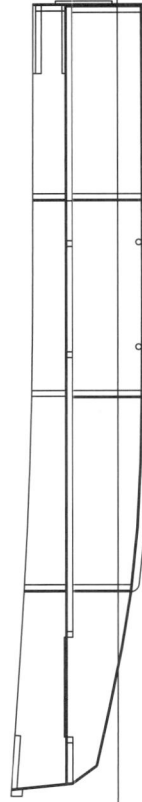

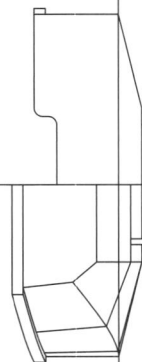

Specific frames

Frame 1

Frame 2

Frame 3

Forward side
of transom

Inches Feet

Dogsbody

650 lb.

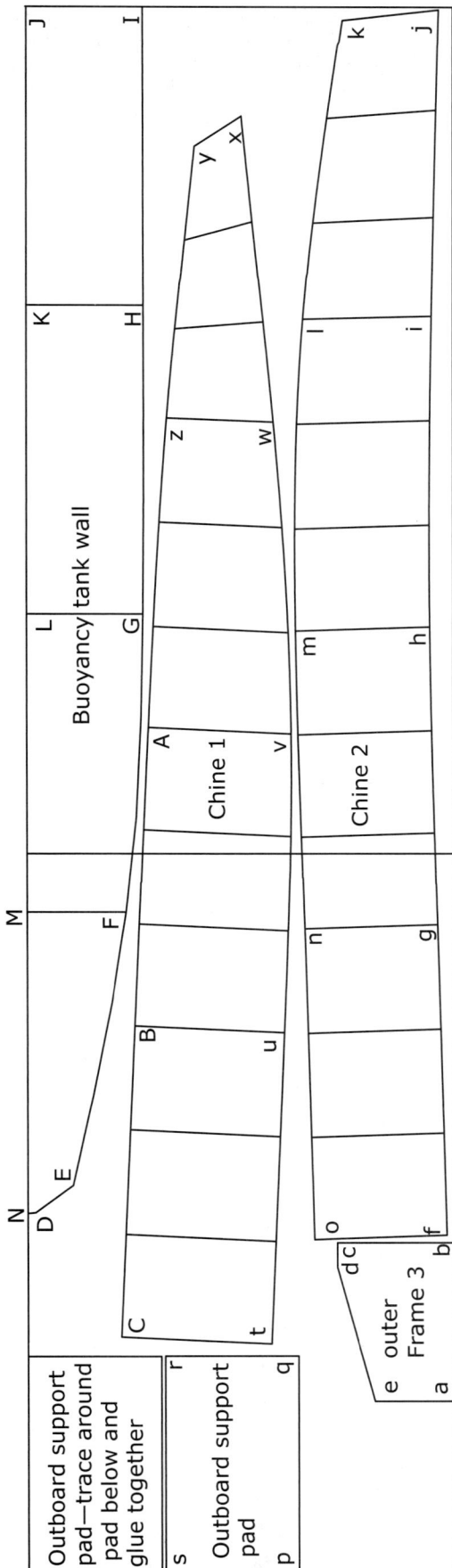

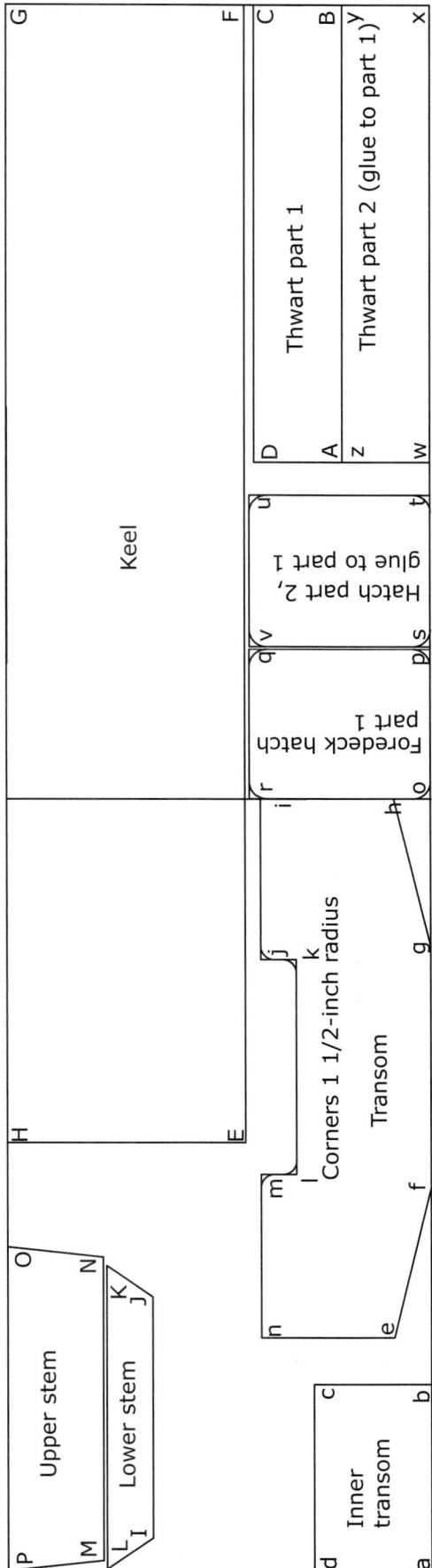

Outboard support pad—trace around pad below and glue together

Outboard support pad

Buoyancy tank wall

Chine 1

Chine 2

outer Frame 3

Keel

Thwart part 1

Thwart part 2 (glue to part 1)

Hatch part 2, glue to part 1

Foredeck hatch part 1

Corners 1 1/2-inch radius

Transom

Upper stem

Lower stem

Inner transom

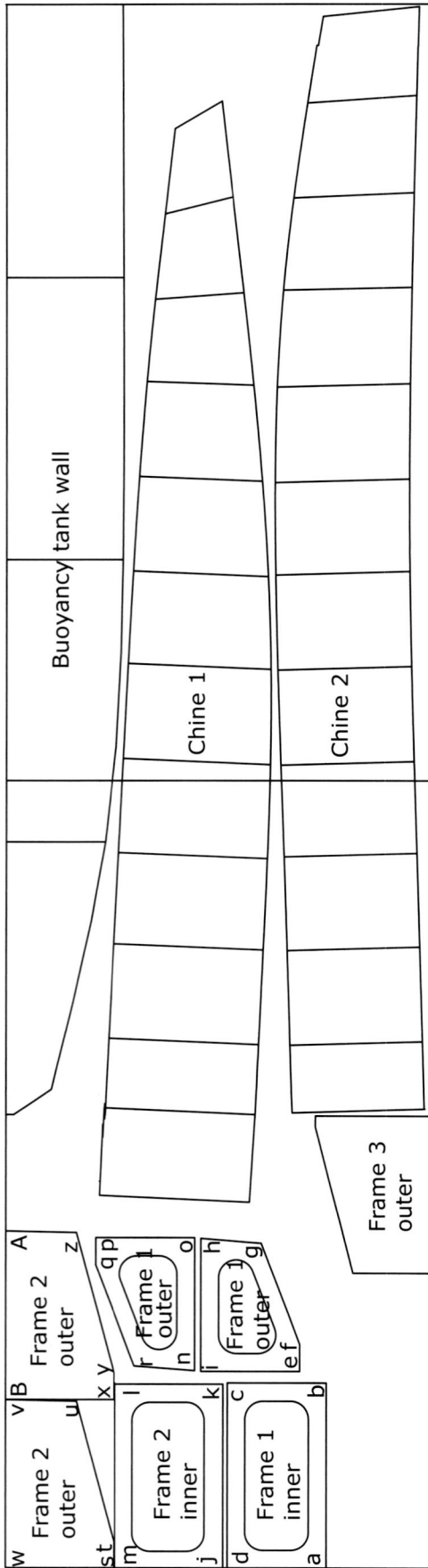

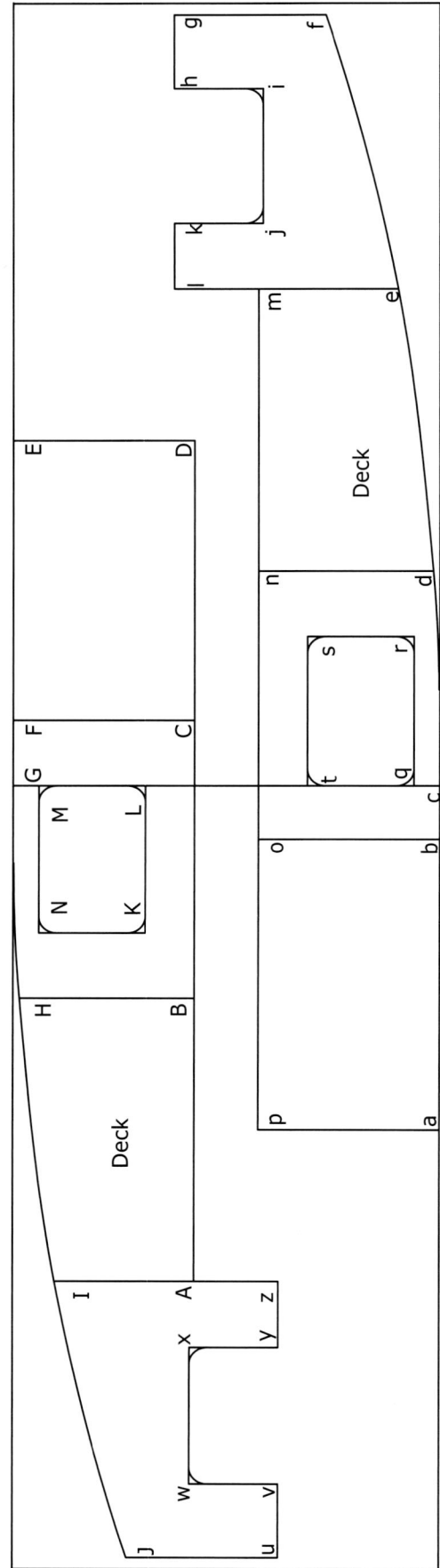

Buoyancy tank wall

Chine 1

Chine 2

Frame 3 outer

Frame 2 outer

Frame 2 outer

Frame 1 outer

Frame 1 outer

Frame 2 inner

Frame 1 inner

Deck

Deck

Dogsbody Coordinates (in inches), Part I

	x	y
Frame 3 outer		
a	$32^5/_8$	0
b	$51^3/_4$	0
c	$51^3/_4$	$13^1/_8$
d	49	$13^1/_8$
e	$32^5/_8$	9
Chine 2		
f	$52^1/_2$	$^7/_8$
g	87	$1^7/_8$
h	$121^1/_2$	$2^5/_8$
i	$156^1/_4$	$2^3/_8$
j	$191^3/_4$	$1^5/_8$
k	$190^1/_2$	$12^3/_8$
l	$155^7/_8$	$16^7/_8$
m	$121^1/_8$	$17^3/_4$
n	$86^5/_8$	$16^3/_4$
o	$52^1/_8$	$15^3/_4$
Outboard support pad		
p	0	$17^1/_2$
q	24	$17^1/_2$
r	24	$32^1/_2$
s	0	$32^1/_2$
Chine I		
t	$40^5/_8$	$20^5/_8$
u	$75^1/_8$	$19^1/_8$
v	$109^5/_8$	$18^1/_4$
w	$144^3/_8$	$20^1/_4$
x	179	$23^3/_4$
y	$175^5/_8$	$29^1/_8$
z	$144^3/_4$	$32^1/_4$
A	$110^1/_4$	$34^1/_4$
B	$75^7/_8$	$35^7/_8$
C	$41^3/_8$	$37^1/_2$
Buoyancy chamber wall		
D	55	$47^1/_8$
E	58	$42^7/_8$
F	$88^1/_2$	$36^7/_8$
G	123	$34^7/_8$
H	$157^1/_2$	$34^7/_8$
I	192	$34^7/_8$
J	192	48
K	$157^1/_2$	48
L	123	48
M	$88^1/_2$	48
N	$54^7/_8$	48

Part 2

	x	y
Frame I inner		
a	0	$11^7/_8$
b	$20^1/_2$	$11^7/_8$
c	$20^1/_2$	23
d	0	23
Frame I outer (first)		
e	$21^3/_4$	$14^7/_8$
f	$24^3/_4$	$14^7/_8$
g	36	$19^1/_4$
h	$36^1/_2$	26
i	$21^3/_4$	26

	x	y
Frame 2 inner		
j	0	$23^1/_2$
k	$20^3/_8$	$23^1/_2$
l	$20^3/_8$	$35^3/_4$
m	0	$35^3/_4$
Frame I outer (second)		
n	$21^7/_8$	$26^3/_4$
o	$36^5/_8$	$26^3/_4$
p	$36^5/_8$	$37^7/_8$
q	$33^5/_8$	$37^7/_8$
r	$22^3/_8$	$33^1/_2$
Frame 2 outer (first)		
s	0	$35^3/_4$
t	3	$35^3/_4$
u	$18^1/_2$	40
v	$18^5/_8$	48
w	0	48
Frame 2 outer (second)		
x	$18^5/_8$	$35^3/_4$
y	$21^5/_8$	$35^3/_4$
z	$37^1/_8$	40
A	$37^3/_8$	48
B	$18^5/_8$	48

Part 3

	x	y
Inner transom		
a	0	0
b	21	0
c	21	$13^3/_8$
d	0	$13^3/_8$
Transom		
e	$35^3/_4$	$4^1/_8$
f	$52^3/_8$	0
g	$79^3/_8$	0
h	96	$4^1/_8$
i	96	$19^1/_4$
j	$77^7/_8$	$19^1/_4$
k	$77^7/_8$	$15^1/_4$
l	$53^7/_8$	$15^1/_4$
m	$53^7/_8$	$19^1/_4$
n	$35^3/_4$	$19^1/_4$
Foredeck hatch part I		
o	96	0
p	113	0
q	113	$20^1/_2$
r	96	$20^1/_2$
Foredeck hatch part 2		
s	$113^1/_4$	0
t	$130^1/_4$	0
u	$130^1/_4$	$20^1/_2$
v	$113^1/_4$	$20^1/_2$
Thwart part 2		
w	134	0
x	192	0
y	192	$9^7/_8$
z	134	$9^7/_8$
Thwart part I		
A	134	$9^7/_8$
B	192	$9^7/_8$
C	192	20
D	134	20

Keel

E	$57^3/_8$	21
F	192	21
G	192	48
H	$57^3/_8$	48

Lower stem

I	$3^3/_4$	$31^5/_8$
J	$30^3/_4$	$31^5/_8$
K	$34^1/_4$	$36^3/_4$
L	$^1/_4$	$36^3/_4$

Upper stem

M	$1^1/_8$	$37^1/_8$
N	$35^1/_8$	$37^1/_8$
O	$36^1/_4$	48
P	0	48

Part 4

	x	y

Deck 1

a	$55^1/_2$	0
b	88	0
c	96	0
d	$122^1/_2$	$^3/_4$
e	157	$4^5/_8$
f	$190^1/_2$	$12^7/_8$
g	$190^1/_2$	$29^7/_8$
h	$181^1/_2$	$29^7/_8$
i	$181^1/_2$	$19^7/_8$
j	165	$19^7/_8$
k	165	$29^7/_8$
l	157	$29^7/_8$
m	157	$20^3/_8$
n	$122^1/_2$	$20^3/_8$
o	88	$20^3/_8$
p	$53^1/_2$	$20^3/_8$
q	96	$2^7/_8$
r	$114^1/_2$	$2^7/_8$
s	$114^1/_2$	$14^7/_8$
t	96	$14^7/_8$

Deck 2

u	$1^1/_2$	$18^1/_8$
v	$10^1/_2$	$18^1/_8$
w	$10^1/_2$	$28^1/_8$
x	27	$28^1/_8$
y	27	$18^1/_8$
z	35	$18^1/_8$
A	35	$27^5/_8$
B	70	$27^5/_8$
C	104	$27^5/_8$
D	$138^1/_2$	$27^5/_8$
E	$138^1/_2$	48
F	104	48
G	96	48
H	$69^1/_2$	48
I	35	$43^3/_8$
J	$1^1/_2$	$35^1/_8$
K	$77^1/_2$	$33^1/_8$
L	96	$33^1/_8$
M	96	$45^1/_8$
N	$77^1/_2$	$45^1/_8$

Dogsbody Coordinates (in millimetes), Part 1

	x	y

Frame 3 outer

a	801	0
b	1267	0
c	1267	323
d	1200	323
e	801	219

Chine 2

f	1287	21
g	2132	45
h	2978	65
i	3828	59
j	4696	39
k	4667	304
l	3820	414
m	2966	435
n	2121	411
o	1276	386

Outboard support pad

p	0	430
q	589	430
r	589	798
s	0	798

Chine 1

t	995	504
u	1839	468
v	2684	447
w	3537	495
x	4386	583
y	4304	712
z	3548	789
A	2702	841
B	1858	880
C	1014	918

Buoyancy chamber wall

D	1348	1154
E	1422	1051
F	2168	903
G	3013	855
H	3858	853
I	4704	855
J	4704	1176
K	3858	1176
L	3013	1176
M	2168	1176
N	1346	1176

Part 2

	x	y

Frame 1 inner

a	0	292
b	502	292
c	502	565
d	0	565

Frame 1 outer (first)

e	534	365
f	607	365
g	882	471
h	895	638
i	534	638

Frame 2 inner

j	0	577
k	499	577
l	499	875
m	0	875

Frame 1 outer (second)

n	536	655
o	897	655
p	897	928
q	823	928
r	549	822

Frame 2 outer (first)

s	0	877
t	73	877
u	453	981
v	457	1176
w	0	1176

Frame 2 outer (second)

x	457	877
y	531	877
z	911	981
A	915	1176
B	457	1176

Part 3

Inner transom

	x	y
a	0	0
b	515	0
c	515	326
d	0	326

Transom

e	875	102
f	1282	0
g	1945	0
h	2352	102
i	2352	472
j	1908	472
k	1908	374
l	1320	374
m	1320	472
n	875	374

Foredeck hatch part 1

o	2352	0
p	2768	0
q	2768	502
r	2352	502

Foredeck hatch part 2

s	2776	0
t	3192	0
u	3192	502
v	2776	502

Thwart part 2

w	3283	0
x	4704	0
y	4704	243
z	3283	243

Thwart part 1

A	3283	243
B	4704	243
C	4704	488
D	3283	488

Keel

E	1406	515
F	4704	515
G	4704	1176
H	1409	1176

Lower stem

I	91	773
J	752	773
K	838	901
L	5	901

Upper stem

M	29	910
N	861	910
O	889	1176
P	0	1176

Part 4

	x	y

Deck 1

	x	y
a	1312	0
b	2157	0
c	2353	0
d	3003	18
e	3848	115
f	4669	315
g	4669	733
h	4448	733
i	4448	488
j	4044	488
k	4044	733
l	3848	733
m	3848	500
n	3003	500
o	2157	500
p	1312	500
q	2353	71
r	2807	71
s	2807	365
t	2353	365

Deck 2

u	38	444
v	259	444
w	259	689
x	663	689
y	663	444
z	859	444
A	853	676
B	1704	676
C	2549	676
D	3395	676
E	3395	1176
F	2549	1176
G	2353	1176
H	1704	1158
I	859	1062
J	38	862
K	1900	811
L	2353	811
M	2353	1105
N	1900	1105

Note: Small discrepancies may exist between the millimeter and inch tables. See page ii.

Dogsbody's "egg crate" structure provides a great deal of stiffness and a lot of enclosed buoyancy. Cleats have been added to the bottom edges of the longitudinal members to provide a secure nailing and gluing surface for the bottom. Note also the vertical cleats that hold the transverse members.

the stitch-and-glue method. Next, flip the boat upright to allow access to the area below the deck so that the chine and side panels can be stitched and glued into place. Don't forget to seal the insides of the buoyancy tanks and stowage areas with paint or epoxy.

Add the cleats along the inside of the hull, across the tops of the frames, and along the line of the seat fronts so that the

The bottom, previously butt-jointed, is attached to the framework. Note the neatly countersunk holes for the fasteners.

decks can be installed. The transom needs a substantial pad of laminated plywood to take the outboard, and because Dogsbody is a workboat there should be a couple of fittings for painters at the bow and stern. Be sure to add sufficient plywood reinforcements for the painter eyes. The rest of the boat will need inwales and gunwales, and the hatch in the bow, where the fuel tank will be installed, needs a lid as shown in the drawings.

The job is just about done. All that's left to do is paint the boat, buy an outboard, and get out on the water.

Gunwales glued and clamped in place. Soon, the decks will go on. The ample volume of the buoyancy tanks is apparent in this shot.

A finished Dogsbody with bimini top. (Ken Newton)

JIGGITY

Length Over All (LOA): 9'4"
Length Waterline (LWL): 7'4½"
Beam: 47½"
Weight: 70–85 lb.
Displacement at Design Waterline: 500 lb.
Crew: 2 adults
Propulsion: Oars, very small outboard
Construction Methods: PU stitch-and-glue
Epoxy stitch-and-glue

I've long admired Philip Bolger's Auray punt, and, based on the frequent discussions about it on the Web, it appears that I'm not alone. There's good reason for the interest in this punt, particularly for use as a dinghy. It rows and tows well, it can handle rough water, and it can carry a huge amount of weight for its waterline length. It can also be used with an outboard of 4 hp or less. Having recently built one of these boats, I found that the design was strikingly simple and quite suitable for the stitch-and-glue method.

It didn't take long for me to start thinking about adding my own touch to the Auray concept and eventually I came up with Jiggity, a 9-foot version of the Auray punt that can be rowed, sculled, or powered by a small outboard. The design is simple and you should be able to build a Jiggity of your own in a couple of weekends. In fact, the build will seem straightforward to anyone who has built in stitch-and-glue. The boat isn't much more complicated than a Mouse, but as simple as the design may be, though, I still suggest building a model first.

To get started on the actual boat, mark out the plywood with 10-inch squares, carefully cutting off the excess where it is slightly longer than the 8 feet it's supposed to be. Plot the curved panels, cut them out, and joint the forward and aft sections of the panels either with a plywood strap or by taping each side, and then stitch-and-glue in the usual way.

You could build from ¼-inch plywood, though I'd suggest using ⅜-inch plywood if the boat is to be used as a tender. Tenders are used hard and often neglected and will, therefore, benefit from a rugged build. If you build the boat as it is drawn and use ⅜-inch plywood, you will have a robust and rigid boat because all the large panels are strongly curved. With this boat, you won't be able to rely on cloth-backed tape alone, so you'll need to use cable ties to hold the boat together while assembling it.

Although I have drawn up a set of panels, bottom and transoms, and a central frame, if you've got some boatbuilding experience you might choose to build this boat even more strongly with, say, nine equally-spaced traditional-style frames, internal chine logs, and long stringers running fore and aft across the traditional frames to support the thwarts (see accompanying illustrations).

If you go this route and intend to apply a traditional-looking workboat-style black finish, make your joints fit well, and work with good metal fasteners and formaldehyde or waterproof polyurethane glue. Don't use epoxy. As it becomes hot, it quickly softens. A black-painted epoxy-built boat left in the hot sun could become fragile.

Jiggity

9'5" by 4' by 500 lbs. rowing and outboard-powered dinghy and tender, based on the Auray punt

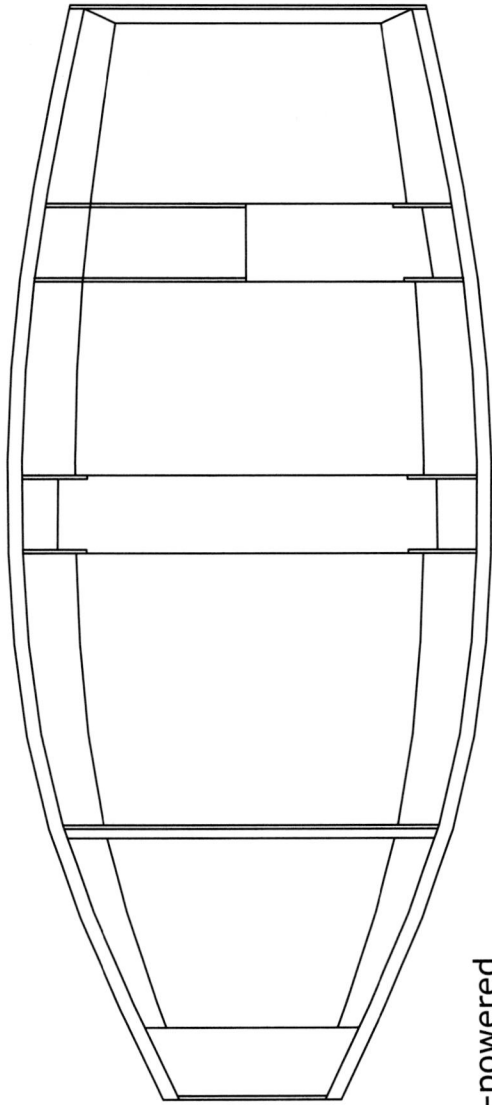

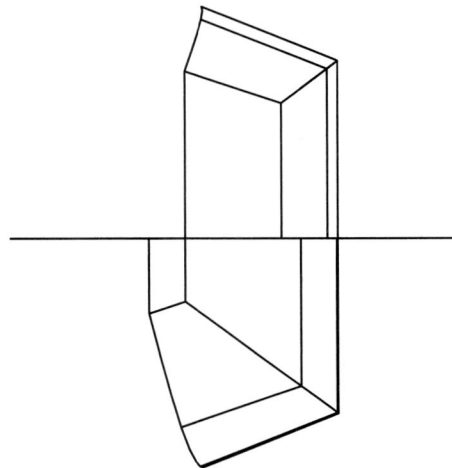

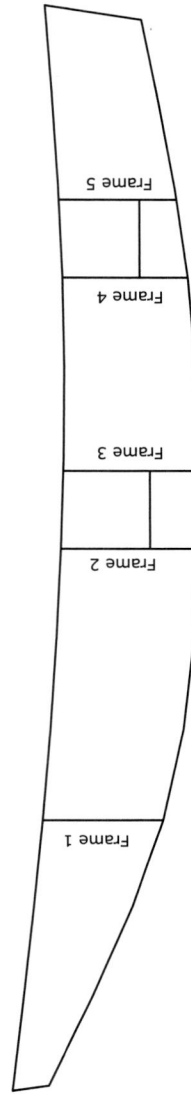

Frame 5

Frame 4

Frame 3

Frame 2

Frame 1

Frame 4

5

34 1/2

13

Frame 5

4 5/8

32 3/4

12 1/4

500 lbs.

2

Frame 2

14 1/4

36

5 1/2

6

Frame 3

14 1/8

36 1/8

5 1/2

6 1/8

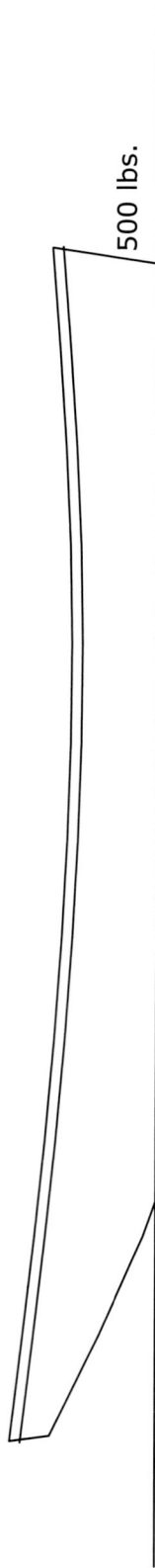

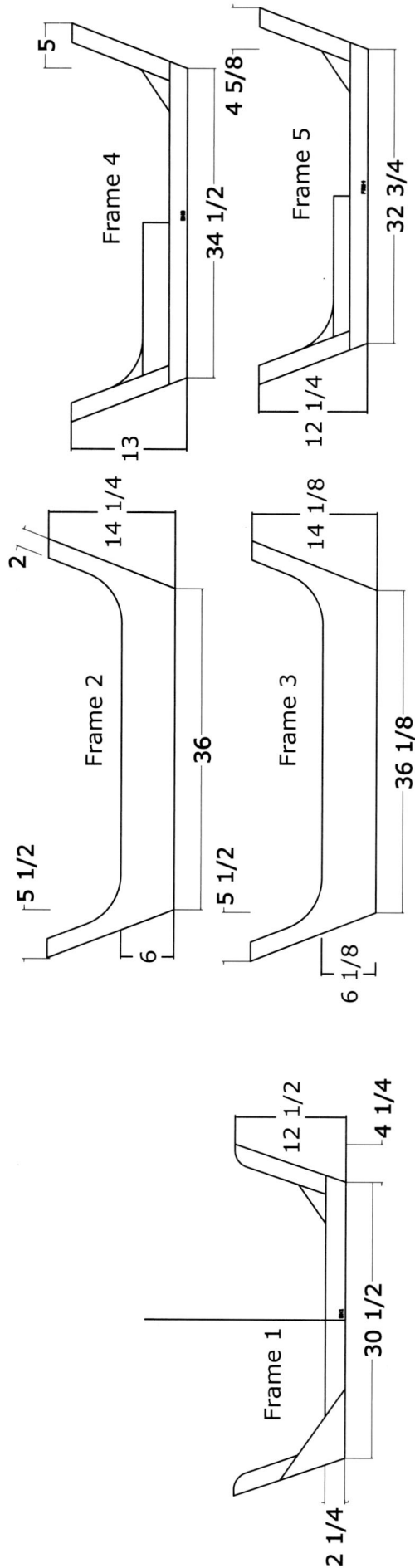

All dimensions in inches

12 1/2

4 1/4

Frame 1

30 1/2

2 1/4

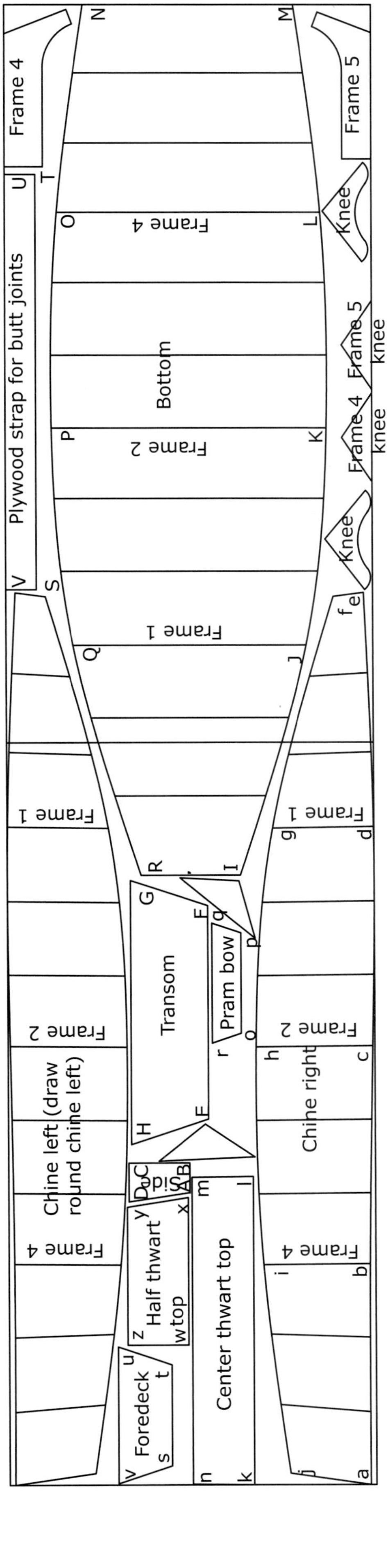

Frame 4

Frame 5

Knee

Plywood strap for butt joints

Frame 4

Frame 5
knee

Frame 4
knee

Bottom

Frame 2

Knee

Frame 1

Frame 1

Frame 1

Transom

Pram bow

Frame 2

Chine left (draw round chine left)

Chine right

Frame 4

Frame 2

Half thwart
top

Foredeck

Center thwart top

Frame 4

Quarter knee
(two are needed)

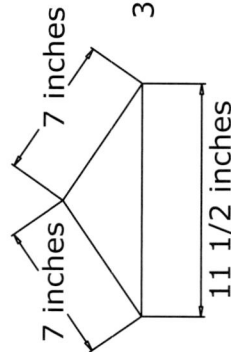

8 1/4 inches

8 3/8 inches

6-inch radius

2-inch radius

13 inches

5 inches

12 7/8 inches

6-inch radius

111 degrees

2 inches

8 3/4 inches

17 1/4 inches

Frame 4 closed-end
knee

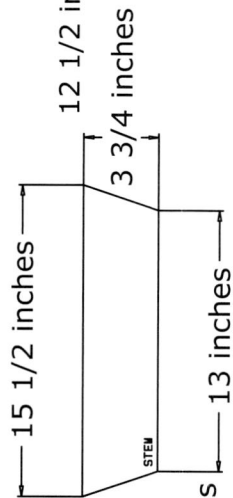

Frame 4 open-end
knee

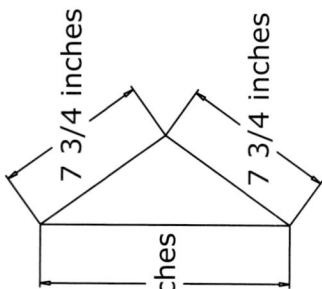

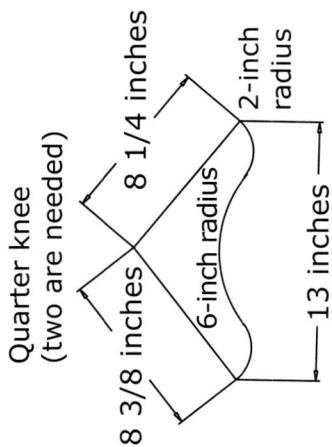

7 3/4 inches

7 3/4 inches

12 1/2 inches

Stem

3 3/4 inches

15 1/2 inches

13 inches

STEM

6-inch radius

2 inches

7 1/2 inches

111 degrees

16 3/8 inches

Frame 5 closed-end
knee

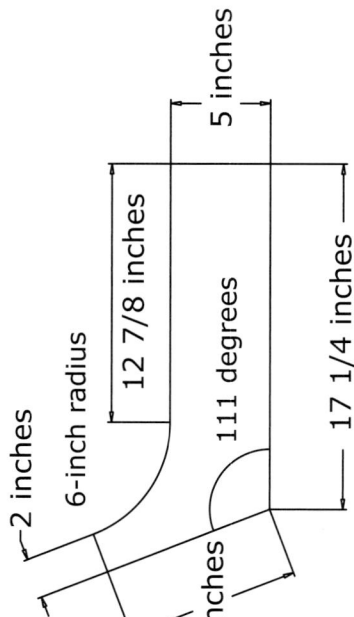

11 1/2 inches

3 3/4 inches

Rear seat side

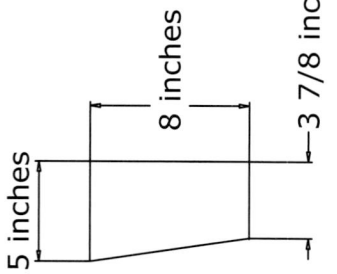

5 inches

8 inches

3 7/8 inches

7 inches

7 inches

7 inches

11 1/2 inches

Frame 5 open-end
knee

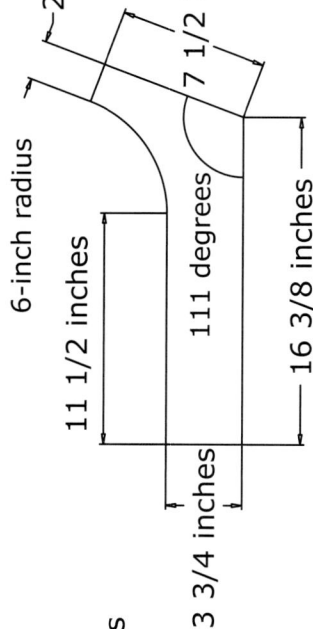

Jiggity Coordinates (in inches)

	x	y
Side panels		
a	0	$\frac{3}{4}$
b	$28\frac{1}{2}$	$\frac{3}{4}$
c	$56\frac{5}{8}$	$\frac{1}{4}$
d	85	0
e	$115\frac{5}{8}$	$1\frac{1}{4}$
f	$115\frac{1}{8}$	$5\frac{1}{8}$
g	$84\frac{7}{8}$	$13\frac{1}{4}$
h	$56\frac{1}{2}$	$15\frac{1}{2}$
i	$28\frac{1}{2}$	$14\frac{5}{8}$
j	$1\frac{1}{2}$	$11\frac{1}{4}$
Thwart		
k	0	16
l	$39\frac{7}{8}$	16
m	$39\frac{3}{4}$	24
n	$\frac{1}{8}$	24
Pram bow transom		
o	$58\frac{1}{4}$	$17\frac{1}{2}$
p	$71\frac{1}{4}$	$17\frac{1}{2}$
q	$72\frac{1}{2}$	$21\frac{1}{4}$
r	57	$21\frac{1}{4}$
Foredeck		
s	$2\frac{1}{2}$	$26\frac{3}{4}$
t	$15\frac{1}{2}$	$26\frac{3}{4}$
u	$17\frac{3}{4}$	$33\frac{3}{4}$
v	$\frac{1}{8}$	$33\frac{3}{4}$
Half thwart		
w	$18\frac{1}{4}$	$24\frac{1}{2}$
x	$37\frac{1}{8}$	$24\frac{1}{2}$
y	$35\frac{7}{8}$	$32\frac{1}{2}$
z	18	$32\frac{1}{2}$
Half thwart side		
A	$37\frac{3}{4}$	$24\frac{1}{4}$
B	$41\frac{5}{8}$	$24\frac{1}{4}$
C	$41\frac{5}{8}$	$32\frac{1}{4}$
D	$36\frac{5}{8}$	$32\frac{1}{4}$
Transom		
E	$47\frac{1}{4}$	$21\frac{3}{4}$
F	$74\frac{3}{4}$	$21\frac{3}{4}$
G	78	$31\frac{7}{8}$
H	44	$31\frac{7}{8}$
Bottom		
I	$78\frac{5}{8}$	$17\frac{1}{2}$
J	$108\frac{3}{4}$	$8\frac{3}{4}$
K	137	6
L	$165\frac{1}{8}$	$6\frac{3}{4}$
M	192	$10\frac{1}{4}$
N	192	$37\frac{3}{4}$
O	$165\frac{1}{8}$	$41\frac{1}{2}$
P	137	42
Q	$108\frac{3}{4}$	$39\frac{1}{4}$
R	$78\frac{5}{8}$	$30\frac{1}{2}$
Butt strap for joints		
S	116	44
T	170	44
U	170	48
V	116	48

Jiggity Coordinates (in millimeters)

	x	y
Side panels		
a	0	18
b	700	18
c	1387	7
d	2083	1
e	2832	30
f	2819	127
g	2079	325
h	1384	381
i	697	359
j	37	276
Thwart		
k	0	392
l	976	392
m	974	588
n	4	588
Pram bow transom		
o	1426	428
p	1745	428
q	1776	521
r	1396	521
Foredeck		
s	61	656
t	379	656
u	435	827
v	4	827
Half thwart		
w	441	600
x	911	600
y	880	796
z	441	796
Half thwart side		
A	926	595
B	1020	595
C	1020	791
D	897	791
Transom		
E	1157	534
F	1831	534
G	1911	782
H	1078	782
Bottom		
I	1927	428
J	2663	214
K	3357	146
L	4044	165
M	4704	250
N	4704	924
O	4044	1010
P	3357	1028
Q	2663	961
R	1927	747
Butt strap for joints		
S	2842	1078
T	4165	1078
U	4165	1176
V	2842	1176

Note: Small discrepancies may exist between the millimeter and inch tables. See page ii.

Jiggity takes its cues from the traditional Auray punt. Fitted with traditional tholepins instead of oarlocks, and painted in tarry-looking workboat black, this little punt seems just right for this piratical-looking Breton sailor in the harbor of Douarnenez. These boats make good tenders because the longish bow keeps the occupants dry, and they are capable of carrying large loads in a short waterline length.

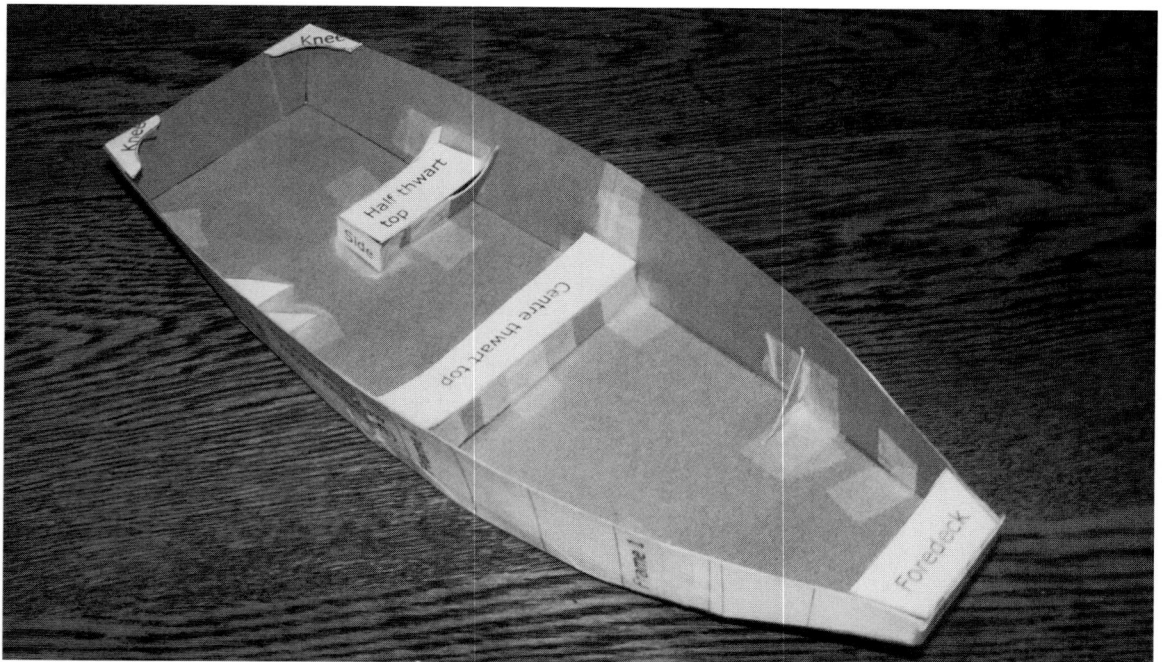

Whether you build in paper or plywood, Jiggity should perform as well as the original Auray punt and will require a good deal less effort.

AURETTE

A SMALL SAILING AURAY PUNT

Length Over All (LOA): 7'7"
Length Waterline (LWL): 5'7"
Beam: 48"
Weight: 65–80 lb.
Displacement at Design Waterline: 380 lb. (author's estimate)
Crew: 2 adults
Propulsion: Oars, very small outboard
Construction Methods: PU stitch-and-glue
Epoxy stitch-and-glue

Here's another interpretation of the Auray punt, this one a sailing version from Australian designer Murray Isles. Aurette is both smaller and more refined than Jiggity (while still being a very straightforward build), so if you like the Auray's hull shape and you want a sailboat, or a boat that's a bit better looking, go ahead and build it. If you're not planning to go sailing, or you want a simpler project, go with Jiggity.

Build Aurette from ¼-inch plywood using the stitch-and-glue method. All the salient building points are covered in the previous chapters, including the last one.

Aurette can be built with either a leeboard, which hooks onto the side and must be shifted across the boat each time you tack or jibe, or with a daggerboard, which slides through a case in the middle of the boat. From the drawing you can see that the daggerboard case is angled so that its top is forward of and higher than the middle thwart. It is always tempting when fitting daggerboard cases to make them low so they don't intrude too much on the limited space inside the boat. When the designers do that, however, they are engineering a low flooding point into the boat. Often, the result is a boat that can't be effectively bailed out because the water comes in through the daggerboard case faster than it can be bailed over the gunwale. Obviously, the leeboard option won't pose any problem in this regard, and the tall daggerboard case in the design addresses the problem of flooding.

The original design does not include built-in buoyancy tanks. For legal and safety reasons, you should glue or strap foam blocks under the thwarts (seats). The blocks should be closed-cell foam and, if necessary, protected from the elements and damage. You'll need about ⅔ of a cubic yard of foam, and you should leave about 1 inch everywhere between the foam and the bottom of the boat to ensure drainage. Make sure you use closed-cell foam; the trashy white stuff used for packaging will eventually soften and soak up a significant amount of water.

A word on the sides: Some boatbuilders have found that the curve in the sides of the boat makes it difficult to fit gunwales made from some types of lumber. Soaking the boards in hot water would help but makes gluing a problem, so I'd suggest a safer option would be to laminate the gunwales from two pieces of thinner stock.

Aurette is sailed with a lug rig that is easy to make with short spars, and there's not a lot of stress on the sail. The sail in a lug rig is shaped more like a square, unlike the triangular sails of a traditional Bermuda rig, and that provides a benefit in that the shape doesn't need careful sheeting to keep it in trim; it's a great sail for novice sailors.

All dimensions in millimeters

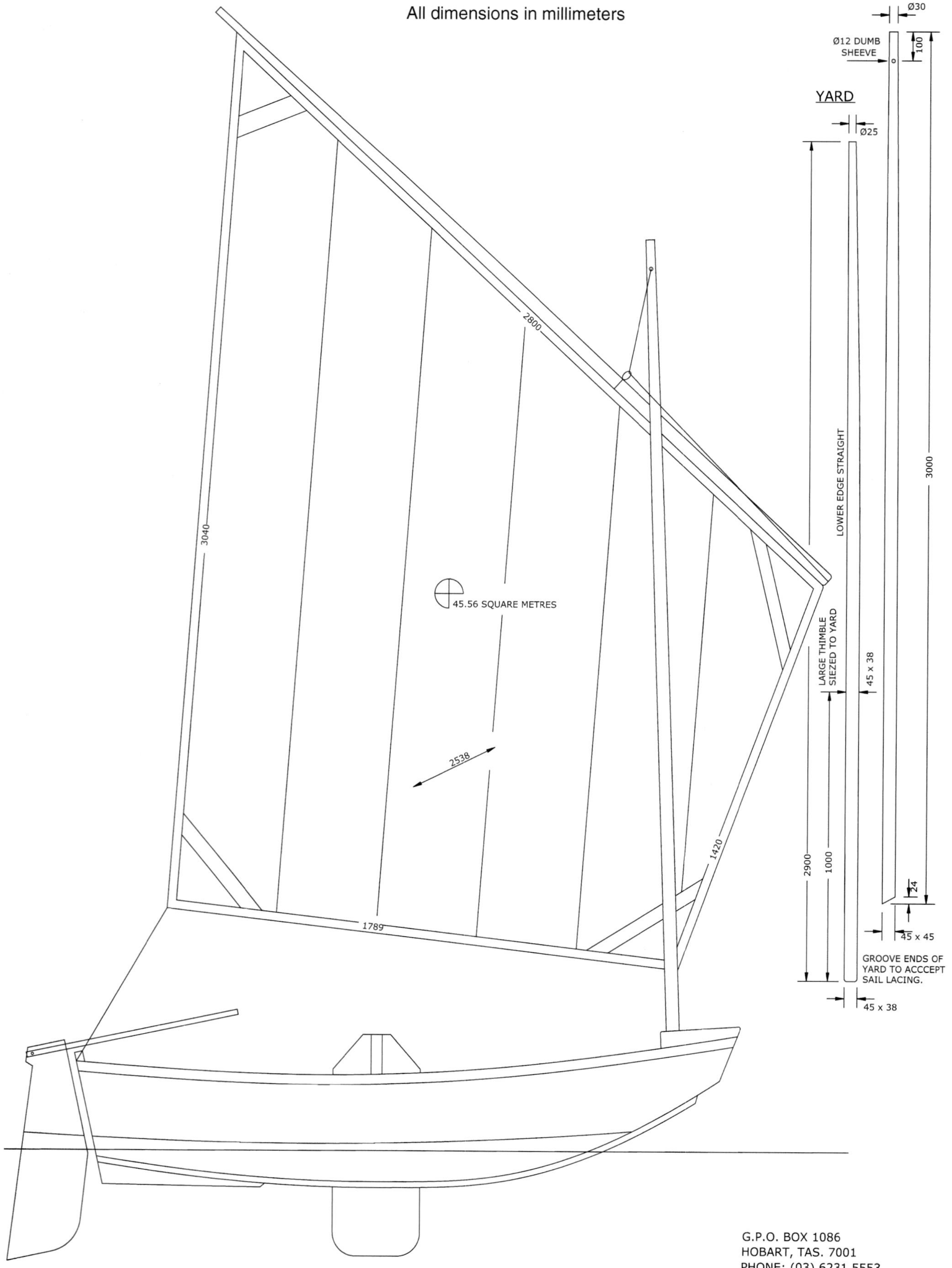

Ø30

Ø12 DUMB
SHEEVE

100

YARD

Ø25

2800

LOWER EDGE STRAIGHT

3000

3040

45.56 SQUARE METRES

LARGE THIMBLE
SIEZED TO YARD

45 x 38

2538

1420

2900

1000

24

1789

45 x 45

GROOVE ENDS OF
YARD TO ACCCEPT
SAIL LACING.

45 x 38

MURRAY ISLES
CONSULTANT IN VESSEL DESIGN & FLEET MANAGEMENT

G.P.O. BOX 1086
HOBART, TAS. 7001
PHONE: (03) 6231 5553
FAX: (03) 6231 5553
MOBILE: 0407 543 941
EMAIL: islesscd@mpx.com.au

All dimensions in millimeters

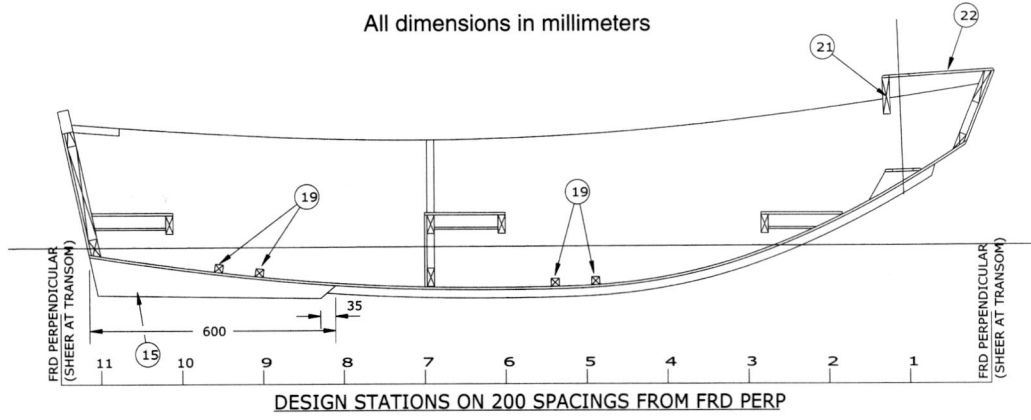

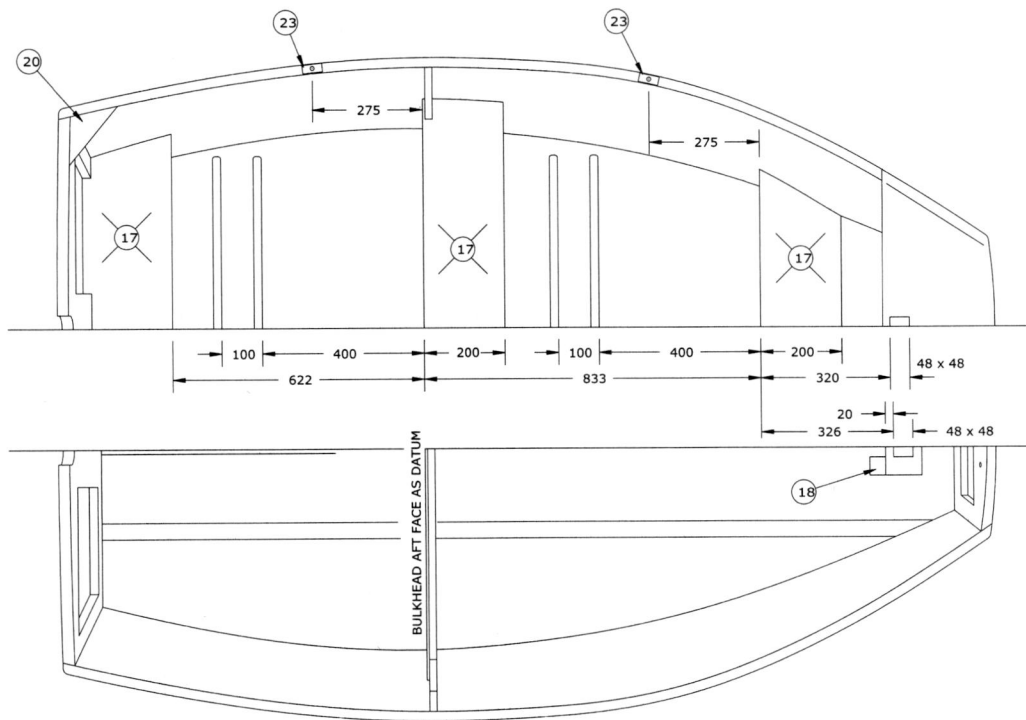

FRD PERPENDICULAR (SHEER AT TRANSOM)

600

35

11 (15) 10   9   8   7   6   5   4   3   2   1

FRD PERPENDICULAR (SHEER AT TRANSOM)

DESIGN STATIONS ON 200 SPACINGS FROM FRD PERP

275

275

100   400   200   100   400   200

622   833   320

48 x 48

20

326   48 x 48

BULKHEAD AFT FACE AS DATUM

139

All dimensions in millimeters

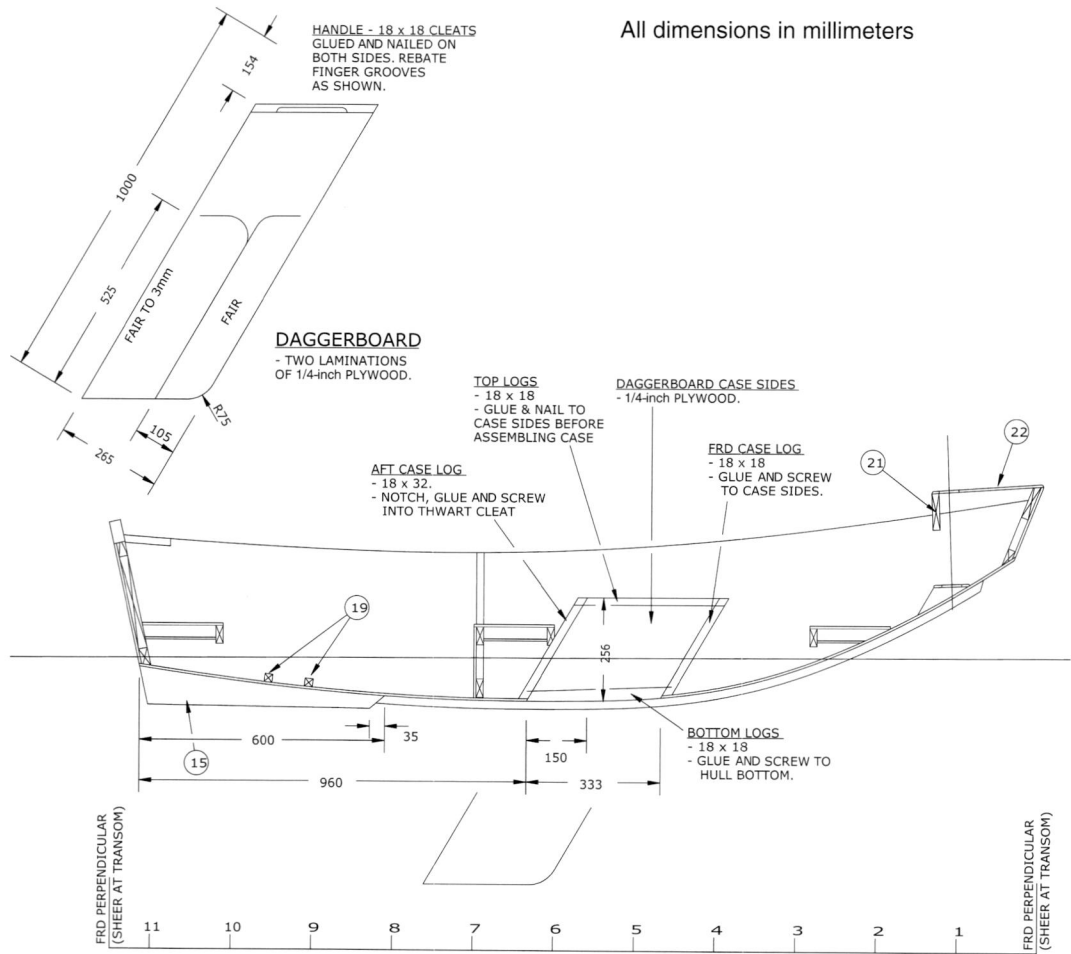

HANDLE - 18 x 18 CLEATS
GLUED AND NAILED ON
BOTH SIDES. REBATE
FINGER GROOVES
AS SHOWN.

154

1000

525

FAIR TO 3mm

FAIR

R75

105

265

DAGGERBOARD
- TWO LAMINATIONS
OF 1/4-inch PLYWOOD.

TOP LOGS
- 18 x 18
- GLUE & NAIL TO
CASE SIDES BEFORE
ASSEMBLING CASE

DAGGERBOARD CASE SIDES
- 1/4-inch PLYWOOD.

FRD CASE LOG
- 18 x 18
- GLUE AND SCREW
TO CASE SIDES.

AFT CASE LOG
- 18 x 32.
- NOTCH, GLUE AND SCREW
INTO THWART CLEAT

22

21

19

256

15

600

35

150

BOTTOM LOGS
- 18 x 18
- GLUE AND SCREW TO
HULL BOTTOM.

960

333

FRD PERPENDICULAR
(SHEER AT TRANSOM)

FRD PERPENDICULAR
(SHEER AT TRANSOM)

11 10 9 8 7 6 5 4 3 2 1

DESIGN STATIONS ON 200 SPACINGS FROM FRD PERP

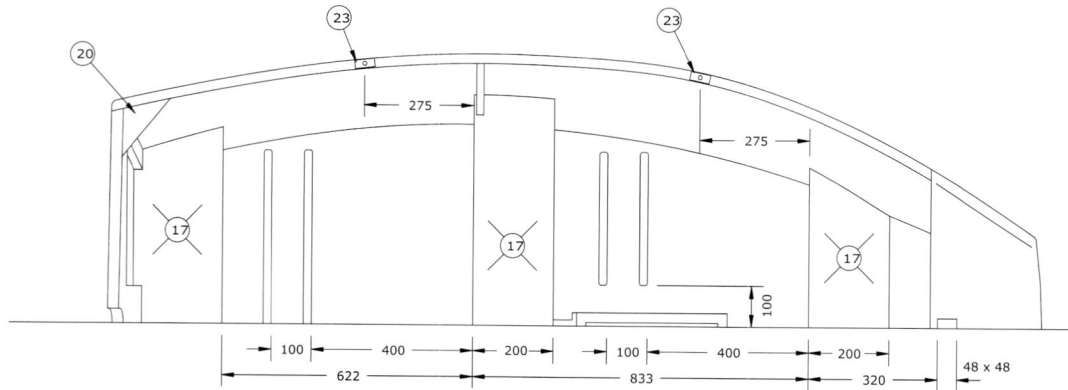

20

23

23

275

275

17

17

17

100

100 400 200 100 400 200 48 x 48

622 833 320

140

All dimensions in millimeters

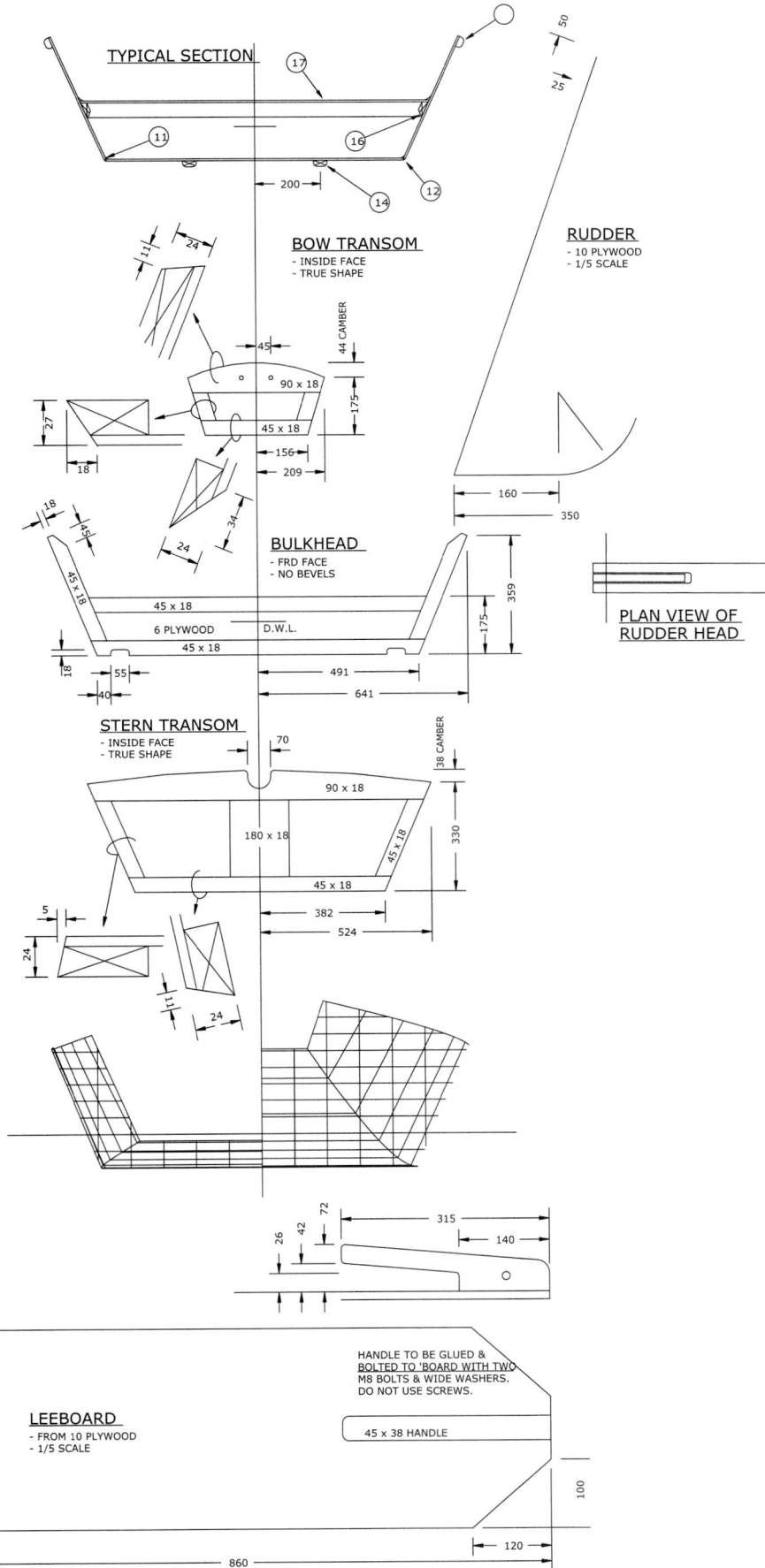

TYPICAL SECTION

17

11   16

200   14   12

RUDDER
- 10 PLYWOOD
- 1/5 SCALE

50

25

BOW TRANSOM
- INSIDE FACE
- TRUE SHAPE

24   11

45   44 CAMBER

90 x 18   175

45 x 18

27   156

18   209

160

350

18   18

45   34

45 x 18   24

BULKHEAD
- FRD FACE
- NO BEVELS

45 x 18   359

6 PLYWOOD   D.W.L.   175

45 x 18

18   55   491

40   641

PLAN VIEW OF
RUDDER HEAD

STERN TRANSOM
- INSIDE FACE
- TRUE SHAPE

70   38 CAMBER

90 x 18

180 x 18   330

45 x 18

45 x 18

5   382

24   524

11   24

72   315

26   42   140

HANDLE TO BE GLUED &
BOLTED TO 'BOARD WITH TWO
M8 BOLTS & WIDE WASHERS.
DO NOT USE SCREWS.

45 x 38 HANDLE

LEEBOARD
- FROM 10 PLYWOOD
- 1/5 SCALE

100

860   120

All dimensions in millimeters

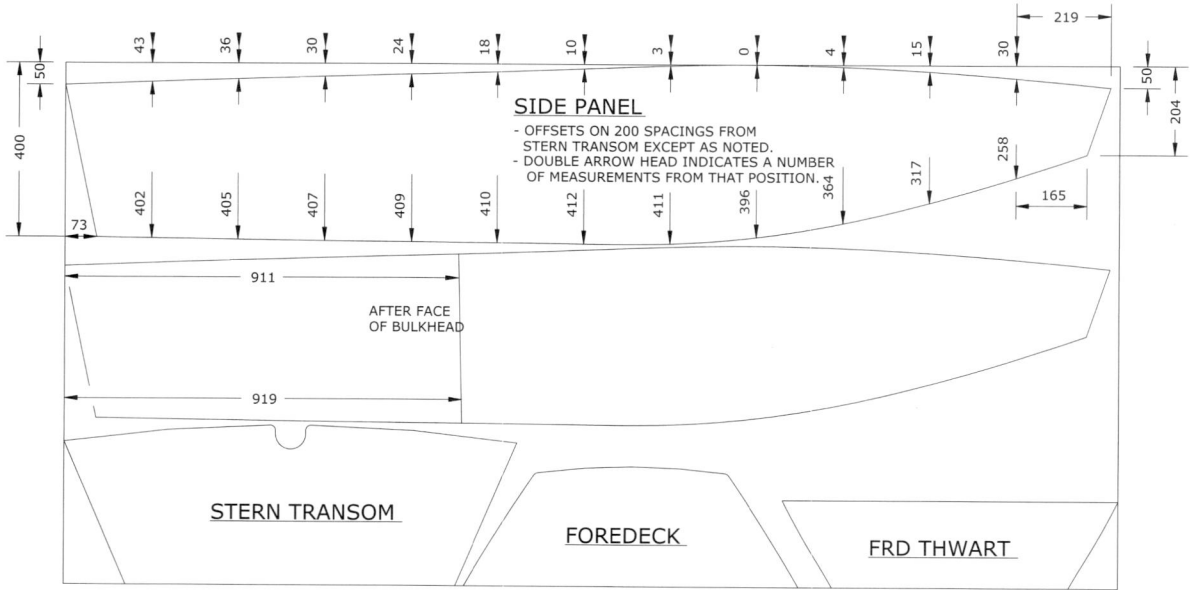

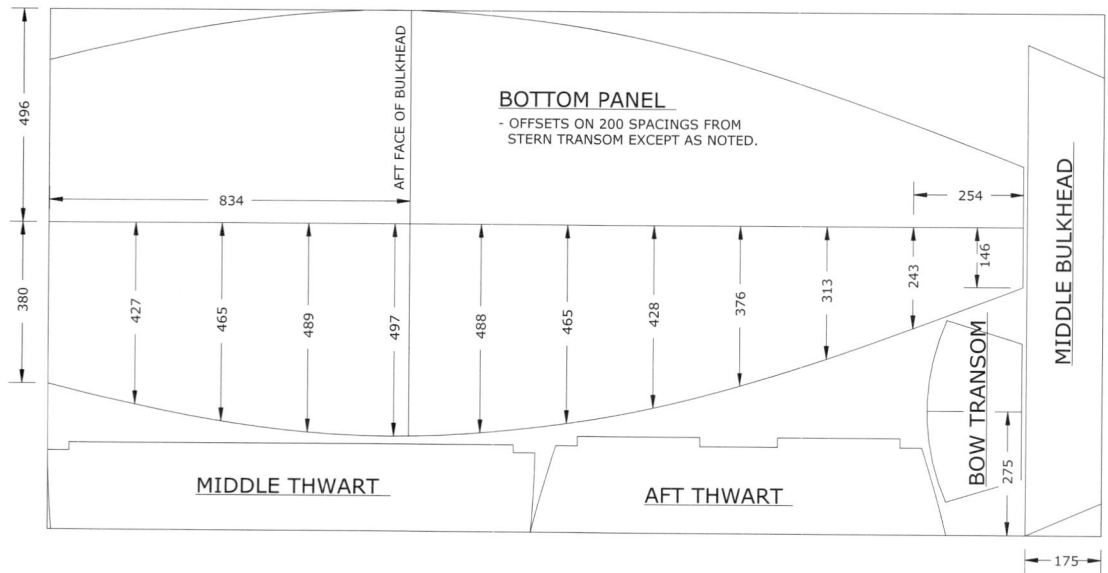

SIDE PANEL

- OFFSETS ON 200 SPACINGS FROM
  STERN TRANSOM EXCEPT AS NOTED.
- DOUBLE ARROW HEAD INDICATES A NUMBER
  OF MEASUREMENTS FROM THAT POSITION.

219

50

43 36 30 24 18 10 3 0 4 15 30

400

204

50

402 405 407 409 410 412 411 396 364 317 258

165

73

911

AFTER FACE
OF BULKHEAD

919

STERN TRANSOM

FOREDECK

FRD THWART

BOTTOM PANEL

- OFFSETS ON 200 SPACINGS FROM
  STERN TRANSOM EXCEPT AS NOTED.

AFT FACE OF BULKHEAD

496

834

254

380

427 465 489 497 488 465 428 376 313 243

146

MIDDLE BULKHEAD

BOW TRANSOM

275

MIDDLE THWART

AFT THWART

175

PIRAGUA

A PIROGUE

Length Over All (LOA): 14'
Length Waterline (LWL): 13'2"
Beam: 30"
Weight: 70 lb.
Displacement at Design Waterline: 300 lb.
Crew: 1 adult
Propulsion: Double or single paddle
Construction Method: Simplified chine log

Jim Michalak's Piragua is an easy-to-build version of the old-fashioned pirogue, which is a very stable platform. These boats are commonly used in swamps.

Jim's version of the pirogue has two features that come up in his boats time after time: a Bolger-style bottom profile that takes the ends of the boat out of the water, and boxed-in buoyancy and storage compartments fore and aft designed to keep clothes and food dry. They also keep the boat afloat and at least a little stable if it swamps.

Piragua, which can be built from two sheets of ¼-inch plywood, is light, only about 70 pounds when empty and dry, making it lighter and more tippy than a pirogue built more heavily using traditional materials. The boat is designed to be double-paddled from a sitting position.

Working from the detailed plans, start the project by marking out the parts for which dimensions and angles are given: that is, the frames and transom (including the angles on the lumber cleats used to attach them to the boat's sides and bottom), one side panel (use the first as a pattern for the opposite side), and the stem.

Make absolutely sure that the positions where the frames are to go are clearly marked on the side panels before cutting out and butt-blocking the fore and aft sections of the

bottom and side panels, and the two halves of the foredeck. The designer favors using butt blocks of ¾-by-3½-inch softwood rather than the plywood straps described earlier in this book.

Assembly is similar to methods used for the flat-bottomed Mouseboats, except for two small differences. The first is that the forward part of the boat comes to a sharp stem rather than a transom and the second is that there is a temporary form in the middle that is removed once it has done its job.

Glue and attach the side panels to the stem, followed by the temporary form (without glue), then the transom, and then the second and third frames. With a boat of this size, you'll need to use a length of rope with a Spanish windlass (see page 39, Chapter 4) or a trucker-style knot (tie a small loop, run the other end around the materials to be held together, then through the loop, and tie off) to help pull the side panels together so the frames and sides can be accurately screwed and glued together in an orderly, well-controlled way. If you haven't done this before, a helper will make it easier.

The chine log is then glued and screwed to the sides as shown and trimmed to accept the bottom, which is glued and screwed into place. Cut the inner gunwale at the correct angle, and glue and screw it in place.

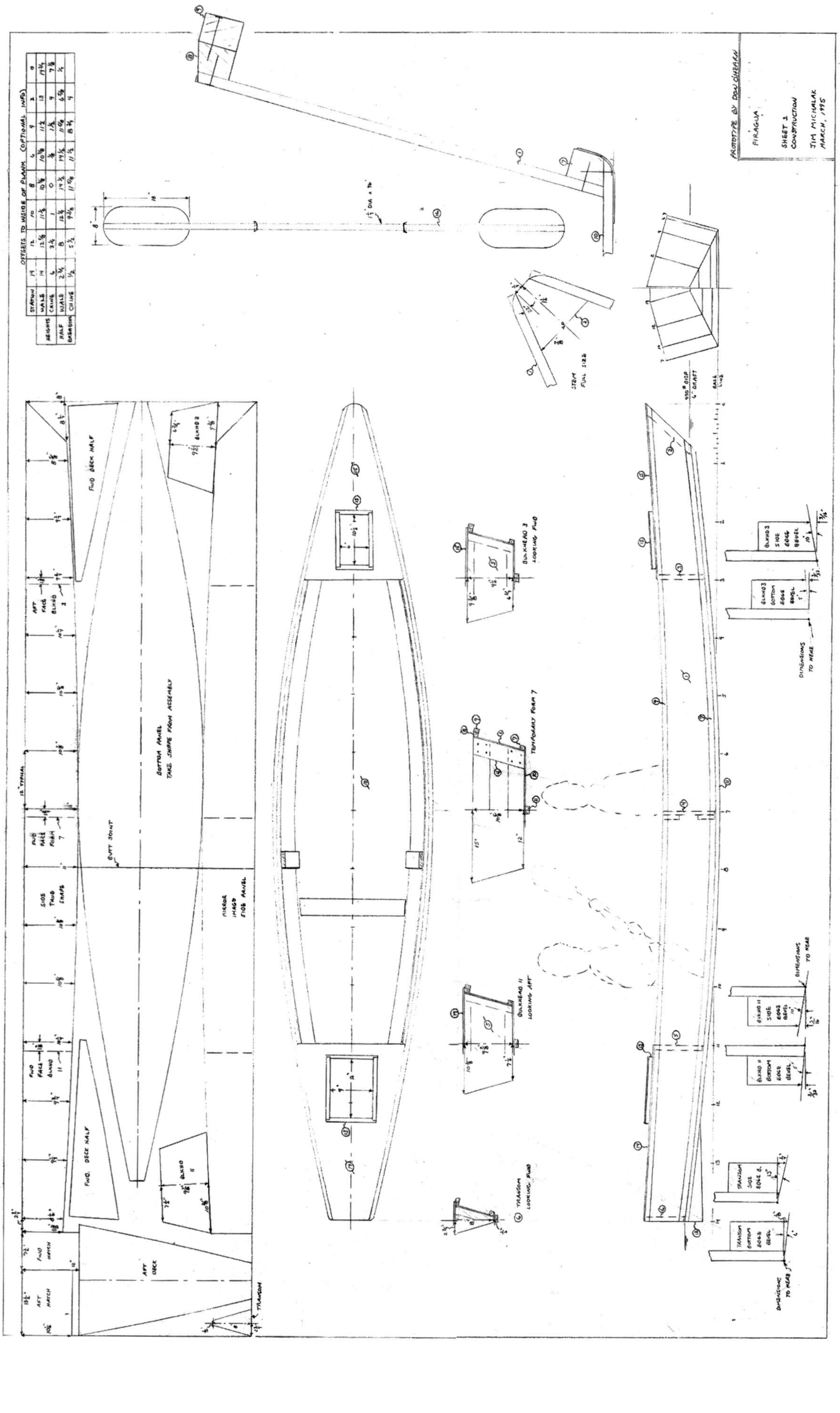

A very capable pirogue, Piragua is one of Jim Michalak's most popular designs. Longer and slimmer than the Cruising Mouse, it doesn't have as much carrying capacity but is faster and more refined. (Garth Battista)

Your next step is to mark the storage chamber and buoyancy tank decks by tracing around the inner gunwale fore and aft. Cut the decks out, and screw and glue them in place, though not before painting the insides of the tanks. Now add the outer gunwales and the storage tank hatch fittings and lids, followed by paint and varnish to finish the job.

CINDERELLA

A DOUBLE-PADDLE CANOE
WITH A "ROUNDISH" BOTTOM

Length Over All (LOA): 12'
Length Waterline (LWL): 11'8"
Beam: 30"
Weight: 40–50 lb.
Displacement at Design Waterline: 325 lb.
Crew: 1 adult and 1 child
Propulsion: Double or single paddle
Construction Methods: PU stitch-and-glue
                           Epoxy stitch-and-glue

Cinderella is a multi-panel, stitch-and-glue, plywood canoe designed to eliminate the need for a strong back. This wooden structure, which has to be built and then discarded accounts for a lot of the hard work usually involved in making boats of this kind.

You could call Cinderella a multi-strake canoe or a soft-chine canoe, but it really is somewhere between.

Begin building Cinderella by temporarily attaching the slender flat-bottom panel to a straight girder. The remaining hull panels (planks or strakes) are then stitched or taped together, and temporary T-shaped frames are inserted into the hull to force it into the proper shape before the strakes are permanently epoxy-and-fiberglass-taped together.

The 12-foot-long girder can be made of almost any straight piece of 2-by-6-inch or 2-by-8-inch lumber. You can even use 2 by 4s, but they're not usually very straight. To address this problem, nail or screw some reasonably straight ones together so the opposing curves help make the lumber straighter.

A better way to fashion a straight girder is to use ¾-inch chipboard or medium-density fiberboard. Cut 12 feet of the stuff 6 inches wide, and 24 feet of it 3 inches wide, and assemble it into an inverted U-shaped channel,

making sure the joints on the top surface don't coincide with the joints on the side pieces.

Cinderella was designed to be built from ⅛-inch plywood, but you can use ¼-inch plywood if you're not concerned about weight. Whichever thickness of material you use, however, it must be at least three-ply marine grade plywood because Cinderella has no internal framing, and its strength comes from the hull panels. Voids found in the cores of cheaper grades of plywood will cause serious weak spots.

Like several other boats in this book, Cinderella is longer than a single sheet of plywood, so a certain amount of butt-jointing is needed. The same technique of epoxy-and-fiberglass taping on both sides of the joint will work perfectly. But where for some designs I suggest cutting out the panel components and then jointing them, this won't work well for Cinderella because the panels are so narrow. It would be difficult to ensure perfect alignment of their very short straight butt edges when gluing them together. Therefore, I recommend cutting the one and a half sheets of plywood you need for the strakes down the middle as shown in the drawing, and then butt-jointing them. Let the joint harden sufficiently, then mark and cut the panels into two 12-by-2-foot

Cinderella

12-foot flat-bottomed, round-sided, elegant but easy to build canoe

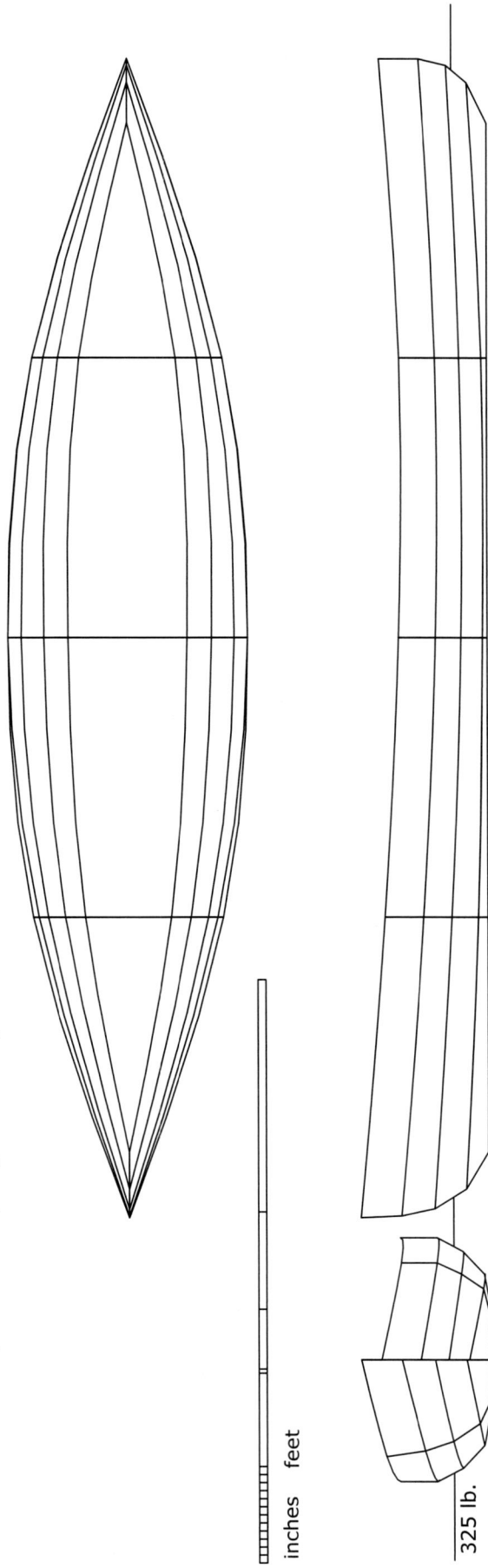

inches   feet

325 lb.

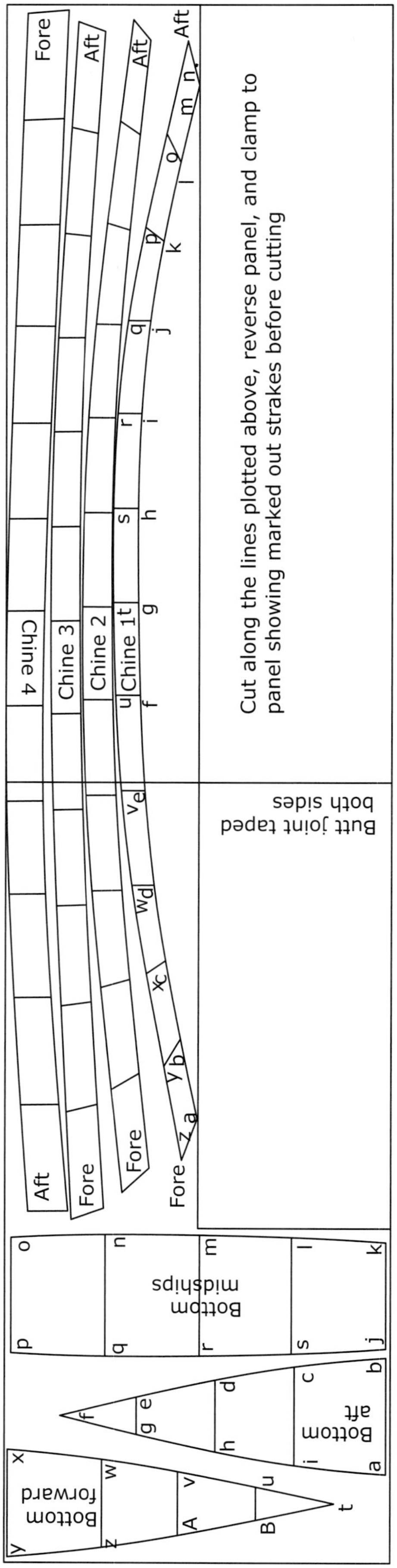

Fore
Aft
Aft
Aft

Chine 4
Chine 3
Chine 2
Chine 1

Cut along the lines plotted above, reverse panel, and clamp to panel showing marked out strakes before cutting

Butt joint taped both sides

Aft
Fore
Fore
Fore

Bottom midships

Bottom aft

Bottom forward

Cinderella Coordinates (in inches)

	x	y
Bottom panels		
a	$10^7/_8$	$^1/_2$
b	25	$^1/_2$
c	24	12
d	$22^3/_8$	$21^3/_4$
e	$20^1/_4$	$31^5/_8$
f	18	$41^1/_8$
g	$15^5/_8$	$31^5/_8$
h	$13^1/_2$	$21^3/_4$
i	$11^7/_8$	12
j	$26^1/_8$	$^5/_8$
k	$39^1/_4$	$^5/_8$
l	$39^7/_8$	$12^3/_8$
m	$40^1/_8$	$23^3/_4$
n	$40^1/_4$	$35^1/_2$
o	$39^7/_8$	$47^1/_4$
p	$25^5/_8$	$47^1/_4$
q	$25^1/_4$	$35^1/_2$
r	$25^1/_4$	$23^3/_4$
s	$25^1/_2$	$12^3/_8$
t	$7^1/_4$	7
u	$9^1/_4$	$16^3/_4$
v	$11^1/_8$	$26^1/_2$
w	$12^5/_8$	$35^7/_8$
x	$13^3/_4$	$47^5/_8$
y	$^3/_4$	$47^5/_8$
x	$1^7/_8$	$35^7/_8$
A	$3^3/_8$	$26^1/_2$
B	$5^1/_4$	$16^3/_4$

	x	y
Chine 1		
a	$54^3/_8$	$24^1/_8$
b	$64^1/_8$	$26^1/_4$
c	$73^3/_4$	$28^1/_8$
d	$83^1/_4$	$29^1/_2$
e	95	$30^3/_4$
f	$106^3/_4$	$31^1/_4$
g	$118^1/_4$	$31^5/_8$
h	$129^7/_8$	$31^5/_8$
i	$141^5/_8$	$31^1/_8$
j	153	$30^1/_8$
k	$162^7/_8$	$28^1/_2$
l	$172^5/_8$	$26^3/_8$
m	$182^1/_8$	24
n	$187^5/_8$	$25^1/_2$
o	$176^1/_4$	$28^1/_8$
p	$164^5/_8$	$30^3/_4$
q	153	$32^7/_8$
r	$141^5/_8$	$34^1/_8$
s	$129^7/_8$	$34^5/_8$
t	$118^1/_4$	$34^5/_8$
u	$106^3/_4$	$34^3/_8$
v	95	$33^5/_8$
w	$83^1/_4$	$32^1/_4$
x	$72^1/_8$	$30^1/_2$
y	$60^7/_8$	$28^1/_4$
z	$49^3/_8$	26

	x	y
Chine 2		
a	$48^3/_8$	30
b	$59^7/_8$	$31^3/_8$
c	$71^1/_8$	$32^1/_2$
d	$82^5/_8$	$33^1/_2$
e	$94^3/_8$	$34^1/_4$
f	$106^1/_8$	$34^3/_4$
g	$117^7/_8$	$34^7/_8$
h	$129^3/_8$	$34^3/_4$
i	$141^1/_8$	$34^1/_2$
j	$152^5/_8$	$33^3/_4$
k	$164^1/_4$	$32^5/_8$
l	176	$31^3/_8$
m	$187^5/_8$	$30^1/_8$
n	$190^1/_8$	$32^1/_4$
o	$177^3/_4$	$33^7/_8$
p	$165^1/_8$	$35^1/_2$
q	$152^5/_8$	$36^7/_8$
r	$141^1/_8$	$37^3/_4$
s	$129^3/_8$	$38^3/_8$
t	$117^5/_8$	$38^1/_2$
u	$106^1/_8$	$38^1/_2$
v	$94^3/_8$	38
w	$82^1/_2$	$37^1/_4$
x	$70^1/_2$	$36^1/_8$
y	58	$34^3/_4$
z	$45^3/_8$	$33^3/_8$

	x	y
Chine 3		
a	$43^3/_8$	$35^3/_4$
b	$56^1/_8$	$36^1/_2$
c	$68^5/_8$	$37^1/_4$
d	$80^3/_4$	38
e	$92^5/_8$	$38^1/_2$
f	$104^1/_2$	$38^3/_4$
g	$115^7/_8$	$38^7/_8$
h	$127^5/_8$	$38^7/_8$
i	$139^3/_8$	$38^5/_8$
j	$150^7/_8$	$38^1/_4$
k	$163^1/_2$	$37^1/_2$
l	$176^1/_4$	$36^5/_8$
m	$188^3/_4$	$35^3/_4$
n	$189^7/_8$	$39^1/_8$
o	177	$39^7/_8$
p	$163^3/_4$	$40^3/_4$
q	$150^7/_8$	$41^1/_2$
r	$139^3/_8$	$42^1/_8$
s	$127^5/_8$	$42^3/_8$
t	$115^7/_8$	$42^1/_2$
u	$104^3/_8$	$42^3/_8$
v	$92^5/_8$	$42^1/_8$
w	$80^5/_8$	$41^5/_8$
x	$68^1/_8$	$41^1/_8$
y	$55^1/_8$	$40^3/_8$
z	42	$39^3/_4$

	x	y
Chine 4		
a	$43^1/_8$	$40^1/_4$
b	$56^1/_8$	$41^1/_4$
c	$69^1/_4$	$42^3/_8$
d	$82^1/_8$	43
e	$93^5/_8$	$43^3/_8$
f	$105^3/_8$	$43^3/_8$
g	$117^1/_8$	$43^1/_4$
h	$128^5/_8$	$42^7/_8$
i	$140^3/_8$	$42^1/_2$
j	$152^3/_8$	$41^7/_8$
k	$164^7/_8$	$41^3/_8$
l	178	$40^7/_8$
m	$191^3/_8$	$40^1/_4$
n	191	$45^3/_8$
o	$177^7/_8$	$45^7/_8$
p	$164^7/_8$	$46^3/_8$
q	$152^3/_8$	$46^7/_8$
r	$140^3/_8$	$47^1/_4$
s	$128^5/_8$	$47^5/_8$
t	$117^1/_8$	$47^7/_8$
u	$105^3/_8$	48
v	$93^5/_8$	$47^7/_8$
w	$82^1/_8$	$47^5/_8$
x	$69^1/_8$	$46^5/_8$
y	56	$46^1/_8$
z	43	$45^1/_8$

Cinderella Coordinates
(in millimeters)

	x	y
Bottom panels		
a	265	14
b	614	13
c	587	293
d	547	533
e	495	774
f	440	1007
g	384	774
h	332	534
i	292	293
j	641	14
k	962	14
l	977	302
m	984	583
n	985	870
o	976	1157
p	627	1157
q	618	870
r	619	583
s	625	302
t	178	171
u	228	409
v	273	648
w	309	879
x	338	1167
y	17	1167
z	46	879
A	82	648
B	127	410
Chine 1		
a	1332	591
b	1570	642
c	1808	688
d	2039	724
e	2327	752
f	2615	767
g	2896	773
h	3182	774
i	3469	764
j	3749	737
k	3990	697
l	4230	646
m	4463	589
n	4596	624
o	4317	688
p	4032	752
q	3749	805
r	3470	836
s	3183	849
t	2896	849
u	2615	842
v	2327	824
w	2039	791
x	1769	747
y	1489	693
z	1209	638
Chine 2		
a	1184	736
b	1468	768
c	1751	797
d	2023	821
e	2313	840
f	2602	851
g	2882	854
h	3169	853
i	3456	844
j	3737	826
k	4023	800
l	4313	769
m	4598	738
n	4659	790
o	4354	830
p	4044	869
q	3739	903
r	3457	925
s	3169	939
t	2882	944
u	2602	942
v	2313	932
w	2021	913
x	1726	885
y	1420	852
z	1111	819
Chine 3		
a	1064	875
b	1374	895
c	1682	914
d	1979	930
e	2270	942
f	2559	949
g	2840	952
h	3127	952
i	3415	947
j	3697	936
k	4005	918
l	4317	897
m	4624	875
n	4653	958
o	4335	978
p	4013	999
q	3697	1018
r	3414	1031
s	3126	1038
t	2839	1040
u	2558	1038
v	2268	1032
w	1975	1021
x	1670	1006
y	1351	990
z	1030	974
Chine 4		
a	1056	985
b	1374	1012
c	1696	1037
d	2012	1055
e	2295	1063
f	2583	1064
g	2870	1060
h	3151	1051
i	3440	1040
j	3732	1028
k	4040	1015
l	4362	1001
m	4688	987
n	4678	1112
o	4357	1124
p	4038	1137
q	3732	1148
r	3440	1158
s	3150	1167
t	2869	1173
u	2582	1176
v	2294	1174
w	2011	1165
x	1695	1149
y	1372	1129
z	1054	1106

Note: Small discrepancies may exist between the millimeter and inch tables. See page ii.

pieces. Stack these one on top of the other, first flipping one of them end-for-end so that the butt lines don't coincide, and clamp them together. The idea here is that only the top sheet has to be marked and cut out, and it is possible to create each pair of matching strakes with a single set of cuts.

The panel labeled "Chine 1" in the nesting diagram is the strake that goes next to the bottom panel, "Chine 2" is the next one up, and so forth. In the interests of clarity, I haven't lettered the coordinates on strakes 2 through 4, but each chine should be treated as if it were labeled "A" through "Z" exactly as shown on Chine 1. In other words, "A" is the lower left-hand corner of the strake, and the lettering goes counterclockwise (anticlockwise) around the strake until it reaches "Z" in the upper left-hand corner. Of course, the blank section of plywood beneath the strakes is intended for the strakes needed to make the other half of the boat.

Label the strakes clearly on both sides with the strake number and an indication of which end is fore and which end is aft, and carefully assemble them in order. In particular, make sure you understand the final strake (Chine 4) is labeled the way it is—that is, upside down—because on the cutting plan it *is* upside down compared to the rest of the strakes. It is only arranged that way on the nesting diagram because it better fits on the plywood sheet. This, of course, means that it *must* be turned upside down before being stitched or taped to the rest of the boat.

Get this wrong, and people have, and when you stitch or tape the boat together you'll wonder why the final strake appears to be too long. It isn't, but it *is* turned the wrong way round! When you have it right, it will fit perfectly.

This is a boat that's best stitched together using cable ties and sealed on the outside using duct tape before doing the inside seams with epoxy and fiberglass tape. It may be possible to tape the boat together using duct tape alone, but no one I know of has tried it and I don't recommend your doing it either. But, if stitching can be avoided, it will save time and eliminate some unevenness in the internal epoxy-taping as a result of using the ties.

With the duct tape holding the side strakes in place, add the central T-mold dead on the centerline of the bottom panel, with the cross of the T exactly athwartship—that is, across the boat at a right angle to the centerline.

Then add the fore and aft T-molds in the same way. Push and jiggle the taped-together hull until the tops of the uppermost strakes are perfectly aligned with the top edges of the T-molds, and nail or tape them in place.

Once everything is firmly in place, epoxy and tape the inside seams (including at least two layers in the bow and stern). Before letting the epoxy harden, check to be sure everything is straight and where it should be. You may need to jiggle or even prop parts of the stem and stern to make them line up vertically and along the centerline, but this should not be difficult. At this stage, it's a good idea to keep everyone away from the work area until the epoxy hardens; you don't want anything disturbing the hull. After the epoxy hardens, remove the duct tape, then relieve all the external strake edges with coarse sandpaper before epoxy-taping the exterior.

You may be wondering why the structure is so flexible and wobbly, but there's no need for concern. Much of the rigidity of these boats comes from adding breasthooks and

All dimensions in inches

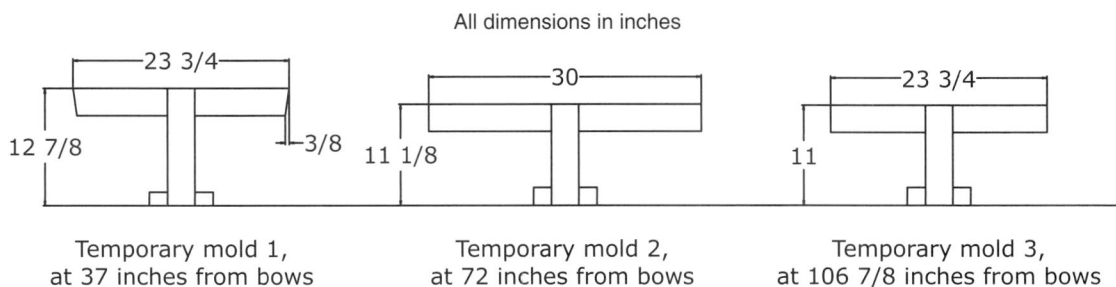

Temporary mold 1,
at 37 inches from bows

Temporary mold 2,
at 72 inches from bows

Temporary mold 3,
at 106 7/8 inches from bows

The T-molds determine Cinderella's shape by establishing the height and width of the uppermost strakes.

Hull - 3/8-inch marine plywood

Gunwale—3/4-inch material

Inwale gap—3/4 inch (or same width as inwale and gunwale)

Inwale - 3/4-inch pine or spruce

Gap filled at bow and stern using either gunwale or inwale material, cut and tried to fit neatly (it will be visible!)

Gapped inwale blocks at 11.5-inch centers

Filler made from doubled hull material, cut and shaped to fit under inwale, then clamped and glued into place

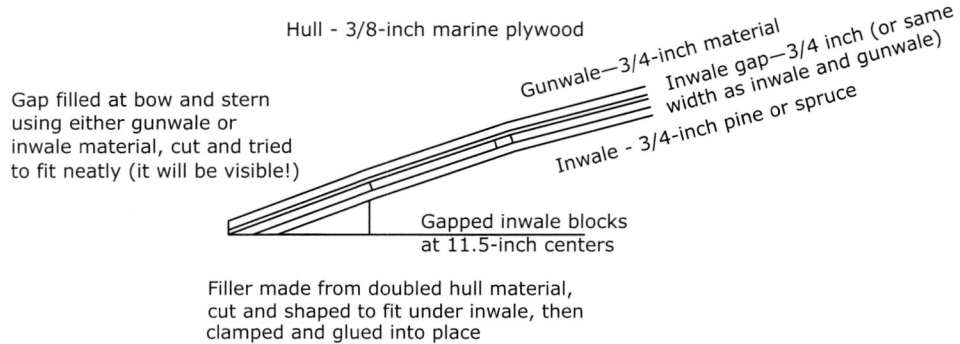

Gapped inwale details.

gapped inwales and gunwales, these features don't just look good, they provide shape and strength, adding remarkable rigidity.

General directions for building these features appear in Chapter 4, so I'll only list the specifics as they apply to Cinderella:

- Inwales: straight-grained, knot-free, rot-resistant softwood 1¼ to 1½ inches deep by ⅝ to ¾ inch thick (spruce or pine). You'll need about 24 feet for the inwales (longer if there are knots to cut around), plus another 5 feet or so for the gapping blocks.

- Gunwales: same dimensions as inwales; use a nice hardwood if available.

- Stretcher and handle: same material as inwales; about 6 feet (enough to laminate a piece 3 feet long to double thickness).

- Breasthooks: ⅛-inch plywood, doubled to make two thicknesses

- Spacing of inwale gapping blocks: 10 inches

- Length of breasthooks: bow, 10 inches along the strake; stern, 8½ to 9 inches along the strake

In Chapter 4, I suggested tapering the inwale gapping blocks and cutting a graceful arc in the breasthooks to add a bit of elegance. If any of the boats in this book deserve these touches, surely it's Cinderella.

Varnish the inwales and their gapping blocks, the breasthooks, and the gunwales, and paint everything else. Before you begin applying finishes, relieve the edges of the inwales and gunwales with coarse sandpaper, then sand, prime, and undercoat. Then undercoat and sand again, and apply a couple of final topcoats.

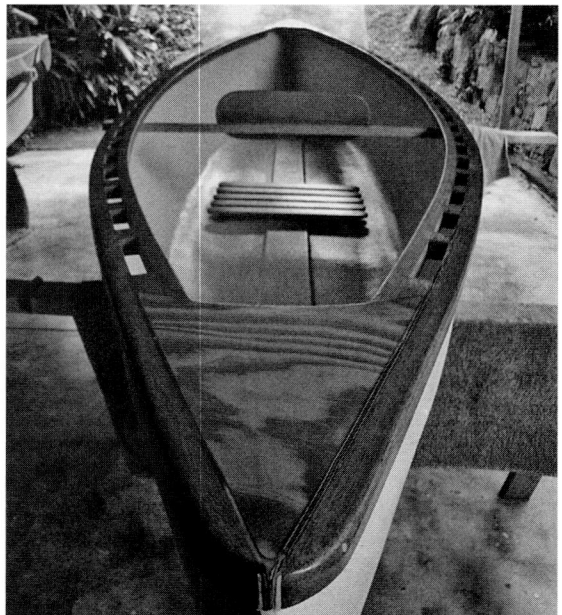

Details of gunwales, gapped inwales, and breasthook, in process (top) and nearly complete (bottom). Note also the seat added by the builder.

Australian Al Burke did a beautiful job building the first Cinderella. He was very sad the day he sold it to a neighbor but still manages to derive some satisfaction from seeing it used regularly for fishing. I need hardly mention that builders like Al make designers like me smile the broadest of smiles!

FLYING MOUSE

A CHILD'S SAILING PRAM

Length Over All (LOA): 7'10"
Length Waterline (LWL): 7'4"
Beam: 36"
Weight: 75–85 lb.
Displacement at Design Waterline: 300 lb.
Crew: 1 child or small adult
Propulsion: Sail
Construction methods: PU stitch-and-glue
Epoxy stitch-and-glue

The Flying Mouse is a sailboat for kids just starting to sail for the first time. Because it was developed from the standard V-bottomed Mouse, its hull is similarly effective; certainly more effective than a flat-bottomed boat of similar length, beam, and displacement.

Its geometry allows the very young skipper to go about from one tack to another without having to try to negotiate the tiny 12-by-12-inch space found between the boom, tiller, and daggerboard of, say, an Optimist. The Flying Mouse has no proper boom but instead has a boom sprit that gives much more headroom, and being low on the aft deck the tiller doesn't get in the way at all. In fact, once the boat's going it's OK to sit on the tiller for a moment while sorting out your sheet!

The benefit of this arrangement, however, is more than just convenience—it also eliminates injuries from being hit by the boom.

The Flying Mouse is rigged with first-time sailors in mind. It takes a lot of wind to blow it over, and the standard 35 square-foot spritsail is small enough that it can be left on the mast and stored in a corner of a garage without getting in the way.

The Spectra or polyester forestay and shrouds are strong, cause no more drag than wire stays, and are easy to set up. Another advantage of having mast stays on this low-decked hull is that it allows a large and watertight forward buoyancy tank. In

this boat, a through-deck mast going down to a mast step in the bottom would reduce the boat's buoyancy in a capsize because it would flood.

The hulls of the Flying Mouse and the original Mouse are not quite the same. While the original Mouse was designed to paddle, the Flying Mouse is definitely a sailer and has a few more inches of beam to make it stable under sail. The cockpits are different too, as the Flying Mouse has side decks that allow the boat to heel a long way before the footwell starts to fill.

Another difference is that the Flying Mouse has a daggerboard and a rudder, and a mast, sail, and various bits of rigging that have to be attached to the hull. It also doesn't have a skeg—if the Flying Mouse had one the sailing rig would have had to be moved aft, reducing the amount of space available in the center of the boat.

As the drawings show, however, the Flying Mouse build is still very much like that of the original Mouse. It should be built using the stitch-and-glue method, although you could use the simplified chine log method. Because of the deck, you'll need two full sheets of plywood, only half a sheet more than the original Mouse.

The hull has to be reinforced with a circle or square of scrap plywood where the shrouds and forestay attach to the hull, where the painter attaches to the deck, under the fairleads

~O8>

Flotation tanks
under foredeck

C/b slot 12 1/2 inches
by 3/4 inch

12 inches

Flotation tank
under aft deck

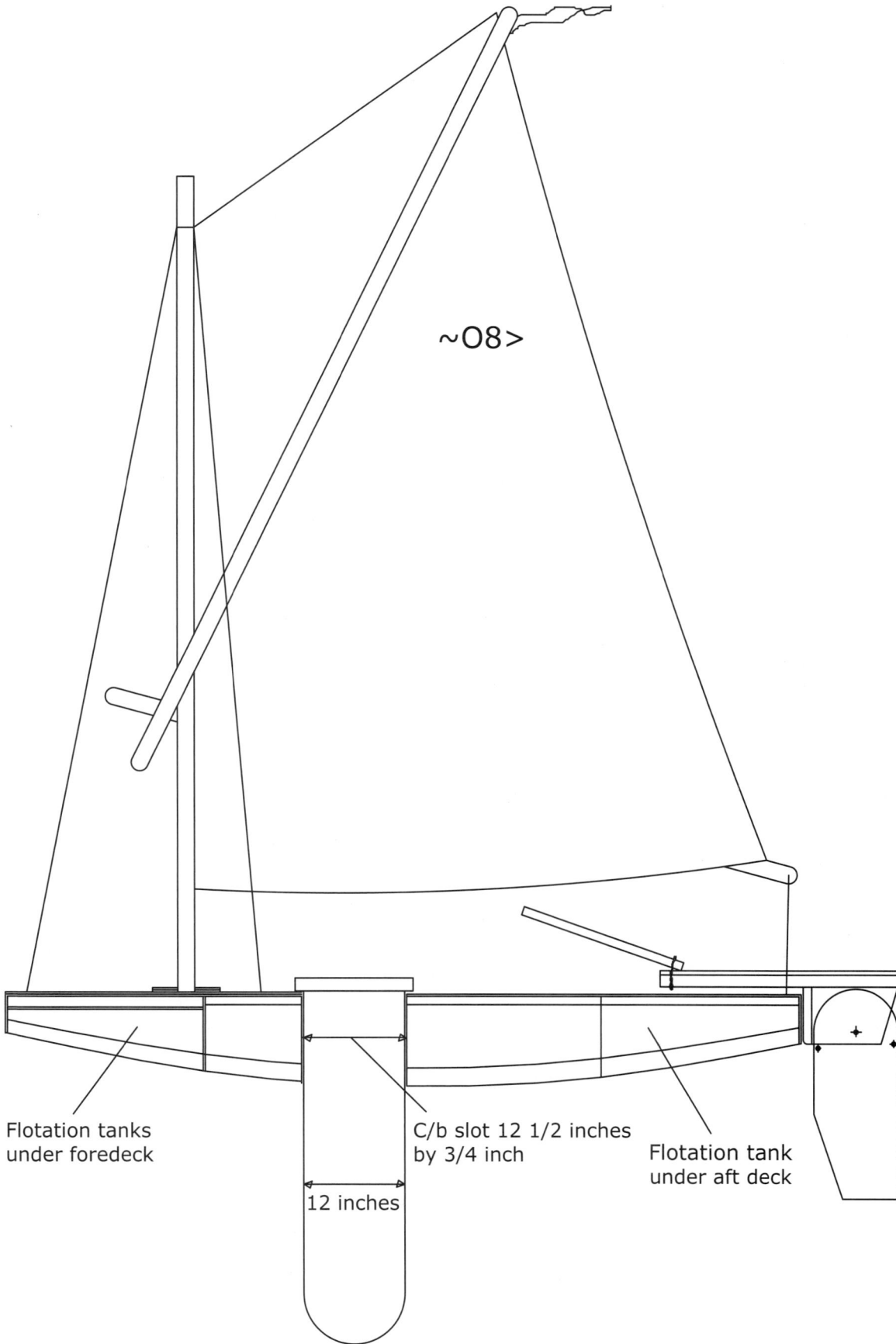

Flying Mouse. (The sail logo "~08>" indicates the boat is a mouseboat. The characters [from left to right] represent the tail, body, ears, and nose.)

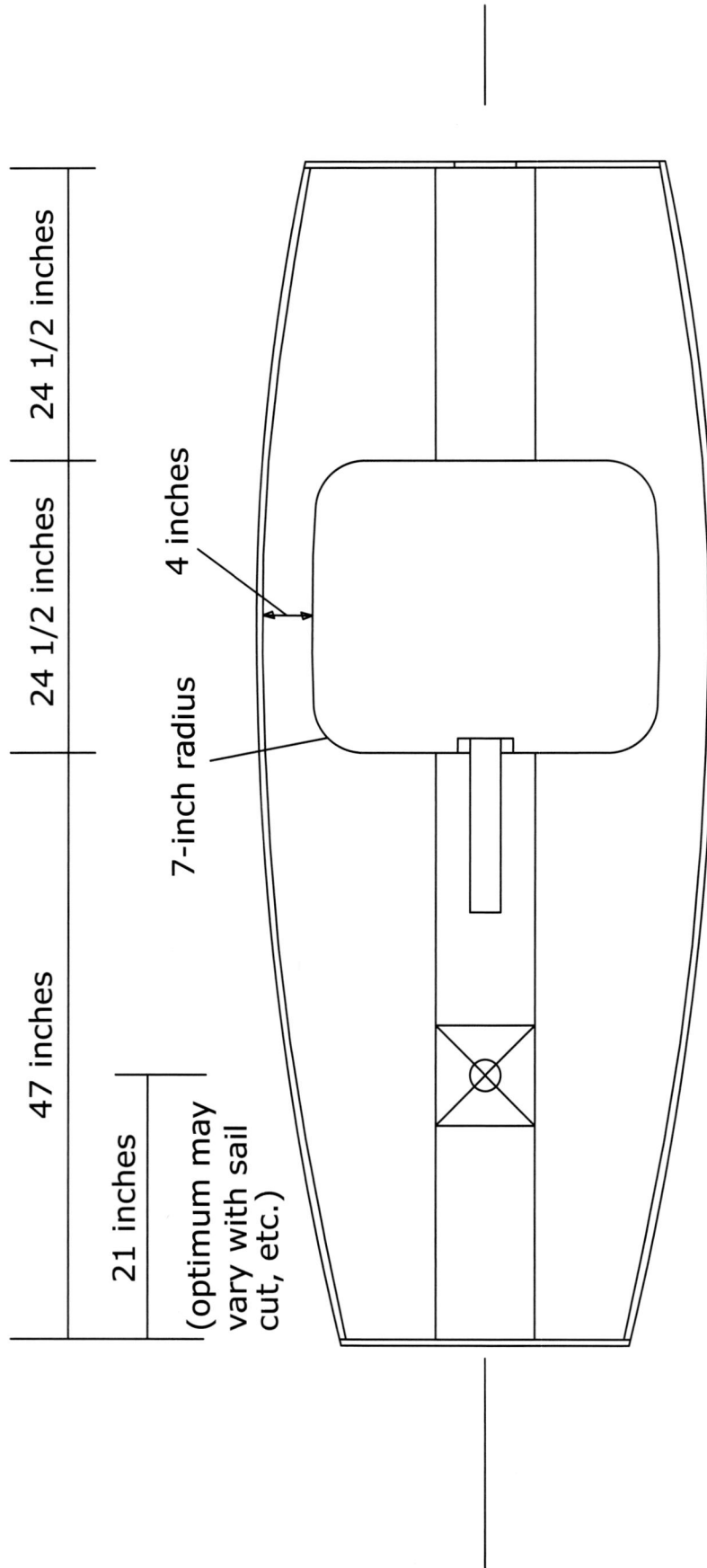

24 1/2 inches

24 1/2 inches

4 inches

7-inch radius

47 inches

21 inches

(optimum may
vary with sail
cut, etc.)

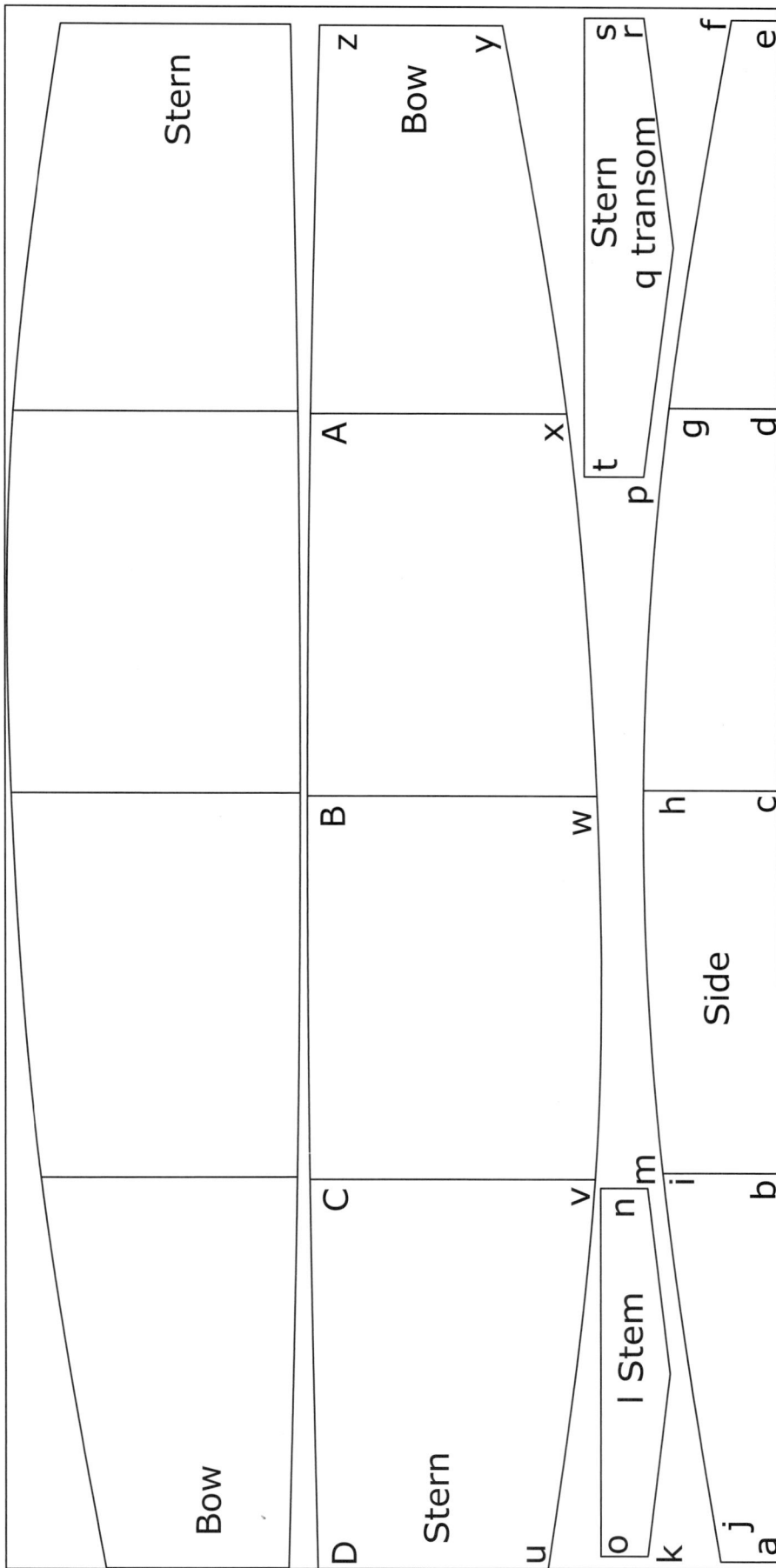

Stern

Bow

Bow

Stern

z

Bow

y

A

x

B

w

C

v

D

Stern

u

s
r

Stern
q transom

t
p

n
m

l Stem

o
k

f
e

g
d

Side

h
c

i
b

j
a

4 inches

7 inches

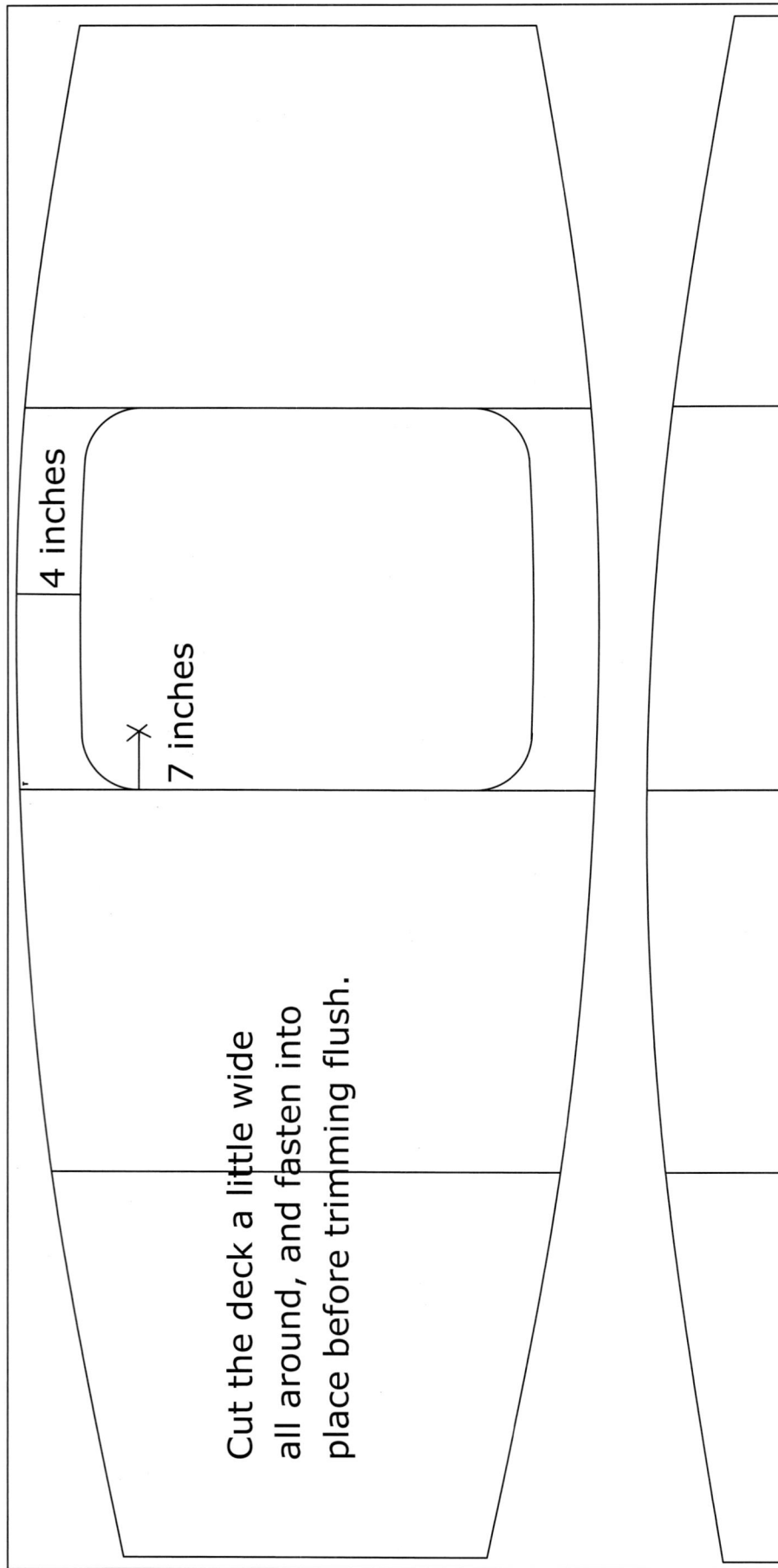

Cut the deck a little wide
all around, and fasten into
place before trimming flush.

```
┌──────────────────────────────────────────────────────────────┐
│ S                              R │                 │          │
│ ⌐  Forward bulkhead  ¬          │                 │          │
│ O                              Q │                 │          │
│       P                          │                 │          │
│ ┌──────────────────────────┐     │                 │          │
│ N                          M │   │                 │          │
│    Cockpit bulkhead          │   │    "           │          │
│ J                          L │   │    Material for "stripe"  detail on deck │
│       K                      │   │                 │          │
│ ┌──────────────────────────┐     │                 │          │
│ I                          H │   │                 │          │
│    Aft bulkhead              │   │                 │          │
│ E                          G │   │                 │          │
│       F                      │   │                 │          │
│                                  │                 │          │
│ Remainder for centerboard,       │                 │          │
│ rudder blade, etc.               │                 │          │
└──────────────────────────────────────────────────────────────┘
```

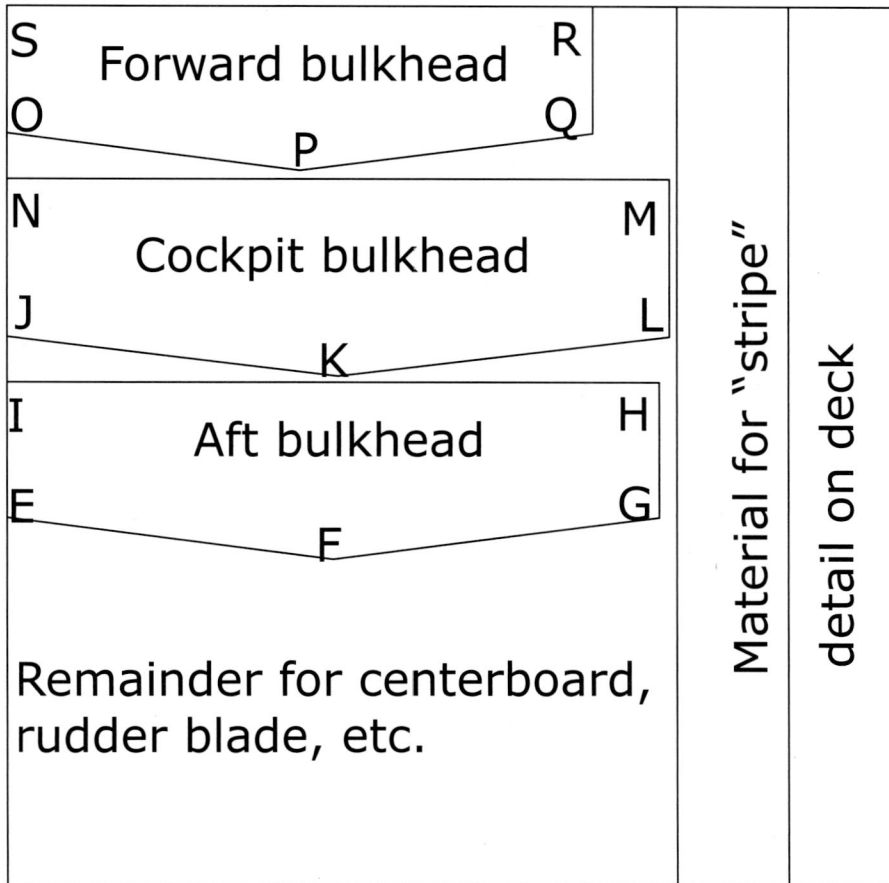

that make the traveler (horse), and where the rudder gudgeons attach to the stern transom.

The foredeck needs to be strong enough to take the downward pressure exerted by the mast. This issue is dealt with simply and effectively by building in a kind of girder under the foredeck made up of two pieces of lumber framing that transfer the pressure onto a sheet of scrap plywood 6 inches wide, extending from the forward frame to the bow transom, as shown in the plans. Another, slightly simpler reinforcement is used on the aft deck, where the deck might occasionally have to take the weight of an adult, though this boat is designed for kids.

Make up the mast foot on the foredeck. As shown, this is made by laminating two or three squares of plywood together and cutting a hole through them just large enough for the mast. The hole can be cut with a hole saw in an electric drill, or, almost as easily, with a hacksaw or fretsaw. The mast step is screwed to the foredeck using four screws—one at each corner—chosen to be just the right length to go through

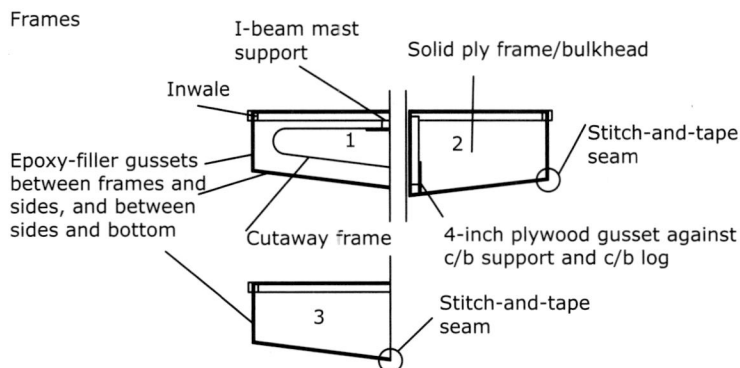

Frames — I-beam mast support — Inwale — Solid ply frame/bulkhead — Epoxy-filler gussets between frames and sides, and between sides and bottom — Cutaway frame — 1 — 2 — Stitch-and-tape seam — 4-inch plywood gusset against c/b support and c/b log — 3 — Stitch-and-tape seam

Rudder components for Flying Mouse

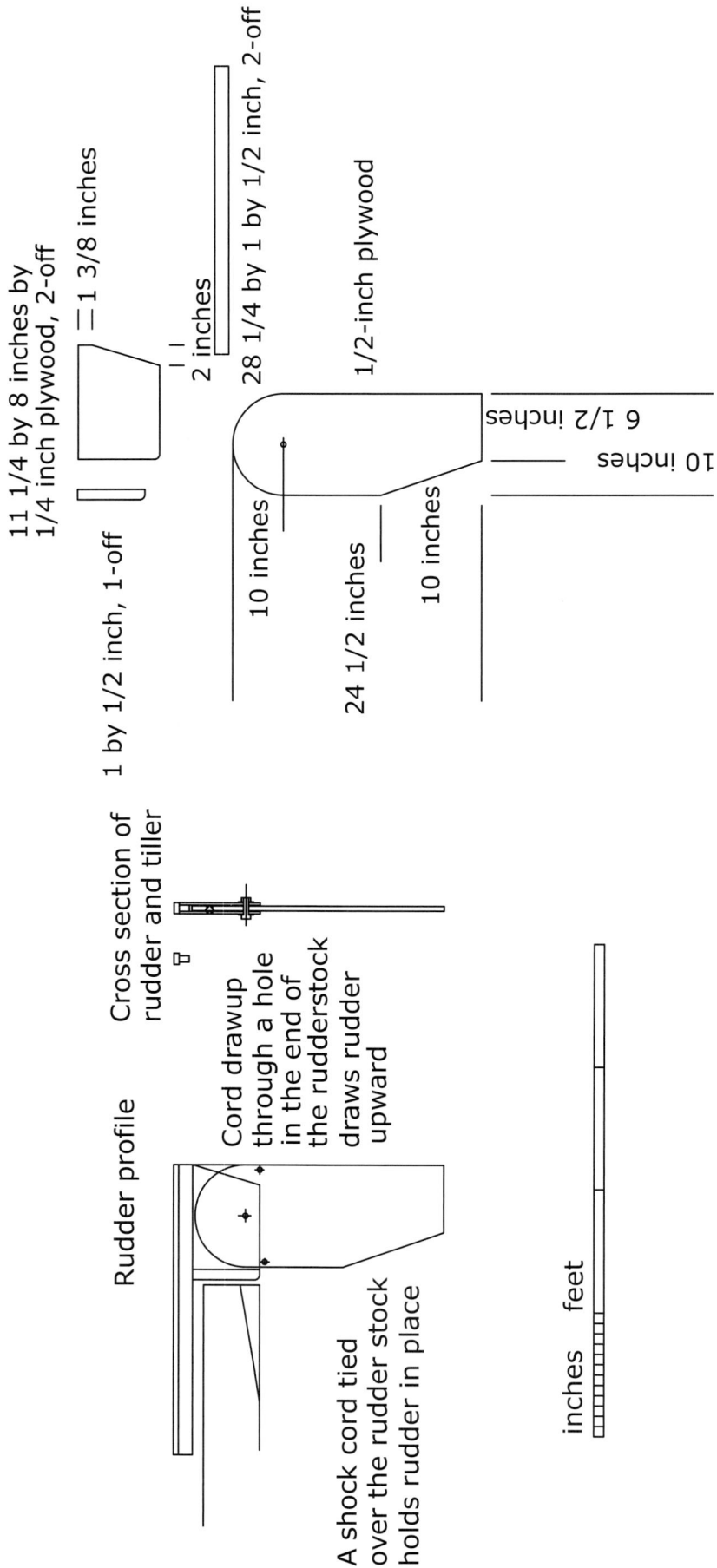

Rudder profile

Cross section of rudder and tiller

11 1/4 by 8 inches by 1/4 inch plywood, 2-off

= 1 3/8 inches

1 by 1/2 inch, 1-off

2 inches

28 1/4 by 1 by 1/2 inch, 2-off

1/2-inch plywood

6 1/2 inches

10 inches

10 inches

24 1/2 inches

10 inches

Cord drawup through a hole in the end of the rudderstock draws rudder upward

A shock cord tied over the rudder stock holds rudder in place

inches   feet

inches

160

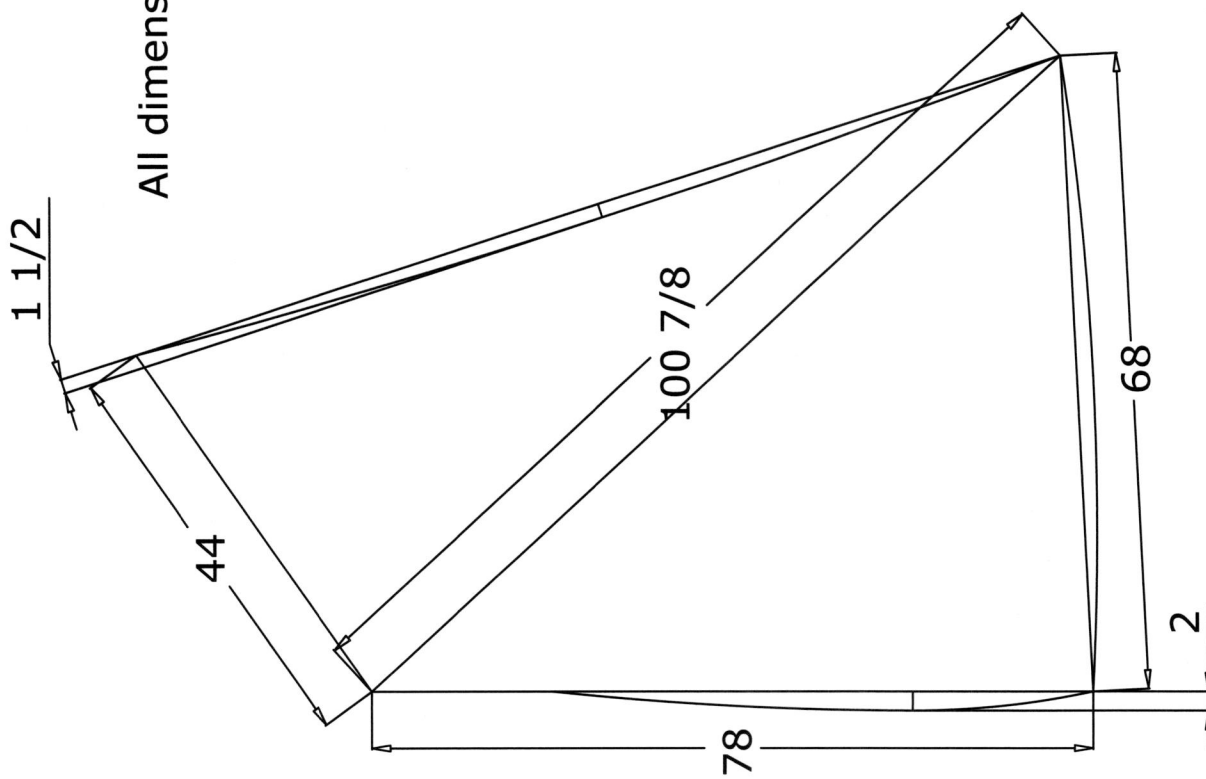

All dimensions in inches

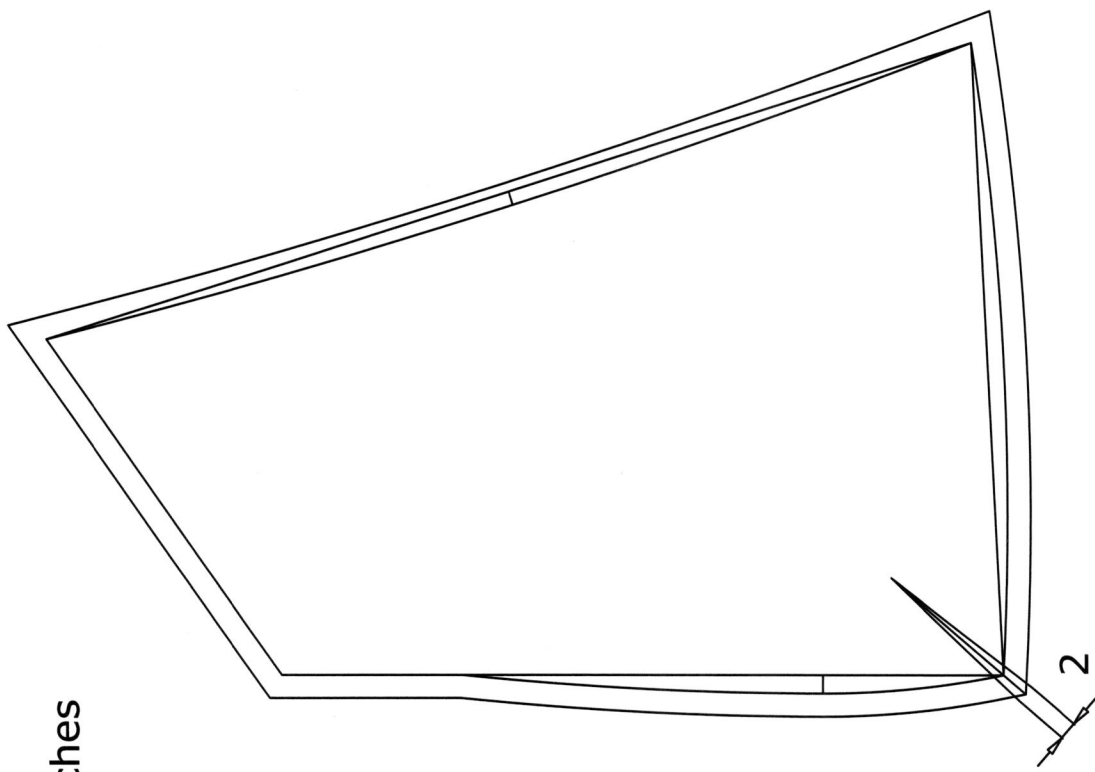

1 1/2

100 7/8

44

78

68

2

2

Flying Mouse Coordinates (in inches)

	x	y
Side		
a	$\frac{1}{2}$	0
b	$24\frac{1}{4}$	0
c	$47\frac{3}{4}$	0
d	$71\frac{3}{8}$	0
e	$95\frac{1}{4}$	0
f	$95\frac{1}{4}$	3
g	$71\frac{3}{8}$	$6\frac{7}{8}$
h	$47\frac{3}{4}$	$8\frac{1}{2}$
i	$24\frac{1}{4}$	$7\frac{1}{4}$
j	$\frac{1}{2}$	$3\frac{3}{4}$
Stem transom		
k	$\frac{3}{4}$	$8\frac{1}{4}$
l	12	$6\frac{7}{8}$
m	$23\frac{1}{4}$	$8\frac{1}{4}$
n	$23\frac{1}{4}$	$11\frac{1}{8}$
o	$\frac{3}{4}$	$11\frac{1}{8}$
Stern transom		
p	$67\frac{1}{8}$	$8\frac{3}{8}$
q	$81\frac{1}{4}$	$6\frac{5}{8}$
r	$95\frac{3}{8}$	$8\frac{3}{8}$
s	$95\frac{3}{8}$	$12\frac{1}{8}$
t	$67\frac{1}{8}$	$12\frac{1}{8}$
Bottom		
u	0	$14\frac{3}{8}$
v	$23\frac{7}{8}$	$11\frac{1}{2}$
w	$47\frac{3}{8}$	$11\frac{3}{8}$
x	71	$13\frac{1}{8}$
y	$94\frac{7}{8}$	$17\frac{1}{8}$
z	$94\frac{7}{8}$	$28\frac{1}{2}$
A	71	29
B	$47\frac{3}{8}$	$29\frac{1}{4}$
C	$23\frac{7}{8}$	$29\frac{1}{8}$
D	0	$28\frac{5}{8}$
Aft bulkhead		
E	0	$20\frac{3}{4}$
F	$17\frac{1}{2}$	$18\frac{1}{2}$
G	35	$20\frac{3}{4}$
H	35	28
I	0	28
Cockpit bulkhead		
J	0	$30\frac{3}{8}$
K	$17\frac{3}{4}$	$28\frac{1}{4}$
L	$35\frac{5}{8}$	$30\frac{3}{8}$
M	$35\frac{5}{8}$	$38\frac{3}{4}$
N	0	$38\frac{3}{4}$
Forward bulkhead		
O	0	$41\frac{1}{4}$
P	$15\frac{3}{4}$	$39\frac{1}{4}$
Q	$31\frac{1}{2}$	$41\frac{1}{4}$
R	$31\frac{1}{2}$	48
S	48	48

Flying Mouse Coordinates (in millimeters)

	x	y
Side		
a	11	0
b	593	0
c	1169	0
d	1747	0
e	2334	0
f	2334	73
g	1747	168
h	1169	207
i	593	178
j	11	91
Stem transom		
k	19	201
l	295	167
m	570	201
n	570	273
o	19	273
Stern transom		
p	1643	206
q	1990	161
r	2336	206
s	2336	297
t	1643	297
Bottom		
u	0	353
v	584	281
w	1160	278
x	1738	323
y	2324	421
z	2324	698
A	1738	712
B	1160	717
C	584	714
D	0	702
Aft bulkhead		
E	0	508
F	429	454
G	859	508
H	859	685
I	0	685
Cockpit bulkhead		
J	0	745
K	436	693
L	872	745
M	872	951
N	0	951
Forward bulkhead		
O	0	1010
P	386	962
Q	772	1010
R	772	1176
S	0	1176

Note: Small discrepancies may exist between the millimeter and inch tables. See page ii.

the material and then almost through the deck. It looks best if it's finished to match the lengthwise deck strap on which it sits.

Finally, I should add that all the bits and pieces needed for sailing this boat will take about as long to make as the hull.

Both of the author's children have had a lot of fun with the Flying Mouse. Ella particularly enjoyed the security and excitement of the boat, and Ewan developed into a real adventurer, constantly pushing the boundaries of how far he was allowed to go and in what conditions.

The author's son Ewan and the lad in the Optimist chased each other all over our nearby sailing lake, and neither could throw the other off despite Ewan's cheap and quick polytarp sail.

EEK!

A LARGER, SPORTIER SAILING PRAM FOR MORE EXPERIENCED KIDS AND SMALL ADULTS

Length Over All (LOA): 10'9"
Length Waterline (LWL): 6'9"
Beam: 36"
Weight: 70–80 lb.
Displacement at Design Waterline: 325 lb.
Crew: 1 or 2 children or a small adult
Propulsion: Sail
Construction Methods: PU stitch-and-glue
Epoxy stitch-and-glue
Simplified chine log

Named by my daughter, Ella, and reflecting the idea that simpler is often better than complicated, the Eek-a-Mouse! (or Eek!, for short) is a significant step in the evolution of the Mouseboat idea. It is designed to meet the need for a kid's sailer that's fun and exciting for kids who don't want to sail all the time with their parents, who want their own boat, and who can't yet handle a boat more suited for adults. You can build one in the garage or backyard in a few days or weeks.

Eek! is rigged with a sail known as a bafter or a Bailey, and its shape is ingenious and offers some significant benefits for those who sail very small boats. One of the cleverest features of this sail is that it maintains nearly the same center of effort (center of wind effort on the sail) when reefed. Further, when an extra panel is added to the bottom and the mast is made correspondingly longer to make the boat sail faster, the center of effort remains in almost exactly the same position in the fore-and-aft direction.

At the size and proportions shown here, the sail combines reasonable on-the-wind performance with exciting sailing on a reach or off the wind in a good breeze, which is just what dinghy sailors seem to expect these days. Its biggest fault is the tendency for the top of the sail to fall off a little, but that isn't a wholly bad thing, for it is useful for spilling wind in a gust to help a young skipper keep the boat upright. Another advantage of this type of sail is that as long as it's small (as it is here), it can be made flat—that is, without any sail shaping. Anything that facilitates sailmaking is a good thing in my book, and I fully expect to use it in further small boats in the future.

If you've read the sections on how to build the standard Mouseboat and the Flying Mouse, this one will seem very basic. The bottom is flat, there are no stays, and, because it's made up of small panels, even the sail is flat.

You can build Eek! using the simplified chine log or the stitch-and-glue method. At 10½ feet in length, the boat is longer than a standard sheet of plywood, and so some butt-jointing of the panels will be needed. I suggest using the glass-taped butt joint because the butt joint coincides with one of the boat's internal frames. However, if you're building using polyurethane glue, a plywood butt strap will do fine. Just remember to cut the thickness of the strap off the lower edge of the forward frame, before assembly, and

Eek-a-Mouse!

129 by 36 inches by 325 lb displacement kid's sailing boat

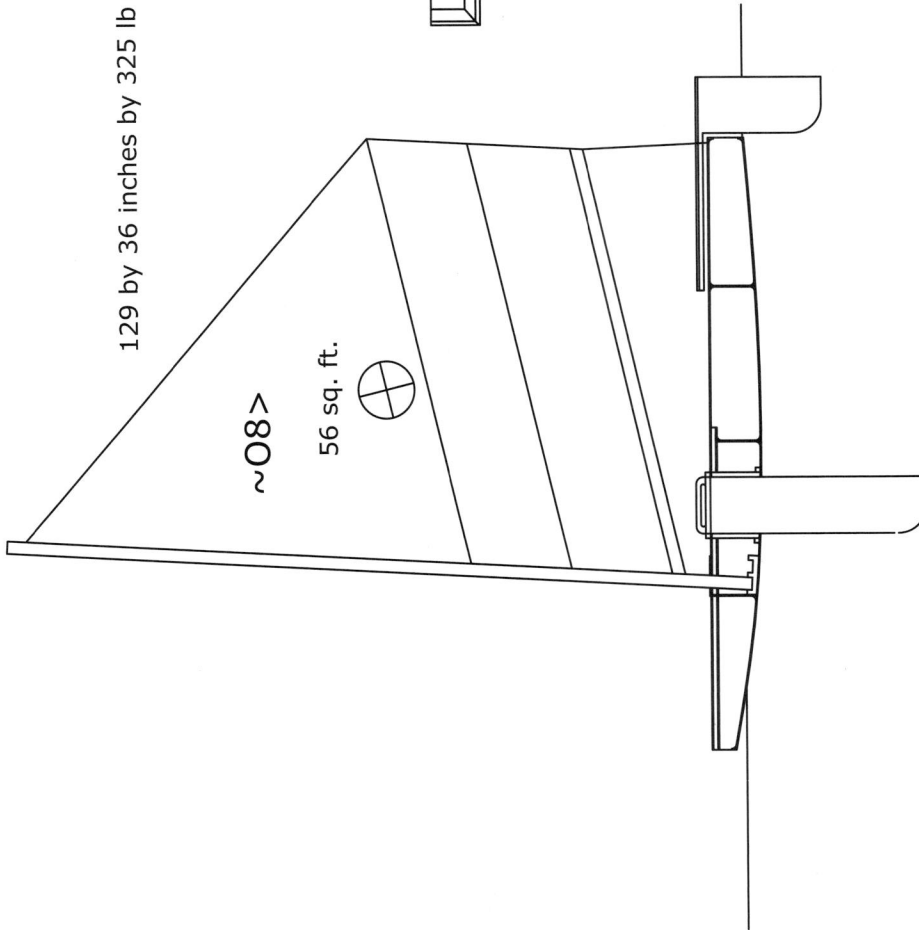

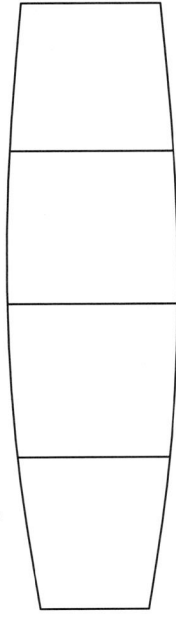

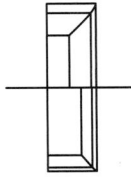

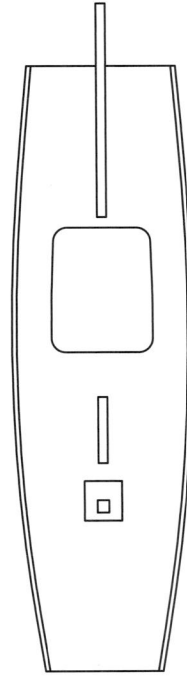

Deck view

~08>

56 sq. ft.

Eek-a-Mouse!
Minor building details

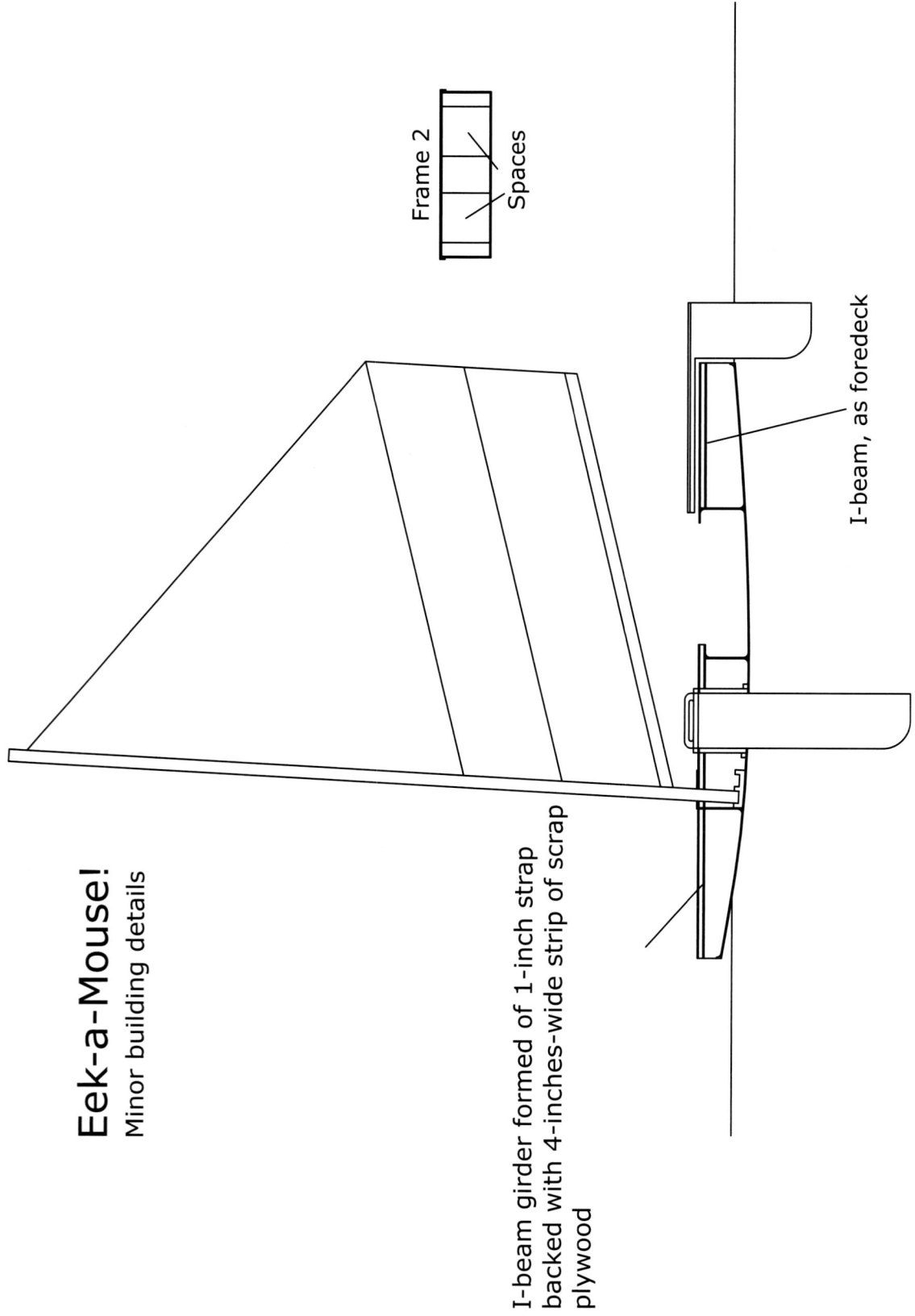

Frame 2

Spaces

I-beam, as foredeck

I-beam girder formed of 1-inch strap
backed with 4-inches-wide strip of scrap
plywood

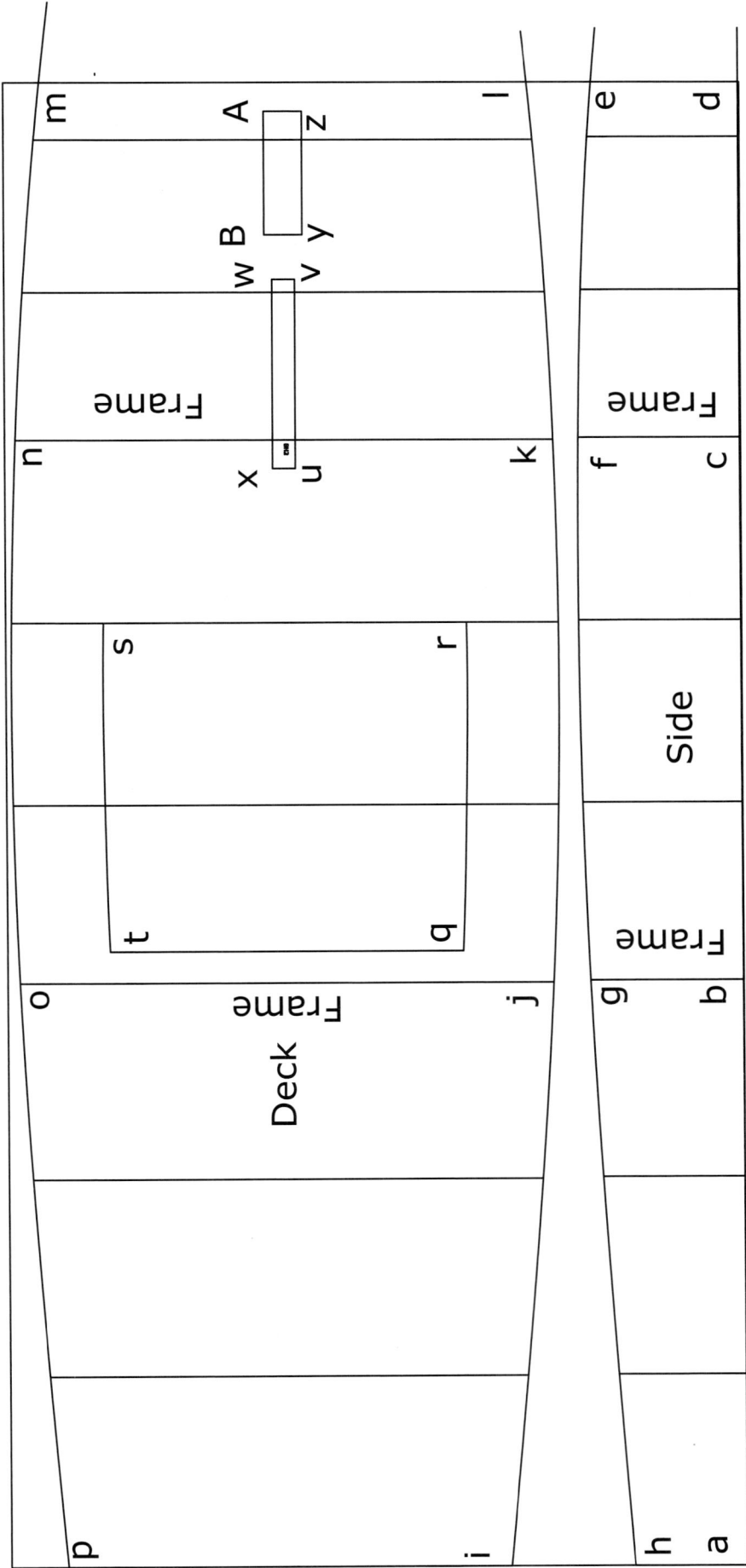

Eek-a-Mouse panels, part 1.

Frame 3
outboard section

Eek-a-Mouse panels, part 2.

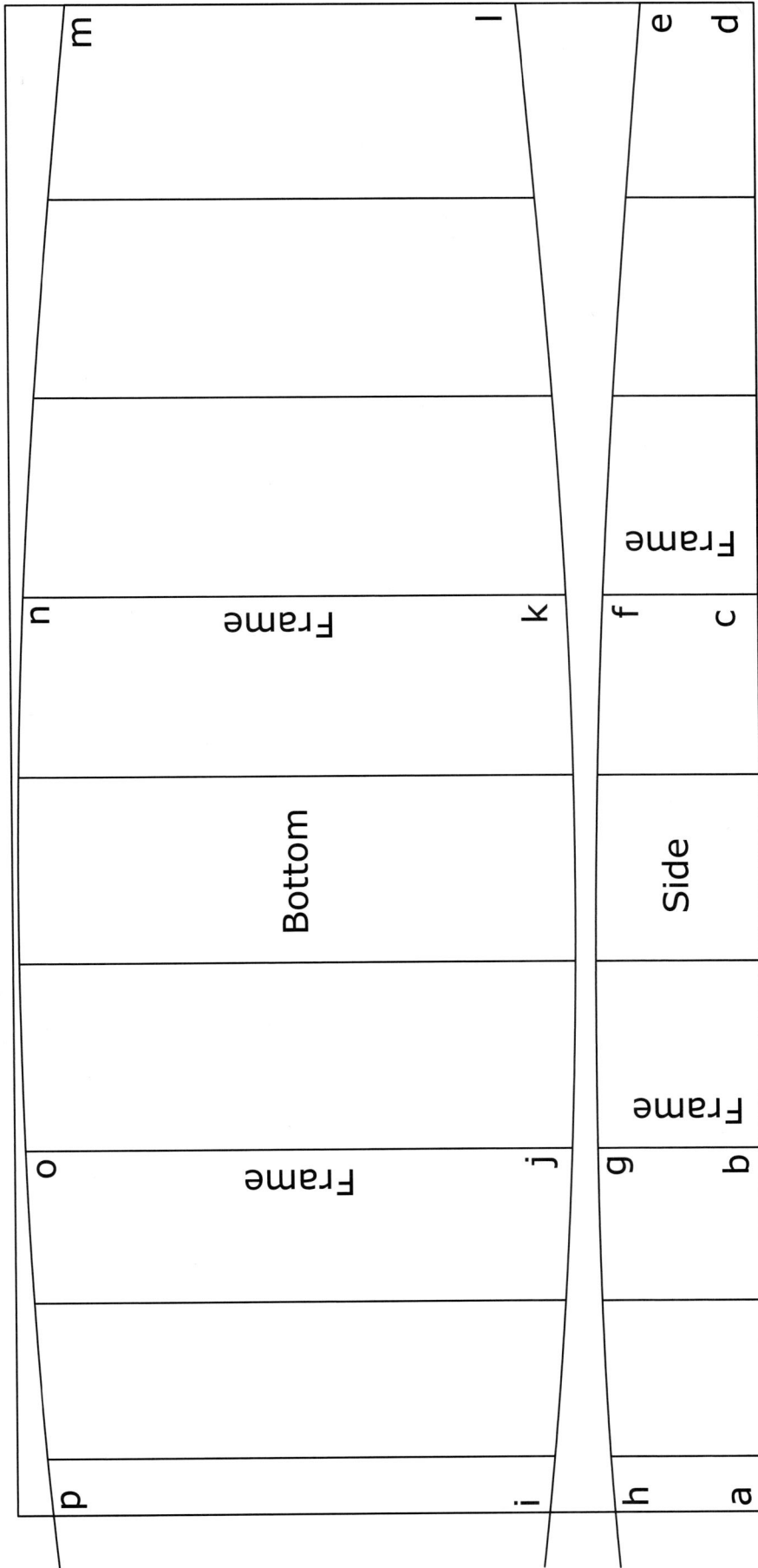

Eek-a-Mouse panels, part 3.

Mast: 2 1/2 by 2 1/2 by 162 inches, square at bottom for
first 11 to 12 inches, then 6-sided and rounded. Top
third may be tapered to 1 1/2 in. diameter.

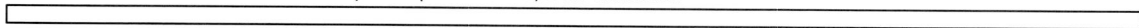

Leading wishbone batten 106 by 2 by 3/4 in. spruce,
two off. Bolted through the sail as indicated, with a leather
or seatbelt webbing cradle for the mast. Webbing is held
in place by a single bolt on either side so that it turns to
conform with the mast.

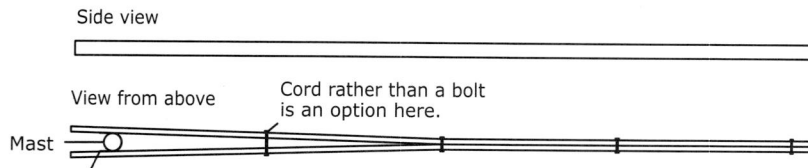

Side view

View from above   Cord rather than a bolt
is an option here.

Mast

then make a polyurethane stitch-and-glue joint
between the frame and the plywood strap.

When using the stitch-and-glue method
with these dual-purpose plan drawings, the bottom should be cut out and stitched in the normal way. However, when using the simplified
chine log method, the bottom shape is marked
out by drawing a pencil around the hull to give
the shape of the bottom. It is then cut out with
a small margin all around, glued and attached,
and then the excess is trimmed off.

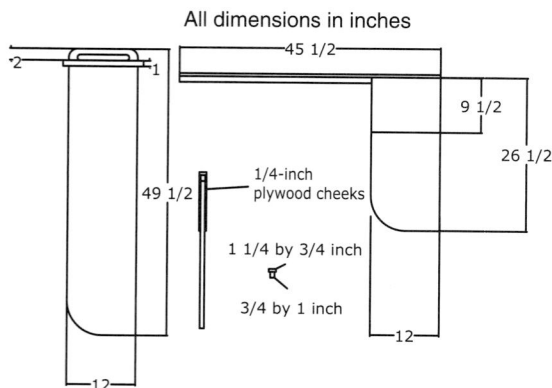

All dimensions in inches

45 1/2

9 1/2

26 1/2

1/4-inch
plywood cheeks

49 1/2

1 1/4 by 3/4 inch

3/4 by 1 inch

12

12

Daggerboard details.

Like the Flying Mouse described earlier,
squares and rectangles of scrap plywood reinforcement will be needed under the aft deck
to support the traveler, on the transom to support the rudder, and in the decking around
the mast to form partners to support the spar.

A detail that doesn't appear in any other
set of plans in this book is the wishbone
battens between the triangular upper section of the sail and the more flexible batten at the bottom of each panel below it.
The flexible panel battens are 92 inches long
and may be made from a variety of materials,
including PVC piping flattened in hot water
at one end and reinforced by wooden dowels
at the other end. You can also use $3/8$-inch plywood, plastic moldings, or solid spruce, particularly if the battens are doubled or even
tripled over the aftermost third or half of the
batten length (whatever curve there is should
be at the forward end). Long pockets, maybe 2
inches wide, should be sewn or taped into the
sail, and the battens should be tied in place to
prevent loss.

Cringles

2in-wide pockets to accept battens

1 1/2-inch convex curve, max depth 41 7/8 inches from tack

1 5/8-inch hollow at center of leach

More sophisticated sail with curved sides designed to give a better sail shape

2-inch-deep convex curve 60 inches up from tack

Eek-a-Mouse!

Weave: Weave of cloth in upper panel parallel to leach; in lower panels parallel to battens

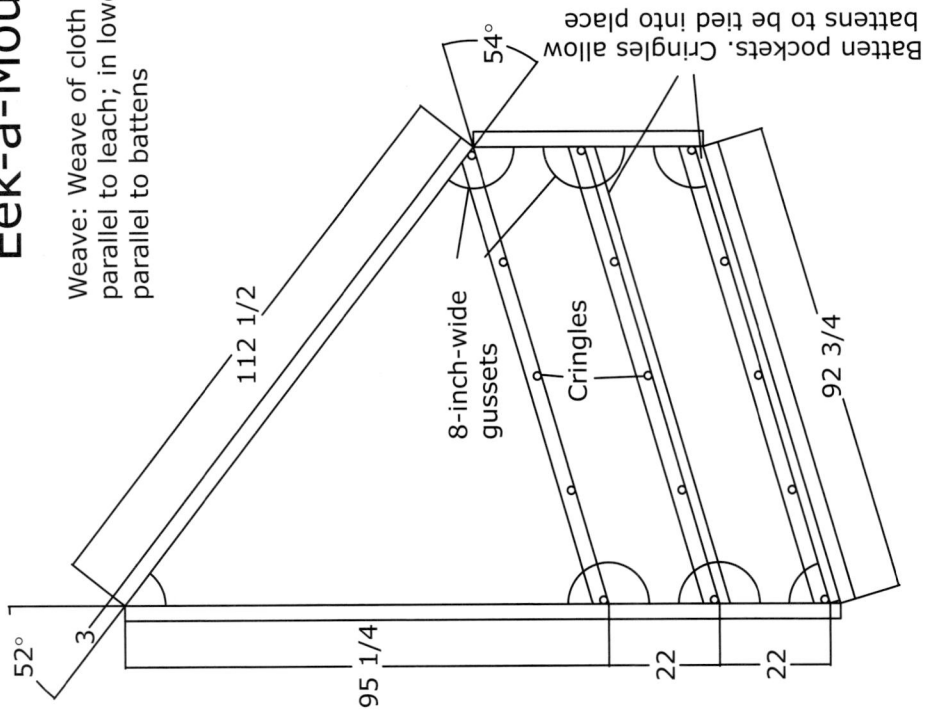

54°

Batten pockets. Cringles allow battens to be tied into place

112 1/2

8-inch-wide gussets

Cringles

92 3/4

52°

3

95 1/4

22

22

Basic flat-cut 56 sq. ft. sail

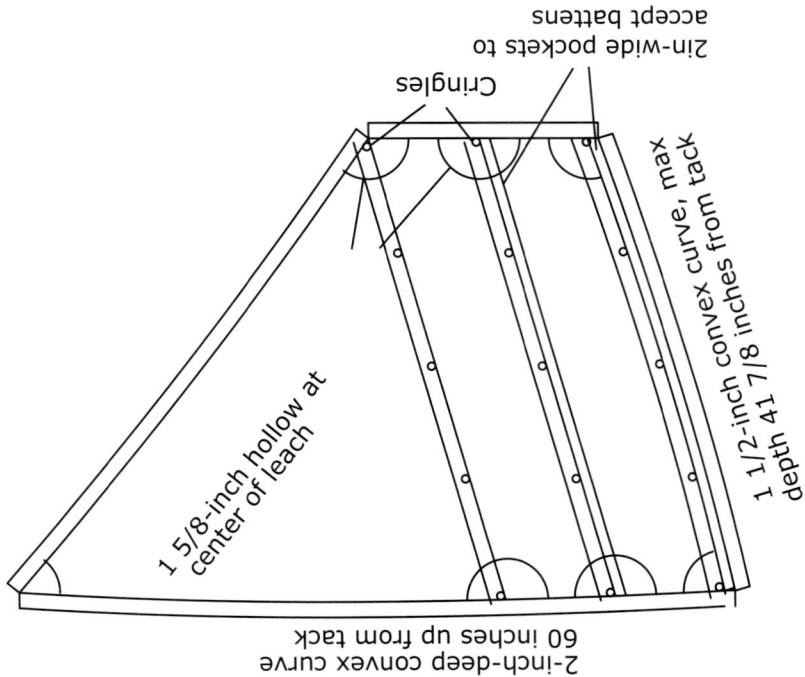

171

Eek-a-Mouse! Coordinates (in inches), Part 1

	x	y
Side 1, aft section		
a	0	0
b	$37^3/_4$	0
c	73	0
d	96	0
e	96	$9^5/_8$
f	73	$10^5/_8$
g	$37^3/_4$	10
h	0	$7^1/_4$
Deck, aft section		
i	0	$15^3/_8$
j	$37^5/_8$	$12^3/_8$
k	$72^7/_8$	$12^1/_4$
l	96	$13^7/_8$
m	96	$45^5/_8$
n	$72^7/_8$	$47^3/_8$
o	$37^5/_8$	$47^1/_4$
p	0	$44^1/_4$
q	$39^5/_8$	$18^1/_4$
r	61	18
s	61	$41^5/_8$
t	$39^5/_8$	$41^1/_4$
u	71	29
v	$83^1/_4$	29
w	$83^1/_4$	$30^1/_2$
x	71	$30^1/_2$
y	$86^1/_8$	$28^1/_2$
z	$94^1/_8$	$28^1/_2$
A	$94^1/_8$	31
B	$86^1/_8$	31

Part 2

	x	y
Side 1, forward section		
a	0	0
b	$6^3/_8$	0
c	$33^1/_2$	0
d	$33^1/_2$	$4^7/_8$
e	$6^3/_8$	9
f	0	$9^5/_8$
Daggerboard case		
g	35	$3/_4$
h	47	$1/_2$
i	47	11
j	35	11
k	49	$3/_4$
l	61	$1/_2$
m	61	11
n	49	11
Side 2, forward section		
o	$62^1/_2$	0
p	$89^5/_8$	0
q	96	0
r	96	$9^5/_8$
s	$89^5/_8$	9
t	$62^1/_2$	$4^7/_8$
Inboard part frame 3		
u	15	9
v	$25^1/_2$	9
w	$25^1/_2$	15
x	15	15

	x	y
Stem transom		
y	$64^1/_2$	$9^5/_8$
z	$86^3/_4$	$9^5/_8$
A	$86^3/_4$	$14^1/_2$
B	$64^1/_2$	$14^1/_2$
Deck, forward section		
C	0	$13^7/_8$
D	$6^1/_8$	$14^1/_2$
E	33	$18^5/_8$
F	33	$40^7/_8$
G	$6^1/_8$	45
H	0	$45^5/_8$
Frame 2		
I	34	$16^3/_8$
J	$43^1/_2$	$16^3/_8$
K	$43^1/_2$	48
L	34	48
Frame 4		
M	44	$13^1/_2$
N	$53^3/_4$	$13^1/_2$
O	$53^3/_4$	48
P	44	48
Transom		
Q	$54^1/_4$	$19^1/_8$
R	$61^1/_2$	$19^1/_8$
S	$61^1/_2$	48
T	$54^1/_4$	48
Bottom, aft section		
U	$61^1/_2$	$18^3/_4$
V	$89^3/_4$	$14^5/_8$
W	96	$13^7/_8$
X	96	$45^3/_4$
Y	$89^3/_4$	45
Z	$62^1/_2$	$40^7/_8$
Outboard sections of frame 3		
aa	$12^3/_8$	45
bb	23	45
cc	23	48
dd	$12^3/_8$	48
ee	$33^1/_2$	45
ff	$33^1/_2$	48

Part 3

	x	y
Side 2, aft section		
a	0	0
b	23	0
c	$58^1/_4$	0
d	96	0
e	96	$7^1/_4$
f	$58^1/_4$	10
g	23	$10^5/_8$
h	0	$9^5/_8$
Bottom, front section		
i	0	$13^7/_8$
j	23	$12^1/_4$
k	$58^3/_8$	$12^3/_8$
l	96	$15^1/_4$
m	96	$44^1/_8$
n	$58^3/_8$	$47^1/_8$
o	$23^1/_8$	$47^1/_4$
p	0	$45^3/_4$

Eek-a-Mouse! Coordinates (in millimeters), Part 1

	x	y

Side 1, aft section

	x	y
a	0	0
b	924	0
c	1788	0
d	2352	0
e	2352	237
f	1788	260
g	924	245
h	0	179

Deck, aft section

	x	y
i	0	376
j	920	303
k	1784	300
l	2352	340
m	2352	1119
n	1785	1159
o	921	1157
p	0	1084
q	970	448
r	1494	440
s	1494	1020
t	970	1012
u	1739	711
v	2039	711
w	2039	748
x	1739	748
y	2110	699
z	2306	699
A	2306	760
B	2110	760

Part 2

	x	y

Side 1, forward section

	x	y
a	0	0
b	156	0
c	822	0
d	822	119
e	156	221
f	0	235

Daggerboard case

	x	y
g	858	18
h	1152	11
i	1152	270
j	858	270
k	1201	18
l	1495	11
m	1495	270
n	1201	270

Side 2, forward section

	x	y
o	1530	0
p	2196	0
q	2352	0
r	2352	235
s	2196	221
t	1530	119

Inboard part frame 3

	x	y
u	368	221
v	626	221
w	626	368
x	368	368

Stem transom

y	1580	237
Z	2124	237
B	2124	355
A	1580	355

Deck, forward section

C	0	340
D	150	357
E	809	457
F	809	1001
G	150	1102
H	0	1119

Frame 2

I	833	402
J	1065	402
K	1065	1176
L	833	1176

Frame 4

M	1078	331
N	1318	331
O	1318	1176
P	1078	1176

Transom

Q	1329	468
R	1508	468
S	1508	1176
T	1329	1176

Bottom, aft section

U	1532	458
V	2198	357
W	2352	340
X	2352	1120
Y	2198	1102
Z	1532	1002

Outboard sections of frame 3

aa	304	1103
bb	563	1103
cc	563	1176
dd	304	1176
ee	822	1103
ff	822	1176

Part 3

	x	y

Side 2, aft section

	x	y
a	0	0
b	564	0
c	1428	0
d	2352	0
e	2352	179
f	1428	245
g	564	260
h	0	237

Bottom, front section

i	0	340
j	565	300
k	1429	302
l	2352	375
m	2352	1083
n	1429	1156
o	565	1159
p	0	1120

Note: Small discrepancies may exist between the millimeter and inch tables. See page ii.

<80°>

69.5 sq. ft.

<80°>

43 sq. ft.

<80°>

56 sq. ft.

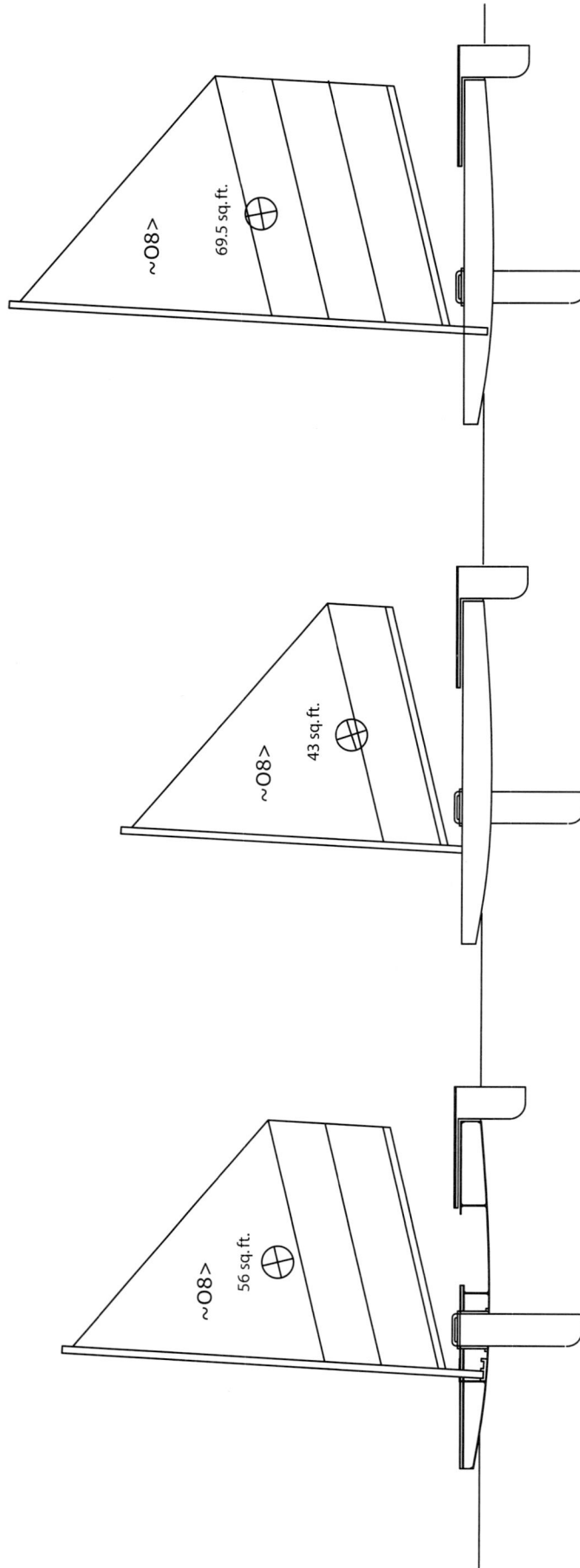

Even when reefed, the bafter sail retains its fore-and-aft center of effort almost unchanged.

It was a nearly windless day when the author launched the first Eek!, but it performed well in the few gusts of wind, and the balance seemed right.

PUDDLE DUCK RACER

THE WORLD'S SIMPLEST RACING CLASS?

Length Over All (LOA): 8'
Length Waterline (LWL): 7'6"
Beam: 48"
Weight: 100 lb.
Displacement at Design Waterline: 650 lb.
Crew: 1–3 adults
Propulsion: Sail
Construction Methods: PU stitch-and-glue
Epoxy stitch-and-glue
Simplified chine log

Not long after I drew up the first of the Mouseboats, I suggested to a group of Mouseboats enthusiasts that it would be fun to create a development racing dinghy class in which the hull would be some standardized version or relative of one of the Mouse hulls, but the builders would have complete freedom in their design of the sailing rig. I argued that racing these boats would often be very wet, but also that it would offer its members challenges and a level of entertainment that you wouldn't find in many dinghy racing classes.

David "Shorty" Routh picked up the idea with an enthusiasm that remains undimmed, and developed a set of rules that defined an iconoclastic 8-by-4-foot straight-sided, flat-bottomed hull that had enough displacement to carry a couple of large adults, and would be very simple and inexpensive to build.

The Puddle Duck Racer was soon under construction, and a new class of small boat racers was born.

The rules of the Puddle Duck Racer class are on the PDR website: www.pdracer.com. They define a boat with flat, parallel sides and flat bow and stern transoms built to a tolerance of 1 inch in all dimensions. The beam must be at least 48 inches. The sides are defined by a drawing that appears on the website. However, the rules state that the boats can be built with much lower sides, as low as 10 inches from the sheer to the deepest part of the a boat, as long as the fore and aft transoms and bottom all follow the defined profiles. One last rule is that the boat is not allowed to have any clever lifting foils that might make it faster.

The first Puddle Duck Racer hull was launched in 2004, and as of August 2007 there were 160 official hull numbers (no doubt more have been built). This is a development class, and its members develop their boats continually, adding their own touches to the design. There's no telling what refinements may arise in the future, but this section presents a version current at the time of writing, based on some boats built by John Wright for visiting Puddle Duck Racers.

Many Puddle Ducks have leeboards, though John Wright's boats feature a centerboard mounted in a trunk on the inside of the port tank, with a lead weight at its tip to make it drop. A pendant line and cleat allow you to control the board so that you can adjust the lateral resistance to match your rig and the point of sail. Melting and pouring lead centerboard ballast is not especially difficult, but it is hazardous and it involves the use of tools and materials that don't appear elsewhere in this book. I recommend against it for beginners looking for an "ultrasimple" approach to boatbuilding, and instead suggest using lead set in epoxy.

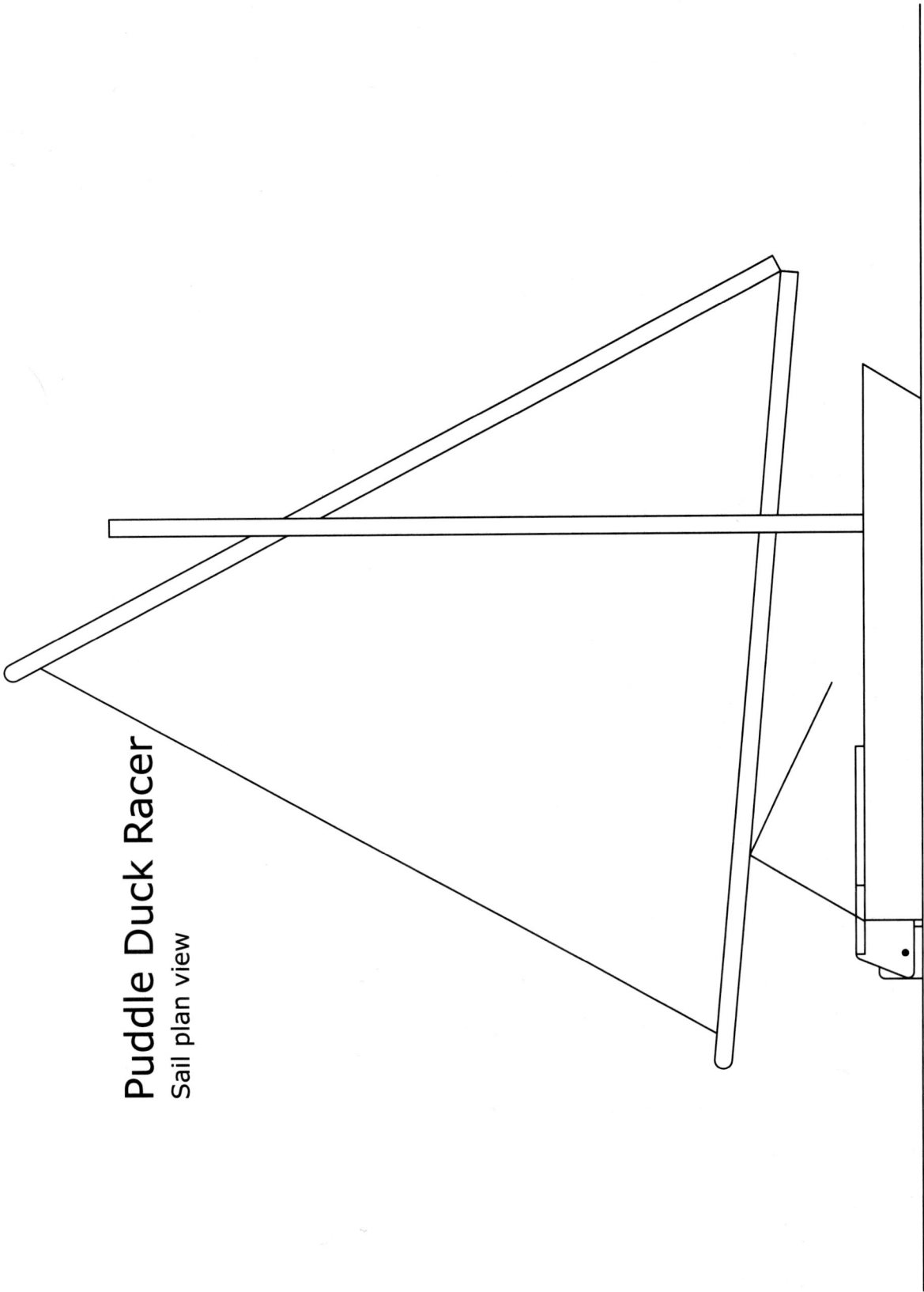

Puddle Duck Racer
Sail plan view

All dimensions in inches

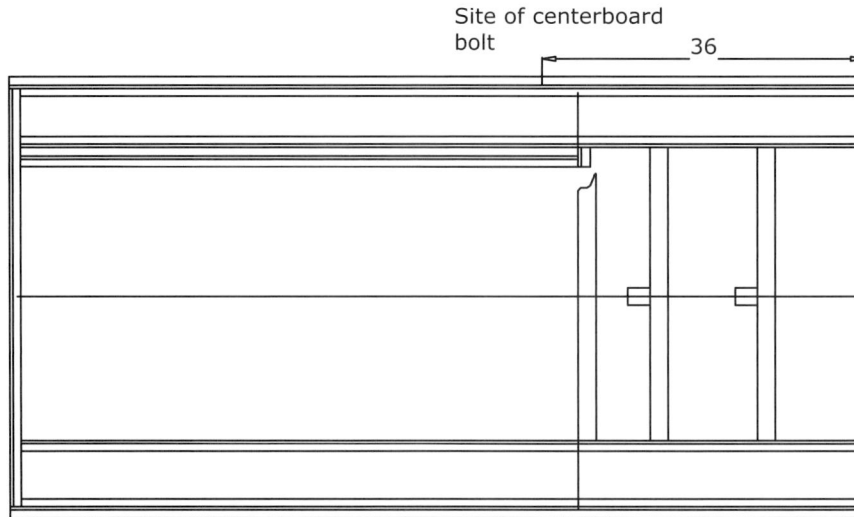

Site of centerboard
bolt

36

Sub-deck structural details.

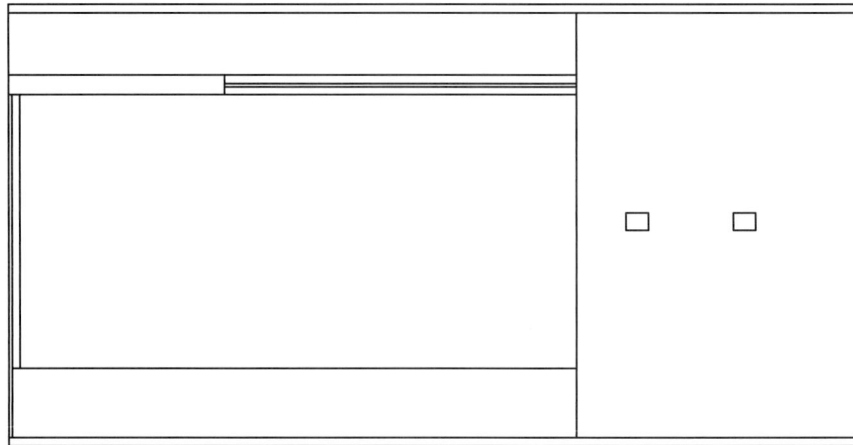

Deck and buoyancy tank tops installed.

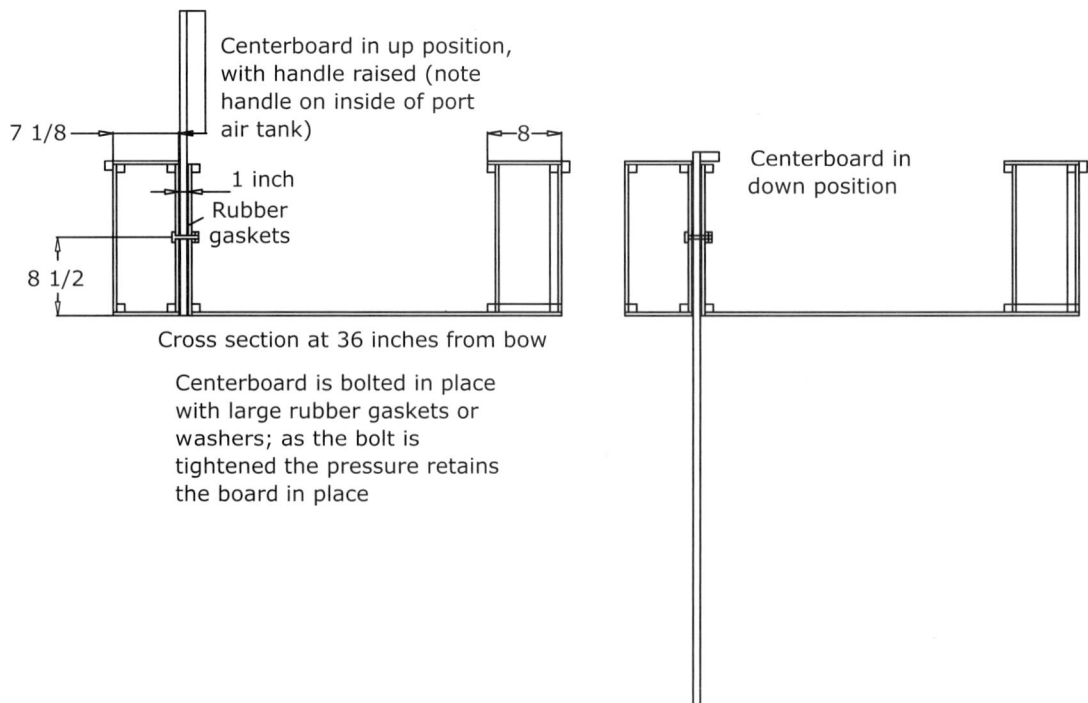

7 1/8

Centerboard in up position,
with handle raised (note
handle on inside of port
air tank)

8

Centerboard in
down position

1 inch
Rubber
gaskets

8 1/2

Cross section at 36 inches from bow

Centerboard is bolted in place
with large rubber gaskets or
washers; as the bolt is
tightened the pressure retains
the board in place

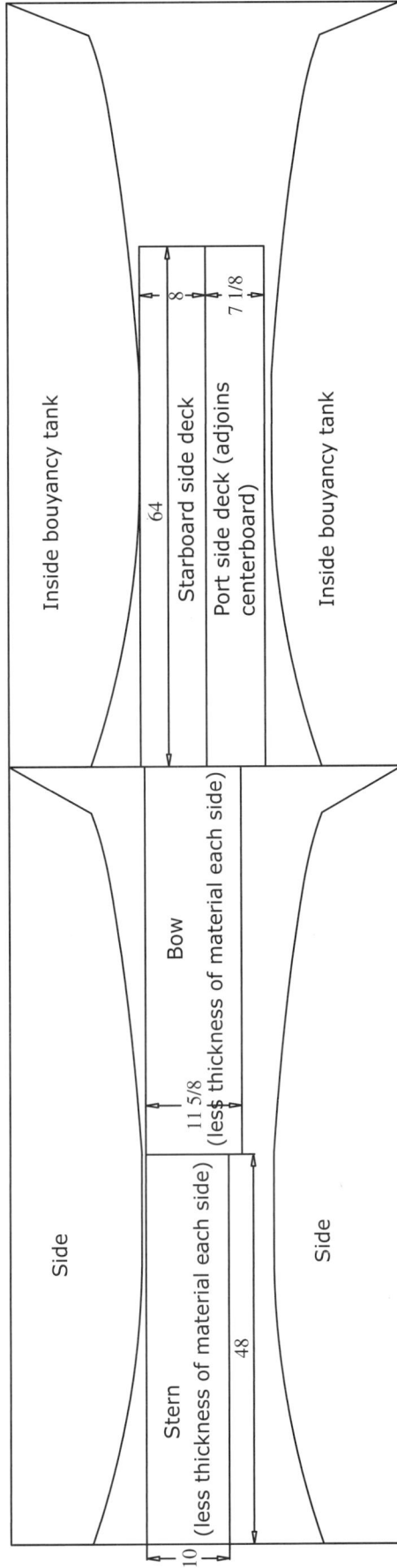

All dimensions in inches

Inside bouyancy tank

Starboard side deck

Port side deck (adjoins centerboard)

64

8

7 1/8

Inside bouyancy tank

Side

Bow

11 5/8
(less thickness of material each side)

Stern
(less thickness of material each side)

Side

48

10

All dimensions in inches

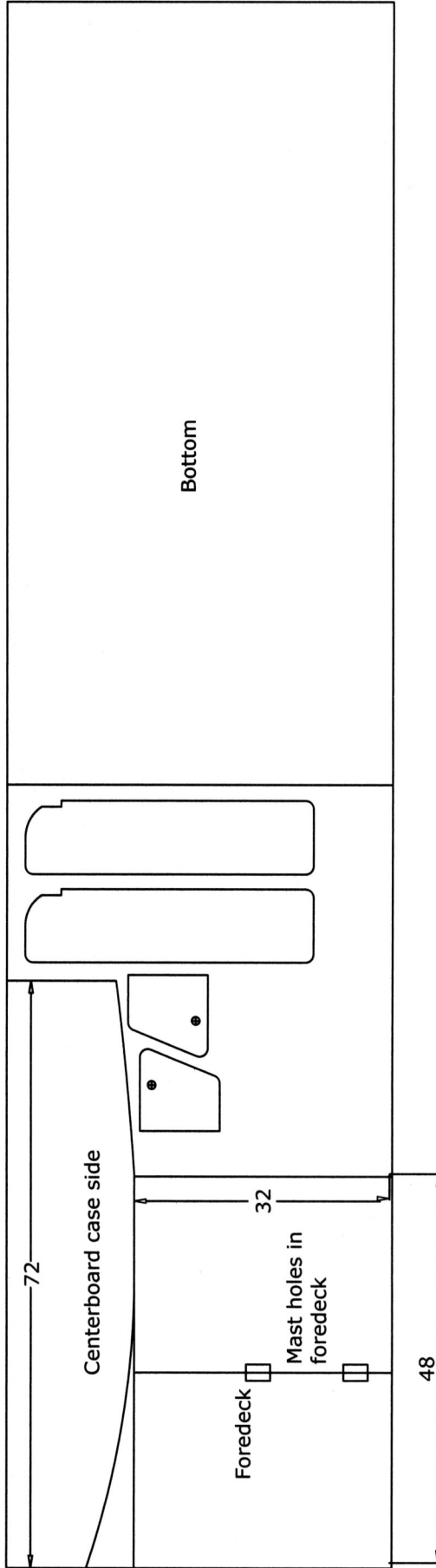

Bottom

Centerboard case side

72

32

Foredeck

Mast holes in foredeck

48

All dimensions in inches

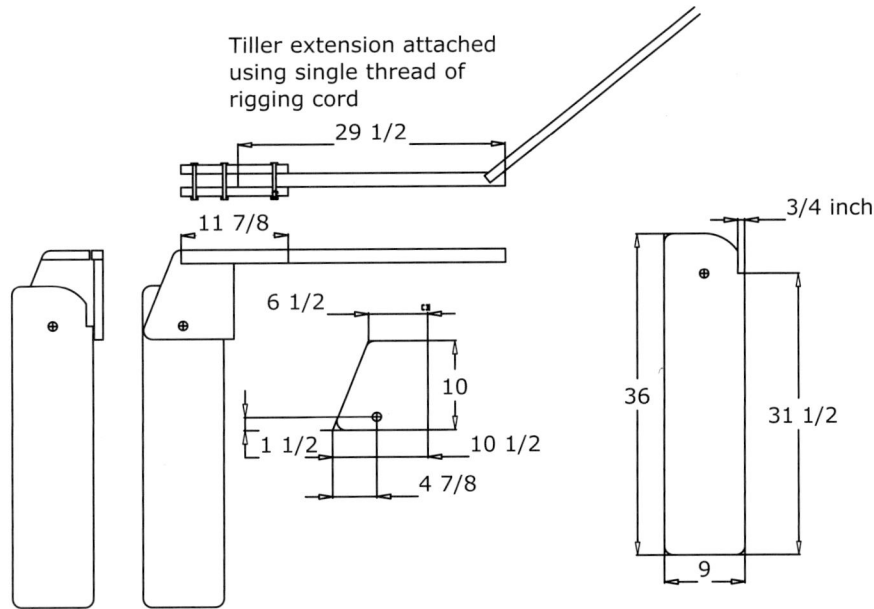

Tiller extension attached
using single thread of
rigging cord

29 1/2

11 7/8

6 1/2

10

1 1/2   10 1/2

4 7/8

3/4 inch

36   31 1/2

9

Rudder details.

10 1/2
Position of centerboard
7 1/2 bolt

Centerboard case with logs and
end cleats, and deck cleats

If you do wish to take the molten lead approach, read up on the subject, and be extremely cautious; molten lead is bad stuff.

This Puddle Duck Racer has sides 16 inches high. This is a good compromise height that provides a large enough bow transom to bounce off waves instead of diving into them, but is low enough to allow competitors to easily sit on the sides and hike out to balance the boat in a blow. It has 7-inch-wide air tanks on either side for emergency buoyancy, which are also good for sitting on when leaning out to balance the boat (a toe strap runs from the foot of the mast to the stern). The boat comes up from a capsize with no water on board. The reason for this is that the buoyancy tanks contain so much air that any water runs out of the cockpit before the boat can be righted.

Spray rails on the bow deck help to keep water out and can be used to mount hardware. A mast sleeve to accommodate a 2½-inch square spar is mounted on the deck 27½ inches from the bow. The sleeve is rectangular in shape, with a drain hole in the bottom that drains any water to the interior of the hull. The Wright boat also has a lifting rudder with a stock with only one strong cheek: it is attached to the rudder blade by a single large bolt, nuts, and a washer.

The lateen sail for the design included in this book was developed by Puddle Duck Racer champion Ken Abraham. It has a yard along the luff and a boom along the foot, and it can be hoisted and secured with a single halyard just like a lug sail. The luff is 12 feet, the foot is 11 feet 4 inches, and the leech is 11 feet, giving a total area of about 62 square feet. The foot and luff are curved outward to give the sail some belly. The curve is a smooth affair running from the tack to the aft end of the boom, with a maximum depth of 3 inches, about 30 percent aft of the luff. The lateen can be easily reefed by running a line of cringles from the tack to the leech and lacing a line through them when necessary.

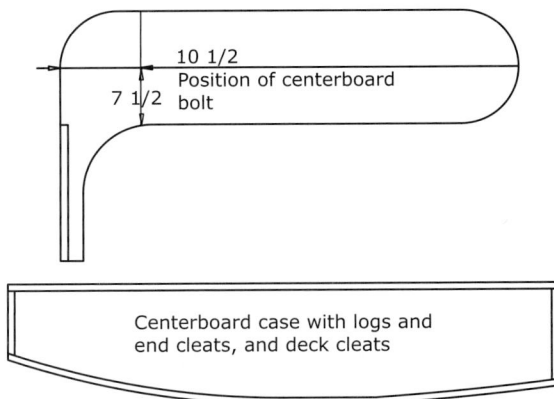

7-1/2 inch radius

Centerboard (i)

9-inch radius

9-inch radius

7 1/2-inch radius

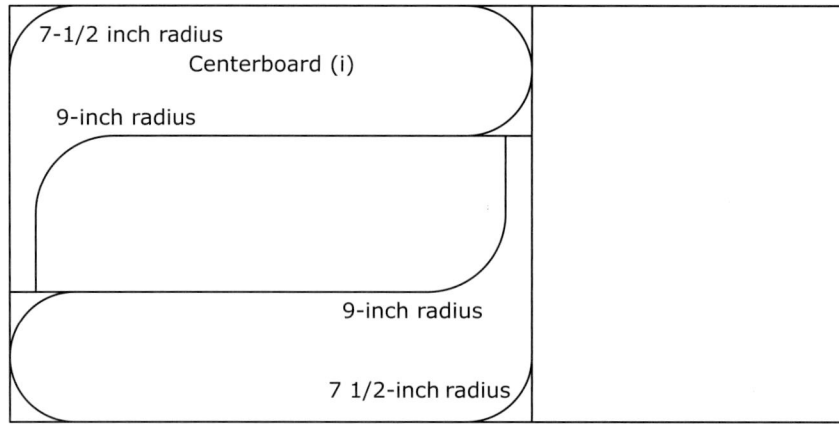

Laminate centre-board pieces together,
then add 2-by-1-by-18-inch handle

Mast 3 by 2 by 143 inches, rounded off to
an oval profile above the deck

Boom 3 by 1 1/2 by 138 inches

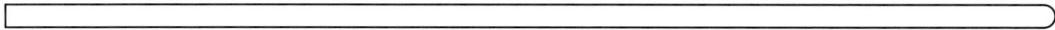

Yard 3 by 1 1/2 by 150 inches

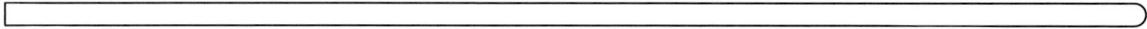

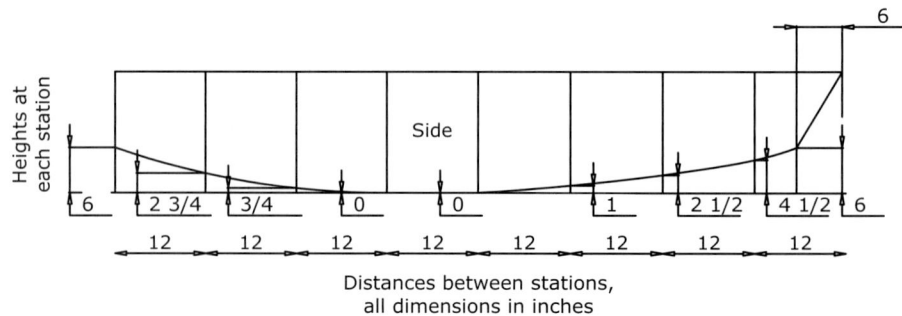

6

Heights at
each station

Side

6   2 3/4   3/4   0   0   1   2 1/2   4 1/2   6

12   12   12   12   12   12   12   12

Distances between stations,
all dimensions in inches

The construction of David "Shorty" Routh's classic PDR *Bucket Ears* could not have been much simpler.

The leeboard arrangement Shorty used could not have been much simpler either.

The PDR cuts a surprising kind of dash on a trailer.

SUMMER BREEZE

A VERSATILE ROWING/SAILING SKIFF

Length Over All (LOA): 11'8"
Length Waterline (LWL): 10'10"
Beam: 48"
Weight: 80–90 lb.
Displacement at Design Waterline: 340 lb.
Crew: 1 adult and 1 child or 2 moderate-sized adults
Propulsion: Oars, sail
Construction Methods: PU stitch-and-glue
Epoxy stitch-and-glue
Simplified chine log

David Beede's Summer Breeze is a great little boat. Although it's simple and cheap to build, it sails and rows well. It's also a very clever design that makes an 11-foot 8-inch boat with a 48-inch beam, 16-inch sides, and a displacement of 340 pounds out of two sheets of plywood, plus a little something extra for the transom.

Those of you who have built any of the boats described in the previous sections will find there's little that's challenging about Summer Breeze, even though it looks a lot closer to everyone's idea of a "real" boat. Because it differs in several of its particulars from most of the boats in this book, I'll present the construction method here in detail. Anyone thinking of building this boat—particularly if this is your first boat—should begin by blowing up the drawing using a photocopier, and building a model using the expanded images glued to thick paper or cereal-box-weight cardboard.

Summer Breeze can be built using either the "instant boat" method (the precursor to my simplified chine log method) that David used for the prototype, or you can go with stitch-and-glue. I suggest using the epoxy stitch-and-glue method because I prefer the look of a well-made taped exterior seam to that of an external chine log, and because stitch-and-glue can be kinder in allowing for small inaccuracies in marking out and cutting. But many of you will prefer handling wood and nails instead of epoxy and fiberglass.

Building the boat using external chine logs cuts down on adhesive costs because cheaper adhesives and sealants can be used, and the amount of epoxy required is either much reduced or not necessary at all, depending on whether you decide to cover the bottom, or the bottom and sides, with fiberglass cloth.

David doesn't favor coordinates when drawing up his panels, but in other respects the process is generally similar to the one used for most of the other boats in this book: draw the fixed dimensions, rule in the straight lines (including the places where the internal panels and bracing go), and cut out using a handsaw.

Just a glance at the drawing showing how the side panels of the boat should be cut out from the first 4-by-8 sheet of plywood reveals just how successful David has been in avoiding waste. In a single sheet of plywood, he somehow managed to fit the panels and still leave something over to make a pair of oar blades, quarter knees, and a skeg. Take care not to throw anything away that you cut from this first plywood sheet—the truth is that the only waste is the dust from cuts and you will find a use for every square inch.

With every cut a certain amount of material, equivalent to the width of the blade, is reduced to dust. This is called the kerf, and if you look at David's drawing you'll see that because of the material lost to the kerf the fore and aft sections of the sides won't be quite the

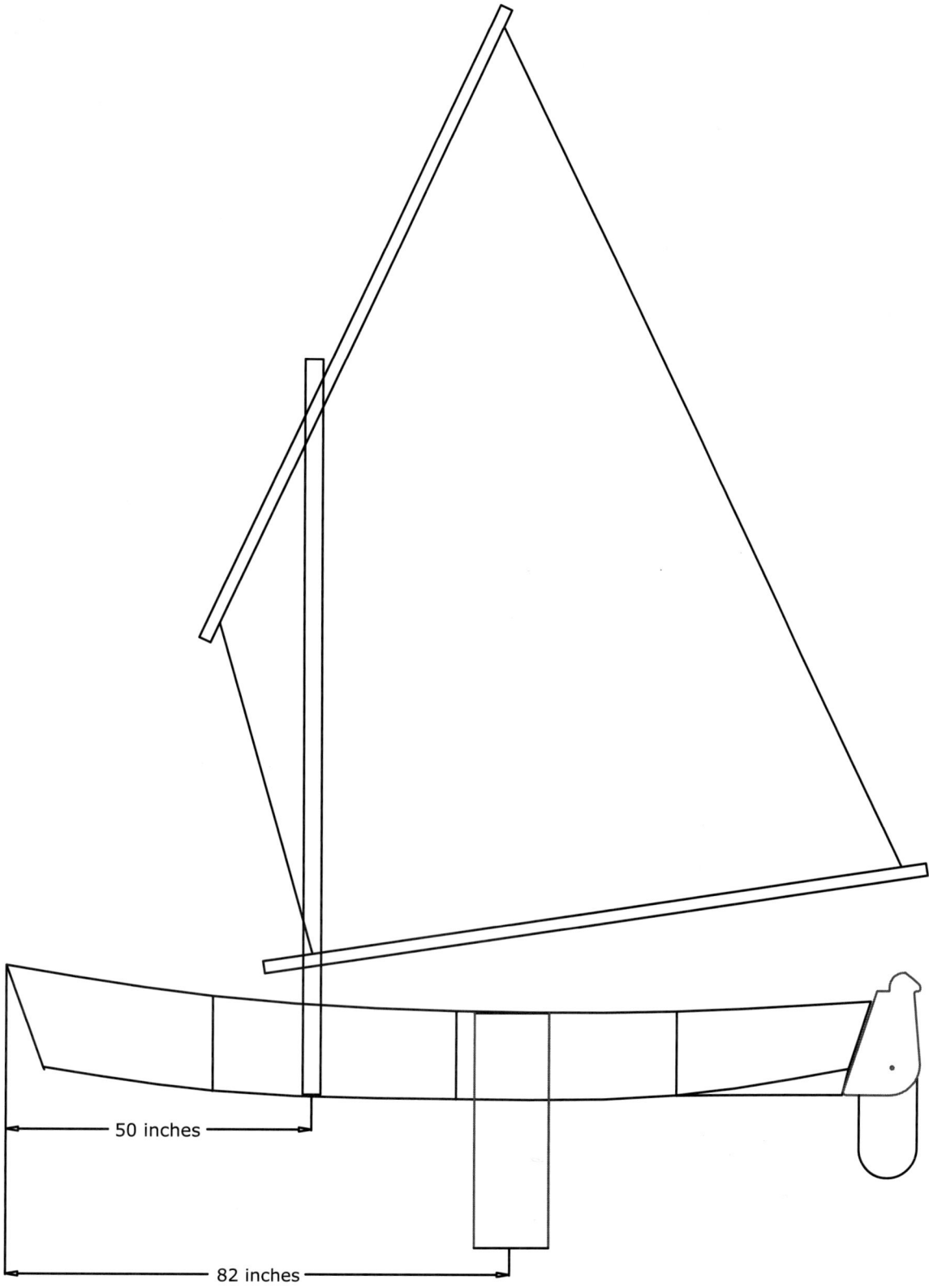

50 inches

82 inches

All dimensions in inches

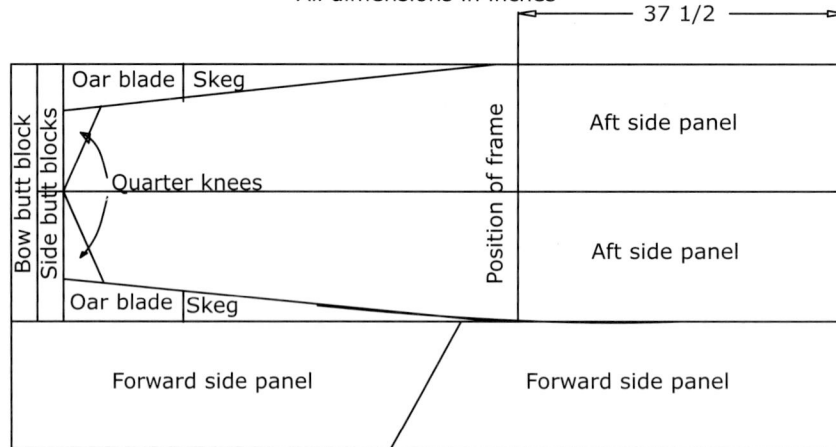

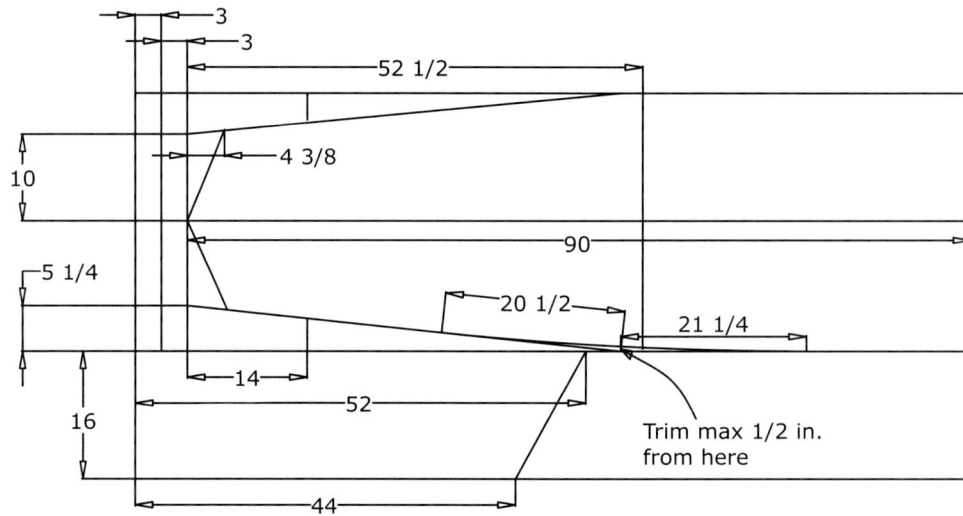

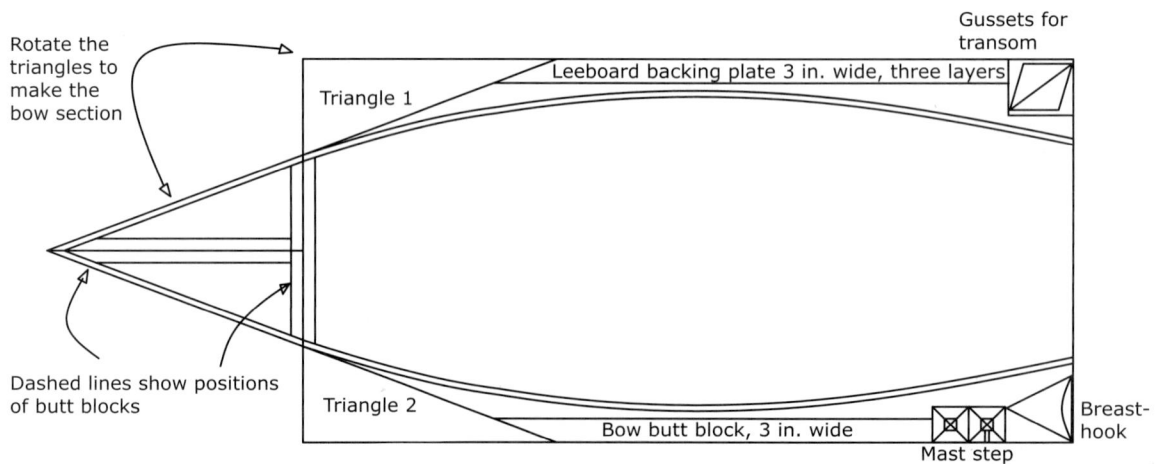

$\leftarrow$ 37 1/2 $\rightarrow$

Oar blade | Skeg

Bow butt block

Side butt blocks

Aft side panel

Quarter knees

Position of frame

Aft side panel

Oar blade | Skeg

Forward side panel

Forward side panel

3

3

52 1/2

4 3/8

10

90

5 1/4

20 1/2

21 1/4

14

16

52

Trim max 1/2 in.
from here

44

Rotate the
triangles to
make the
bow section

Gussets for
transom

Leeboard backing plate 3 in. wide, three layers

Triangle 1

Dashed lines show positions
of butt blocks

Triangle 2

Bow butt block, 3 in. wide

Mast step

Breast-
hook

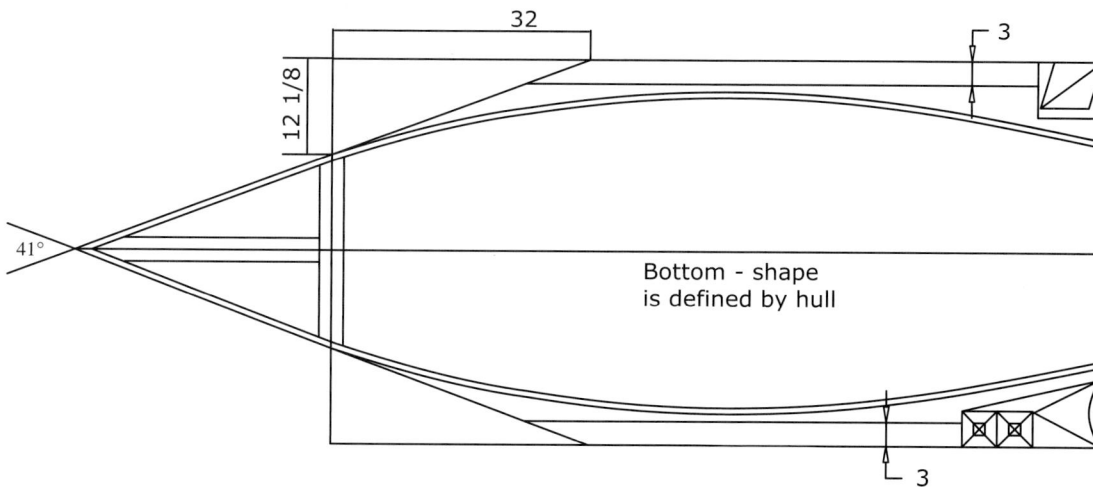

same width. To reduce this difference to a minimum, you have to abandon the usual carpentry principle of always cutting on the wastewood side of the line. These cuts are an exception: when making the long cuts between these panels, cut right on the line so that the sides are exactly equal because the kerf is effectively shared between them.

In these circumstances, it's good to use a Japanese pull saw that leaves a very narrow kerf. Because the blade of a pull-saw is in tension when cutting, it is thinner than the blade of a handsaw that cuts on the push stroke, which is thicker because it's in compression when cutting.

Similarly, cut the forward-most panels along the center of the line, and your panels should be closely matched.

As well as leaving material for the oar blades and skeg in this cutting-out plan, David has carefully included two strips of plywood to

be used as butt blocks for the sides and the forward part of its bottom, and they too should be cut out down the centerline so that the dimensions are closely matched. Cut one of the butt blocks into two, carefully join the side panels, and allow the glue to harden.

The drawing shows a distinct but wide angle along the bottom edge of the cut-out side panels. This has to be rounded off equally for both panels. I'd suggest clamping both sides together, marking a point about half an inch in from the "corner" of the obtuse angle, and rounding off gently with a Stanley Surform or rasp until you achieve a graceful curve that lines up with the mark you have just made.

The second nesting diagram showing the layout of the bottom panels shows two additional long, narrow butt blocks along the outer edges. Between them are the two large bottom panels, with the twin triangular forward bottom panels in the two triangular

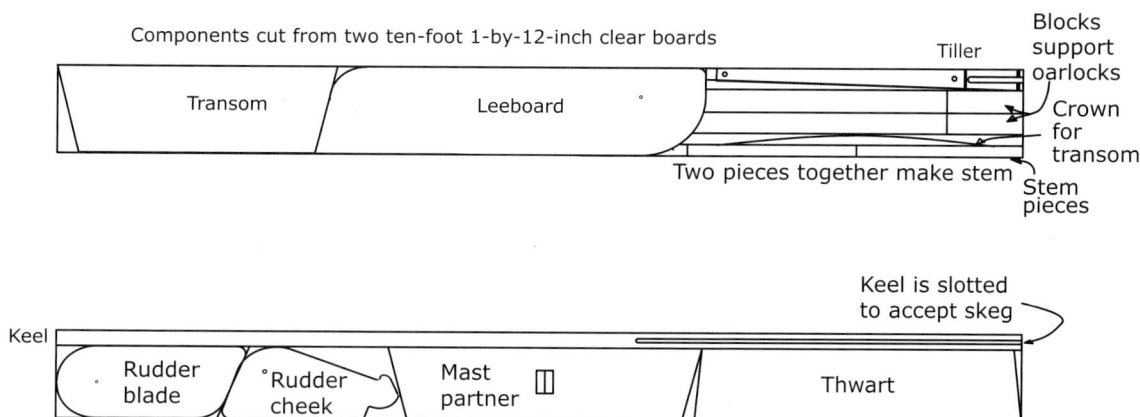

All dimensions in inches

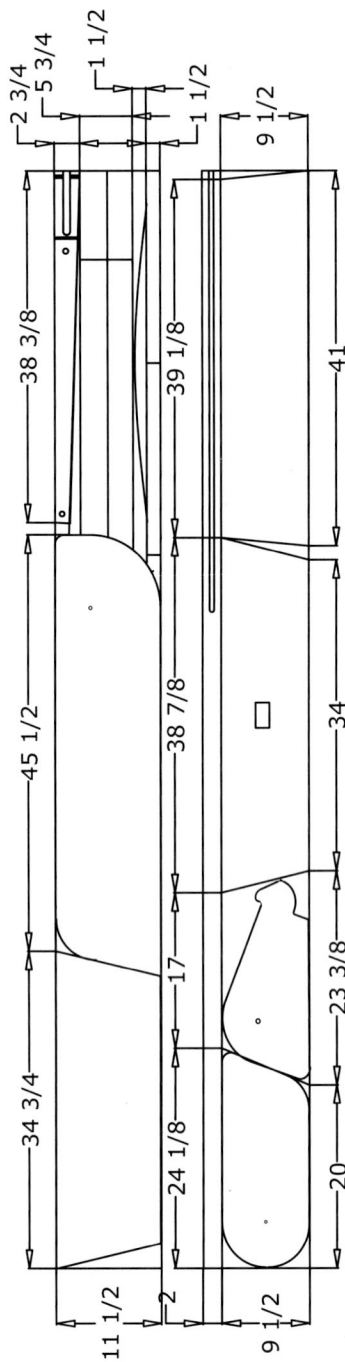

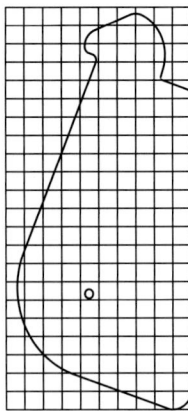

2 3/4
5 3/4
1 1/2
1 1/2

9 1/2

38 3/8

39 1/8

41

45 1/2

38 7/8

34

34 3/4

17

23 3/8

24 1/8

20

11 1/2

2

9 1/2

Not to scale.

7 5/8

1 5/8

5 3/4

48

Plywood gussets between
frame members

15 3/8

39

Center frame

9

8 1/4

1 1/2

4 1/2

4 1/2

Second mast foot has
drain cut into it

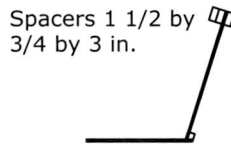

Spacers 1 1/2 by
3/4 by 3 in.

Gunwales and inwales
3/4 by 1 1/2 in.

Chine logs make from 1 by 1 in.
trimmed to accept bottom (or one
piece 6/8 by 1 1/2 in. split at 18 degrees)

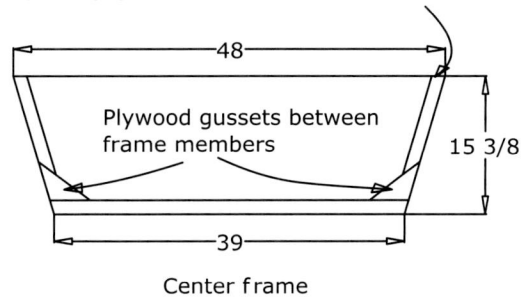

spaces at the left-hand end of the plywood sheet. These have to be butt-jointed together with the long, narrow butt blocks laid in a T-shape as shown.

David also manages to find pieces for the mast step, breasthook, backing for the leeboard, and other small parts from this sheet.

With everything safely cut out, the first job is to butt-joint all the large panel components. This can be done in a variety of ways, but I recommend that you use the full set of butt blocks in the plans. Depending on your building method, you're likely to need to cut some of this butt block material away.

You can use a variety of weights to hold the plywood panels and butt blocks in place while the glue sets. David uses staples for this purpose and removes them after the glue has hardened.

Once the bottom and side butt joints are complete, you have to decide between building using the simplified chine log method or by the stitch-and-glue method. I won't repeat everything I said in earlier chapters about these methods, but only pick out some key points as they apply specifically to Summer Breeze.

DETAILS FOR THE SIMPLIFIED CHINE LOG METHOD

Stem and Transom
The solid lumber stock has to be marked out and cut out carefully. The stem can be cut from a piece of solid stock sawn to create a 41-degree wedge, but another way is to cut a piece of 2-by-1-inch material to an angle of 20.5 degrees. The two halves are then reversed and glued together to create a stem with little or

All dimensions in inches

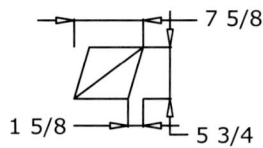

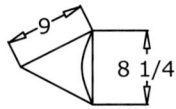

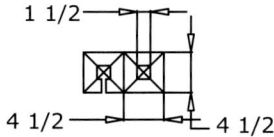

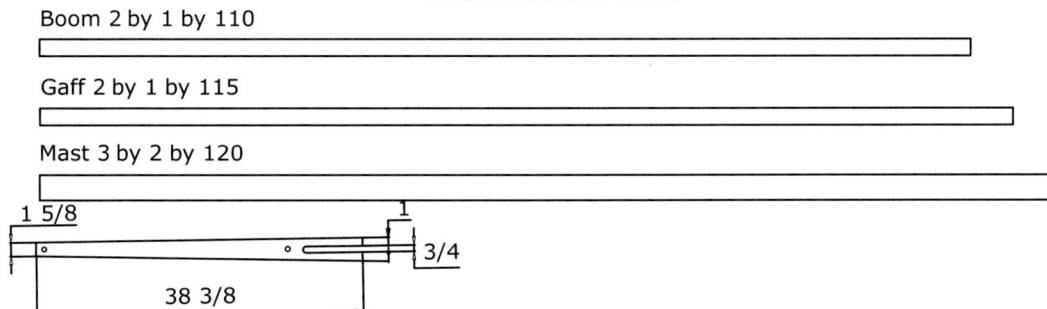

Boom 2 by 1 by 110

Gaff 2 by 1 by 115

Mast 3 by 2 by 120

1 5/8

1

3/4

38 3/8

All dimensions in inches

107 1/4

151 3/8

Midpoint 2

3-inch seams all around

3 Midpoint

56

31 7/8   3

96 3/8

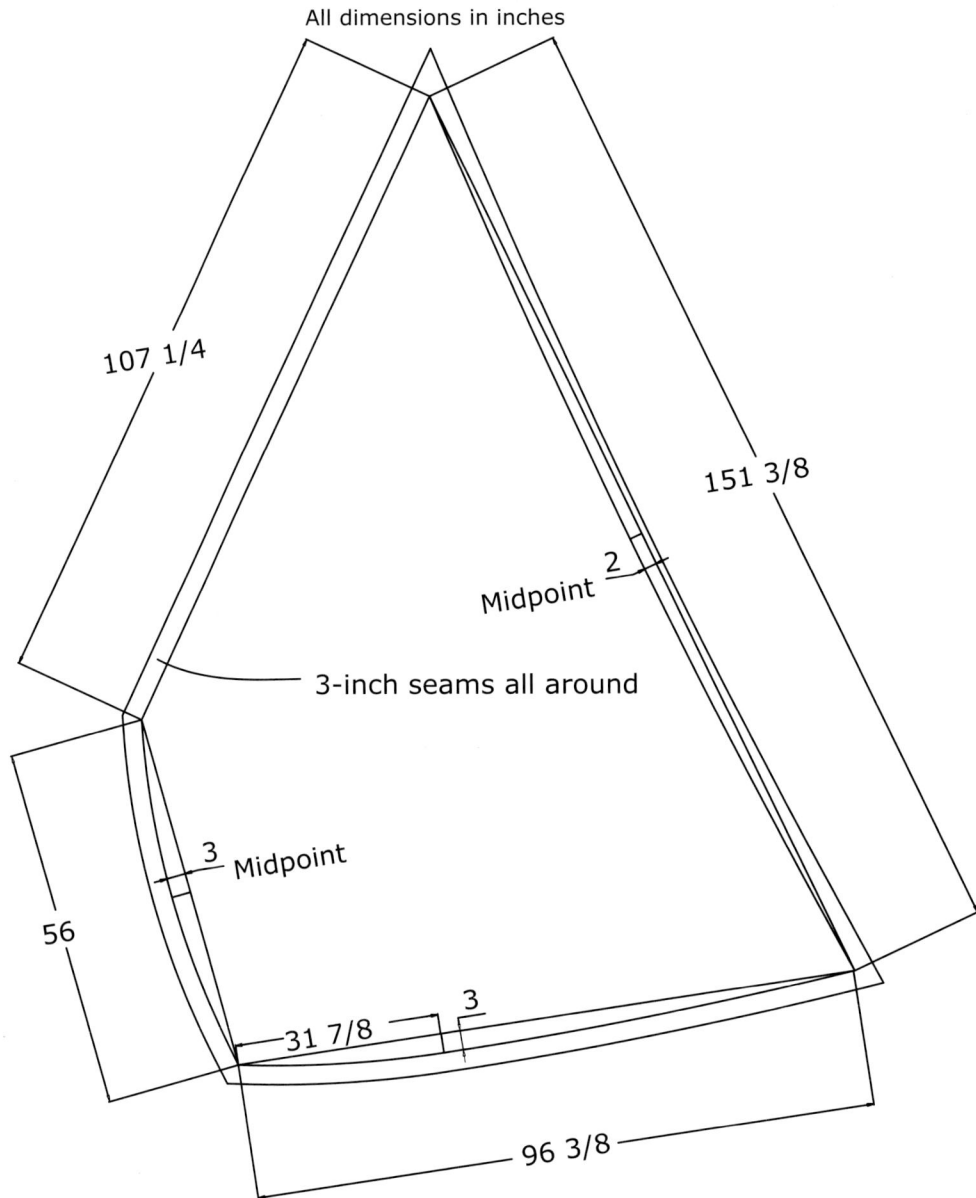

no waste. Use a protractor to measure and mark the angles, and don't worry about the half degree; you'll never be that accurate anyway.

When you come to fit the stern, you will also need to cut the inch-thick transom as shown in the drawing to ensure that it neatly fits the shapes created by the side and bottom panels.

Frame

Summer Breeze's central frame is placed at the widest point of the boat where the bottom is deepest and where the sides and bottom form a 90-degree angle to the frame. Because of this

the central frame needs no further shaping before you assemble the boat.

The best way to make up the frame is to draw up the frame profile full-sized on a large piece of paper (brown paper used to wrap packages is good), and then make up the frame so that it matches the paper profile. Leave the sides 18 or 19 inches long so that they can later be trimmed to exactly the right length.

Mark out and cut out the gussets as shown in the drawing. These are sandwiched between the piece of lumber that makes the "bottom" of the frame and the pieces that make the "sides." Placing the gussets between the bottom and side members helps eliminate some shaping, and the excess where the frame components meet at the chines is easily removed with a rasp or Stanley Surform.

Assembling the Sides

Once the central frame has been screwed and glued and trimmed to match the paper profile, it's time to go three-dimensional. This is always an exciting moment in any boatbuilding project, and it usually goes better with a helper.

The first move into the third dimension is to screw and glue the shaped stem to the bow-end of one of the sides—it doesn't matter which, and we do it this way only because it's easier to do it at this stage than later.

The second is to glue and screw both side panels to the central frame.

The third is to find a line and tie a loop around the aft end of the side panels so that they won't pull outward as you draw the two halves of the bow together. The fourth step is to use a Spanish windlass (see page 39, Chapter 4) to draw the front ends of the panels together. Begin by placing clamps—one each at the sheer-line and at the bottom on each side panel—about a foot from the wobbling ends. Then tighten the loop on the inside of these so that it doesn't get forced off the end.

Tie a loop tied in a second length of line, place it behind the clamps and around the wobbling forward section of the sides, and pass a stick through the loop and crank it around until both sides come together at the stem.

As you recall, you've already glued and screwed the pre-cut stem to one of the side sections. The fifth step is to glue and screw the other side panel to the stem, making sure the tops and bottoms of the two side panels coincide. When you're done, you should have a pointed bow, with the stern sections of the sides only held by their restraining line.

The sixth step is to glue, clamp, and screw the sides so that the frame is exactly where you marked it, making sure that the bottom of the frame and the bottom edges of the side panels correspond exactly. Tighten the line a little so that the gluing surfaces are brought into close contact.

The seventh step is to glue and screw the transom into place, shaping its edges to match the angles made by the sides, while making sure the lower edge is slightly proud so it can be trimmed to accept the bottom.

As well as being very traditional, this is also a great, fast way to build a boat, but if you're working alone it can be a little like wrestling an octopus. It's best to get someone to lend a hand with the Spanish windlass.

As with the Mouseboats, at this point it's good practice to use a tape measure to ensure that the distance from the stem to each stern quarter is the same. This is called "horning." If the two distances aren't identical, jiggle the stern until they are, and you can leave the boat knowing that all is true while the glue hardens. Another way to do this is to mark the center point of the frame, stretch a string from the stem to the center point of the transom, and jiggle the whole thing around until it lines up perfectly.

Installing the Bottom

When everything is good and solid, you can start on the external chine logs and the bottom. The chine logs are 12-foot-long, 1¼-by-¾-inch boards glued and screwed to the outside bottom edges of the sides. These should be single boards and should be attached or clamped to each side simultaneously starting from the stern so that the boat does not become twisted.

As you go screwing, or clamping and gluing, from the stern to the bow, keep the edges of the chine logs flush with the plywood edges. When everything has set use a straightedge and a rasp or Stanley Surform to make the gluing edges horizontal so that they offer a rough but flat gluing surface for the bottom.

The bottom is not properly marked out and does not need to be; it's represented in the drawings as a space on the plywood that's not taken up by any other parts. What you'll have to do is lay the remaining plywood onto the upturned boat, and jiggle it about until the chines, bow, and stern are all covered. Then drive a few temporary nails through the bottom into the chines to keep everything still, and then use a pencil to carefully trace onto the plywood along the chines, bow, and stern, making sure that you leave a small margin all the way around (see the drawing). Hold the flat of the pencil against the boat so that the margin is exactly half the pencil's thickness. This will give you a small but comfortable margin all around for fitting to the hull.

The bottom is then glued down to the chines, transom, frame, and stem, and clamped, nailed, or screwed at intervals of 8 to 10 inches all the way around. To cut down on the amount of sanding required later, wipe up any glue that squeezes out before it hardens. After the glue has well and truly set, get the rasp or Stanley Surform out again and trim the excess plywood all around the bottom.

USING STITCH-AND-GLUE

Summer Breeze's bow has a much more acute angle than any stitch-and-glue joint you have seen in the Mouseboats, but that should cause no problems—the stitch-and-glue procedure for making the stem is the same for any other stitch-and-glue joint, and the resulting joint will be very strong. Nevertheless, because the bow of a boat often receives a lot of punishment, it should be reinforced with several layers of fiberglass and epoxy to make it strong and durable.

The central frame is glued and screwed into place as we've seen, but this time the transom is attached to the sides and bottom just like any other stitch-and-glue joint. It's perfectly okay to cut the transom from something other than the 1-inch-thick solid board that the designer shows in the plans. You can use ⅜-inch or ½-inch plywood for the transom, though I'd use two or possibly even three layers to make a pad for the outboard clamp, if you plan to use an outboard.

As you saw above, the area of plywood set aside for the bottom in the original plans is intended to cover a boat built with external chines, and it is, therefore, a little larger than is strictly needed for the stitch-and-glue version. You will still need to lay the bottom panel on the upturned boat and trace around the hull using a pencil, but due to the

absence of chines, you probably won't be able to use small nails to temporarily hold the bottom firmly to the rest of the hull while doing the marking out. Instead you'll need to use a variety of weights from around the house and the workshop area.

When tracing the bottom, rather than leaving a margin for error, as you did with the simplified chine log method, place the pencil point right up against the boat's sides and transom to scribe the exact shape and size needed for stitch-and-glue assembly. If you don't, you'll end up with an awkward ridge of epoxy that will require a lot of unnecessary work to remove from the exterior seams in preparation for taping.

FINISHING THE HULL

The nesting drawing that includes the bottom panel also shows suitable areas of scrap plywood for the breasthook and quarter knees (labeled "gussets for transom"). I think these may be a little slight for the purpose, so I recommend finding some extra bits of plywood to make up the quarter knees double-thick. I would recommend making them in a more graceful shape than the simple triangles shown—something that gives it some curve but without weakening the structure would be nice, but not essential.

The breasthook and quarter knees can be made up several ways. You can take the approach used on the Cinderella canoe; it would work very well for Summer Breeze, as the gunwale would go on last and would neatly cover the seams. But since this is, in some ways, a simpler boat than Cinderella and needs blocks for oarlocks, you might choose to dispense with inwales altogether. In their stead, rely on the frames, the bent panel shapes of the boat, substantial gunwales laminated from two 1½-by-¾-inch strips to a finished dimension of 1½ by 1½ inches, and the breasthooks and quarter knees to keep everything rigid. Make up simple softwood cleats and attach them beneath the quarter knees against the boat's sides, shaping the ends with a little taper and cutting them an inch or so shorter than the knees to make them as unobtrusive as possible. Shape them to fit the boat's transom and sides by the good old cut and try method, using a vise or a clamping workbench and a rasp or Stanley Surform. Once the parts have been trimmed to fit, screw and glue them into place. Alternatively, if you have built the hull by stitch and glue, you may prefer to use the same method for the breasthook and quarter knees.

Make the gunwales from 12-foot lengths of whatever rot-resistant hardwood you can find at a good price.

Regardless of which construction method you used, Summer Breeze can easily be covered in fiberglass and epoxy, as it is made up of large flat surfaces. This is a worthwhile, but by no means essential, upgrade to the basic design of the boat.

With or without the glass and epoxy, the flat bottom is wide enough to require the reinforcement of a simple plank keel running from bow to stern. This should be made from 2½-by-¾-inch lumber. Start by drawing a centerline from the bow to the transom using a straight piece of lumber or molding. Then draw two parallel lines on either side of the centerline to indicate where the edges of the keel should lie: the distance between them should be the same as the breadth of the keel material.

Before you fasten the keel plank down, you have to cut a slot in its aft end to accommodate the skeg. David's scheme is to laminate the skeg from two pieces, which means that the slot's width must be twice the thickness of the plywood. Drill a hole in the plank keel, of the same diameter as the thickness of the skeg, where the forward end of the skeg should go, then mark out lines from the hole to the back end of the board and make straight cuts. Cut out the skeg pieces as shown on the plywood cutting diagram, and laminate them together using either epoxy or polyurethane.

Once the keel has been cut to length and the slot has been cut out, it is nearly ready to be glued and screwed into place. First, however, you need to mark and then drill the screw holes. Make the first tick on the centerline, 3 inches or so aft from the bow, then

continue making marks at 12-inch intervals back to the point where the skeg begins. At that point, continue at 12-inch intervals on both sides of the centerline so that they don't coincide with the slot itself, but drive into the material of the keel.

Screwing the keel into place is one of those jobs where it's useful to have someone help you, but, failing that, it's good to have the boat propped up securely in such a way that you can easily get underneath it. However, if you're going to be underneath even a small boat it's good to have someone around for obvious safety reasons. It's best to enlist a helper for this procedure.

From above the upside-down hull, use a hand drill to put the screw holes through the plywood bottom at the points you have marked along the centerline.

Then measure, mark, and check that all is as it should be, and drill the first screw hole in the keel at the bow end. Loosely drive a screw from inside the boat up into the hole in the keel. With this done, line up the keel along the centerline, and either weight it in place, or ask a helper to hold it in place while you scramble under the boat and mark where the rest of the screw holes should go with a pencil.

Unscrew the first loose screw, remove the keel, turn it over, and drill the remaining small screw holes. Finally, roughen the gluing side of the keel, apply the glue, lay it back into position on the bottom, crawl back underneath the boat, and drive all the screws home. Again, this really requires a helper to hold the keel in place while you work from underneath.

An electric screwdriver and a cushion to sit or lie on underneath the boat will both make life easier for you.

As the glue sets, take care to wipe and trim any excess away, partly in order to do a neat job, and partly so that fitting the skeg will be a simple task that won't have to begin with chiseling out a lot of hardened adhesive. So long as you do this, and assuming you have made the slot for the skeg wide enough, fitting it should be easier and more comfortable than fitting the keel!

Fitting the skeg is a matter of cut and try, or rather try and then cut. The process can be summed up as follows: try putting the skeg in place and assess how it needs to be trimmed first to go into the slot and second to seat home stably. The particular issue I see here is that while the bottom of the boat is curved in profile, the side of the skeg is straight. This means that the profile of the skeg will have to be hollowed slightly first by spiling, as described in Chapter 4, and then cut to match the curve of the bottom. Because the curve will be slight, you can probably do this pretty quickly with a rasp; however, if you have some epoxy handy, even this may not be necessary.

Once the skeg seats properly, it can be glued in place. As with the Mouseboats, however, I would nail a little scrap plywood to the transom to use as a brace to keep the skeg perfectly perpendicular to the bottom until the glue sets. If you do this, of course, you'll need to use a little polyethylene or plastic shopping bag material to prevent the support from being permanently stuck to the boat.

ROWING ACCOUTREMENTS

If the boat is only for rowing, all you need to add are a thwart to sit on, a stretcher for your feet, and a pair of oarlocks.

Further, if your boat will only be used for single-handed rowing, I recommend installing a fixed thwart as shown, so that its front edge lines up with the central frame. This will align the boat's center of buoyancy roughly with the rower's center of gravity, which I reckon will be somewhere around the front of the stomach, if his or her legs are spread out aft in the usual way. This is an approximation that has worked very well for me in several small boat designs, and I'm confident that it's quite close enough.

The drawings show the thwart supports. Support the seat using the central frame and a horizontal cleat of framing lumber on both sides of the boat, running from the frame to a point about 14 inches forward of the frame. Take two pieces of 1-by-2-inch lumber cut to size, and cut a taper into the forward 5 inches or so. This taper isn't essential, but it will help distribute pressure exerted by the thwart on the plywood sides.

Dry-screw the cleats temporarily into the position shown, about 9½ inches down from the sheer on each side of the boat. With the boat on a horizontal surface, use a level to mark how the cleats should be trimmed to present a wide horizontal surface to support the thwart. Trim them, then glue and screw the supports permanently in place.

Make the thwart from two pieces of ¼-inch plywood, 8 to 9 inches wide and laminated together so that the exterior grain runs from one side of the boat to the other. An athwartships orientation is stronger and looks good. A couple of knot-free lengths of 1-by-2-inch lumber beneath the thwart will make it as strong as it needs to be.

The specified width is important because it's just the right size for most people to "hook" their bottom onto when rowing—and with the comfort of the rower in mind, do please round the edges with a rasp and sandpaper. The length should be measured and cut as in the drawing so that it bears on the supports, but does not press directly onto the sides of the boat.

If you plan to only use the boat as a single-handed rower, and you don't intend to sail, entertain passengers, or carry anything heavy in your boat, you can complete your hull by gluing and screwing the thwart in place to create a strong, permanent structure.

Most of us, however, will have passengers on board from time to time. In a small boat, a passenger in the stern combined with a rower placed in the middle can cause a serious imbalance: the bow will point at the sky and the transom will drag in the water, making the boat slow and difficult to steer and row.

If the boat is strictly for use as a rowboat (or with an outboard) that will carry passengers, add a second similar rowing thwart a foot or a foot and a half farther toward the bow, so that the rower can move forward to balance the weight of a passenger. (The extra thwart will conflict with the mast in the sailing version.) The new thwart should be built like the central thwart, but needs to be fitted using two cleats on each side since there is no frame in this position, and the cleats should be tapered fore and aft.

To make room for the rower's knees in this forward position, the central thwart must be left loose so that it can be removed when required. A piece of elastic cord and some deadeyes screwed to the frame and cleat on each side can be used to hold it in place when in use.

On a passenger-carrying version of this boat, you might also add a thwart near the stern, although this isn't strictly necessary because non-rowing passengers in a boat this size are often happier sitting in the bottom.

Stretchers for your feet at either rowing position can be made up using the method described in Chapter 6. You'll need to experiment with the position of this stretcher in order to find the most comfortable position.

Oarlocks need to be placed just the right distance from each of the thwarts. A good traditional way to do this is to make the horizontal distance along the sheer, from directly above the aft edge of the thwart to the center of the oarlock socket, the same as the length of the rower's forearm. If you're adding an extra rowing position, add a second set of oarlocks using the same measurements.

The oarlock itself must be well supported. Make up two blocks, each laminated from three 4-by-8-inch pieces of ¼-inch plywood, and then glue and screw them to the hull sides (driving the screws from the outside of the hull). Make sure they are flush with the sheerline and that the fittings of each oarlock sit in the middle of each block. The oarlocks can be standard models. I think a pair of oars of 6 to 7 feet will be perfect for Summer Breeze.

SAILING RIG

Sailing rigs usually add a lot of complexity to a boat, but David has kept this rig about as simple as it can be.

The leeboard should be made as shown in the drawings. It's a pretty easy piece of work and should be attached on the port side and bolted to the side of the boat using a large stainless steel bolt, big washers with holes that match the bolts, and two nuts (the second is a locknut). The bolt should pass through a good-sized circular piece of scrap plywood glued to the side.

If you built your boat using the stitch-and-glue method, there will be no bearing surface along the lower chine, so add two bearing strakes, one along the chine and another balancing strake just under the gunwale or use the gunwale itself. The bearing strakes should be a couple of feet long and should be tapered at the ends to reduce stresses and for a better appearance.

David's rudder design is better looking than the one I drew up for the Flying Mouse. Both follow the same basic idea, with a long, strong tiller and rudder assembly, and a lifting blade, but in this case there is only one strong cheek. A simple stainless steel bolt, with two washers and two nuts (again, one is a locknut), provides a pivot. (See diagrams on page 71 in Chapter 6.) Be sure the installation is not too tight because the rudder must be able to freely lift if it hits an underwater obstruction.

Make the mast from spruce around 2 to 2½ inches square using either a single straight-grained piece of lumber or two pieces laminated together. The lower end of the mast should be square in section, but from about 18 inches up from the bottom it should be planed on its corners to make it 8-sided, then 16-sided, and then it should be sanded with coarse and then

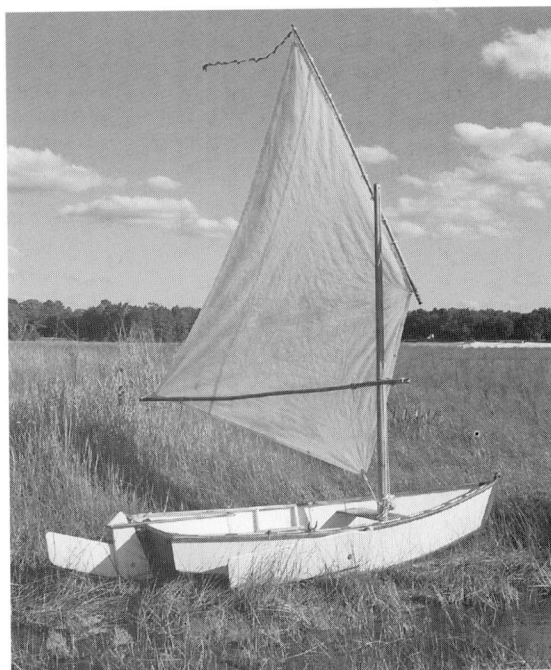

David Beede's Summer Breeze is a classic that uses very little in the way of materials, but is a lot of fun to sail in a sheltered area.

medium and fine sandpaper to make it smooth and round. Don't plane it so that it becomes too thin, and sand lengthwise rather than across the grain for a better appearance after it's varnished.

The mast support partner is like the solid version of the thwart I discussed earlier, but it needs to be sited near the sheer, with the center 50 inches aft of the stem. In the center there should be a square hole, sized to accept the mast, and directly below that should be the mast step. Block the hull so that it is horizontal (the bottom of the stem and the lower edge of the transom will be at exactly the same height above your working surface), and use a level to find the point directly below the hole in the mast partner.

The mast step is made from four 6-inch-square pieces of plywood. The bottom one is left whole, the second one has a square cut in the middle to accept the mast and drain holes cut in the sides, and the third and fourth have the simple square hole. These pieces should be laminated by gluing and clamping them together, and should then be glued and screwed to the bottom of the boat in the position marked.

The yard and sprit-boom can be made from 1-by-2-inch lumber, but I strongly suggest laminating them from two pieces of 1-by-1-inch material chosen to have opposing grain to prevent warping. The ends of the spars should be rounded for safety.

DORIS THE DORY

A ROWING AND SAILING EXPEDITION BOAT

Length Over All (LOA): 15'3"
Length Waterline (LWL): 13'
Beam: 48"
Weight: 100 lb.
Displacement at Design Waterline: 480 lb.
Crew: 2 moderate-sized adults or 1 adult and 1 child
Propulsion: Oars, sail
Construction Methods: PU stitch-and-glue
                                   Epoxy stitch-and-glue

In many ways, flat-sided Banks-style dories are great boats when built from plywood. They're simple to build, exciting to row, and good in moderately lumpy water (particularly if there's something heavy in the boat's bottom), and they usually have a dramatically sweeping sheer and an attractive narrow transom.

Dories are small boats for their length. Although a more mainstream 16-foot boat with, say, a 5-foot beam can be quite roomy, a 16-foot rowing dory just over 2 feet wide on the bottom and 4 feet on the beam is really not much more than a single-handed boat. A dog or a couple of kids (one fore and one aft) would be OK, and a petite adult in one end of the boat may also be OK for a short while if you put a balancing bag of picnic stuff in the other end—but if you need a boat you can share with the family, a dory isn't the answer. By contrast, if you need a boat to get away from the demands of work and home, a dory might be just the thing you need. Although you'll be alone, your boat will certainly have the carrying capacity to accommodate camping or other gear.

Another consideration: the narrow, high ends that give the dory its characteristic looks and easy ride and stability also tend to catch strong winds and can make the boat a little difficult to control. The legendary seaworthiness of the dory comes from heavily-built traditional models loaded with fish and gear, and manned by expert boat handlers who knew just how to look after themselves even when separated from their mother fishing schooner by fog or bad weather. These boats sometimes survived all sorts of storms and dangers, and the legend is that some even made the Atlantic crossing to Ireland.

However, a light plywood dory piloted by a flabby desk jockey (like me, for example) is a very different kind of boat and shouldn't be expected to carry you through the same situations. Using a dory does not mean you can ignore the weather forecast.

The stability characteristics of a dory can seem strange and unsettling. Although a dory may be pretty steady when heeled, when sitting bolt upright they can seem very tippy to someone used to more conventional boats. If you decide to build a dory, you should probably expect to spend a little time getting used to its interesting and unusual ways.

Doris the Dory (in French, the word *doris* means "dory") is enjoyably quick to row. It sails pretty well by traditional boat standards, with a 65-square-foot spritsail setup so that I can lower the whole rig into the boat in an instant. However, it is a *challenging* boat to manage under sail and isn't at all suitable for beginners in sailing. The technique required is very different from dinghy sailing, and it could irritate many sailors familiar with conventional small boats.

197

Doris the Dory

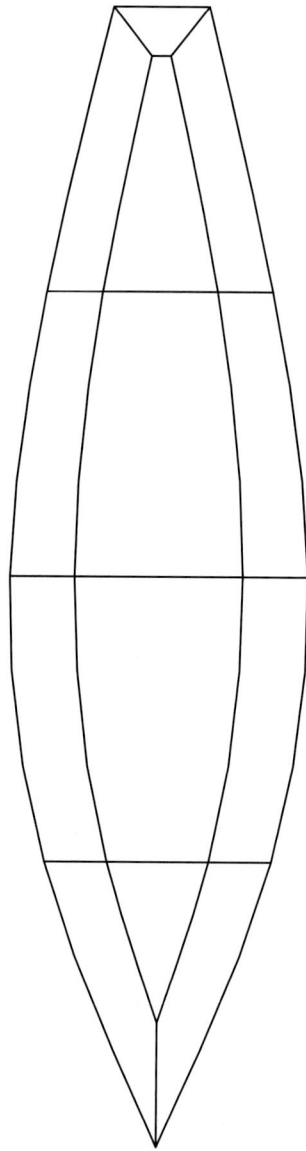

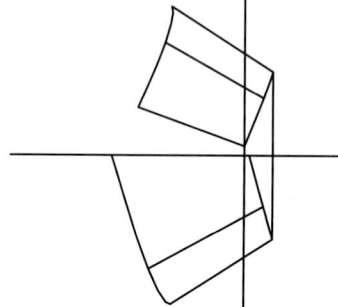

480 lb.

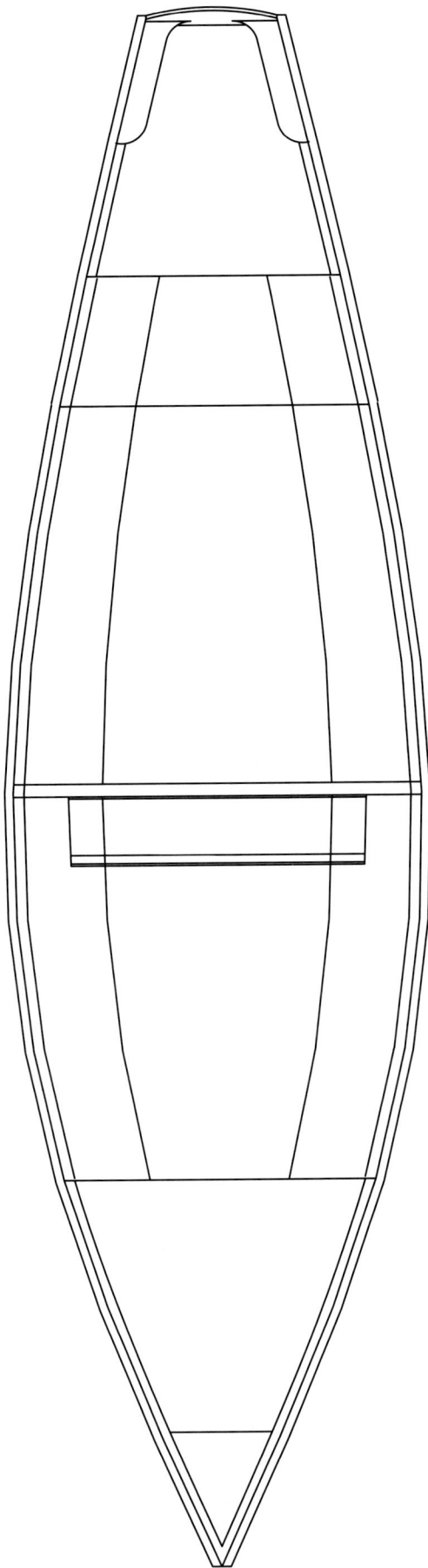

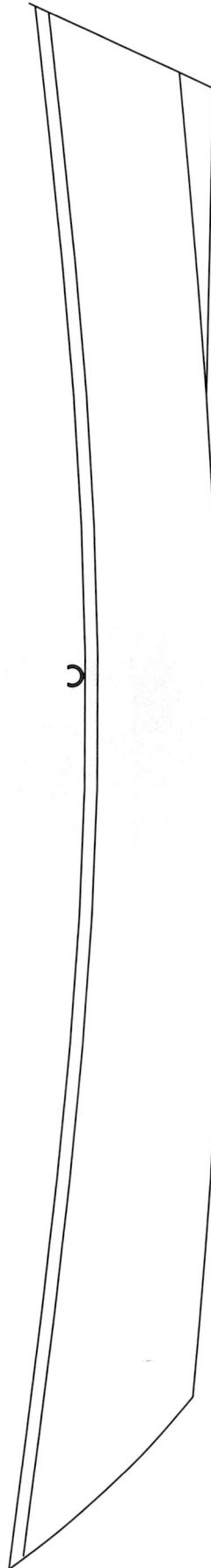

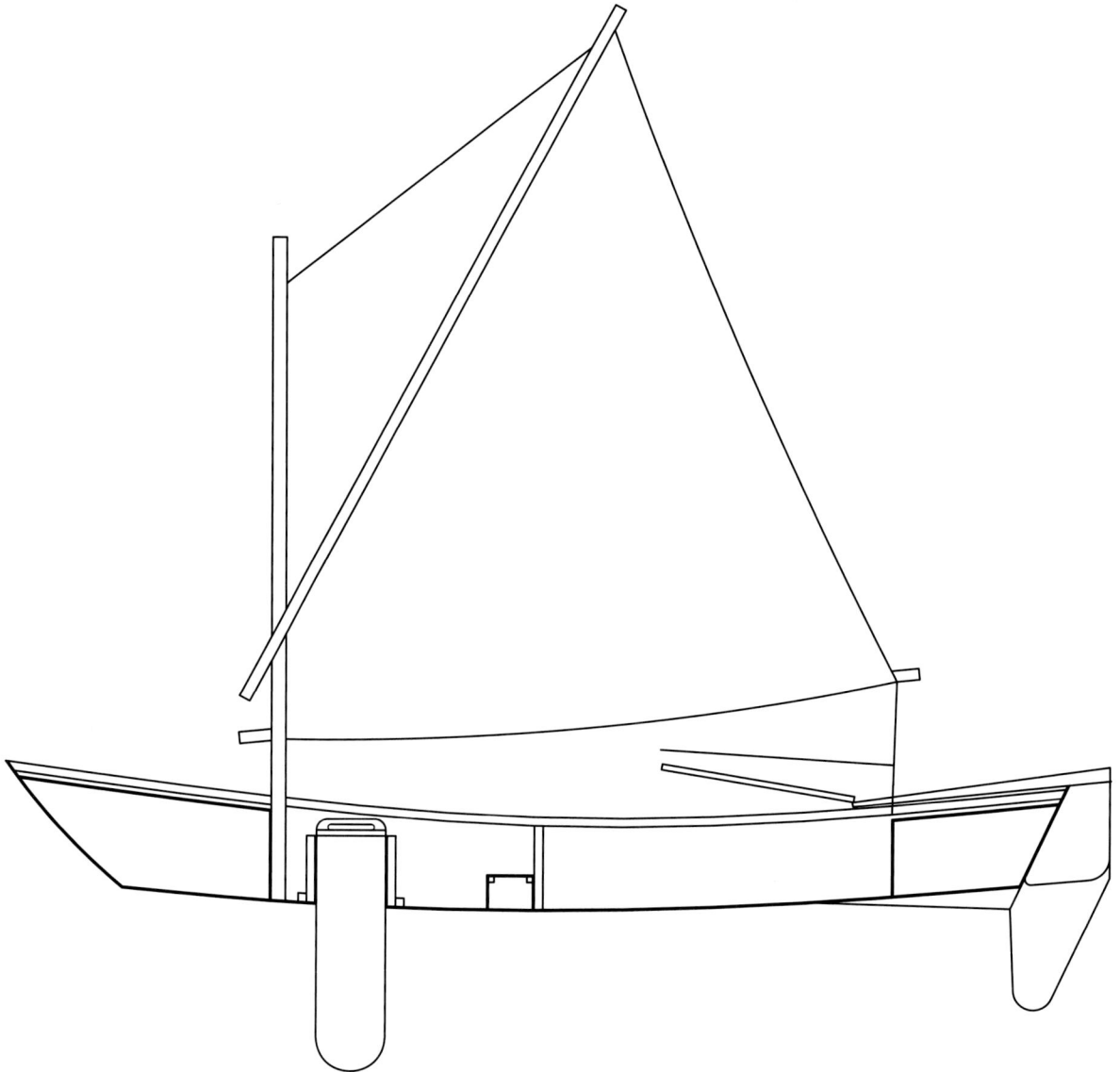

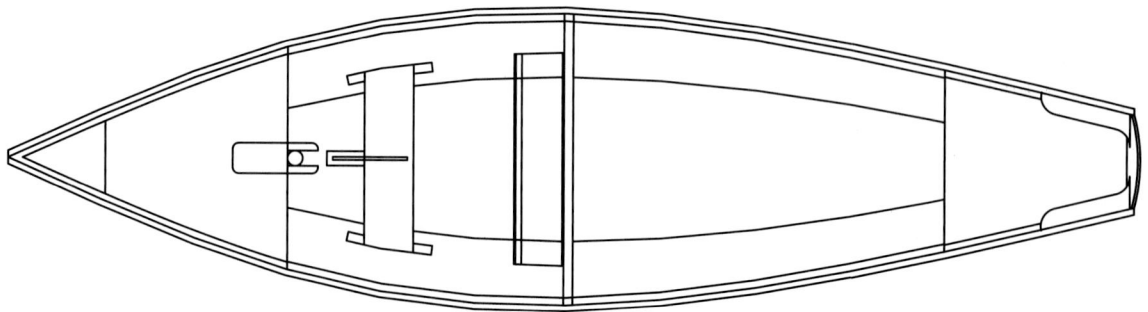

72

3-inch borders

123 5/8

60 sq. ft.

2

126 3/8

79 5/8

2

26 1/2

3

106 3/8

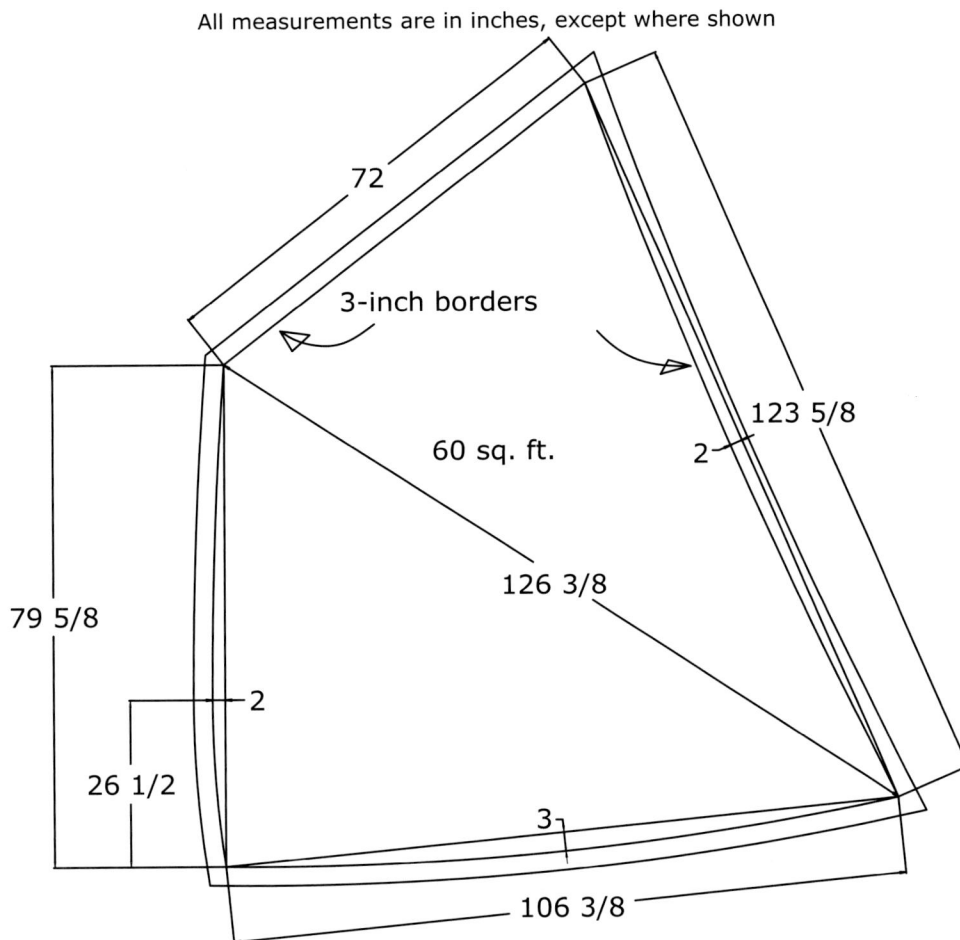

Unlike a conventional dinghy, Doris sails as straight when heeled well over as it does upright, which is useful in a gust—there are no sudden broaches with this boat. This is a good feature, as is the comforting way it becomes steadily more stable the farther it heels, once you're used to it.

However, don't expect to execute any dramatic racing turns when you go about. Like most long and lean boats, Doris needs to be coaxed around using a slightly different technique, which I think many beginning sailors will find difficult, largely because it's different from what the sailing schools and textbooks teach.

If you're on the wind in Doris—or in a sailing canoe, for that matter—it's normal to bear off from the wind for a moment, get some extra speed up, turn smoothly but smartly into the wind, pulling smoothly on the sheet and moving your weight forward to help the boat into and through the eye of the wind. Once you're through the wind you have to move aft to let the bow fall off on the other side.

In common with other narrow boats, Doris is also difficult to right and re-enter after a capsize, though not impossible. In order to avoid capsize as far as possible, I devised a quick-release mechanism that lets the mast down into the boat in a hurry when necessary. It is an essential part of the gear and should be used as soon as there's the slightest hint of trouble.

The quick-release is simplicity itself. The mast is supported by two open partners (supports). What prevents it from falling into the boat when in use is a line looped around it that runs through a ring and back to a cleat I can reach from the central thwart. When I need the mast to come down, I slacken the line with one hand and lower the mast into my other arm before rolling the whole thing up and putting it to one side in the boat. Fortunately, I rarely have to do this because Doris can stand up to a strong breeze.

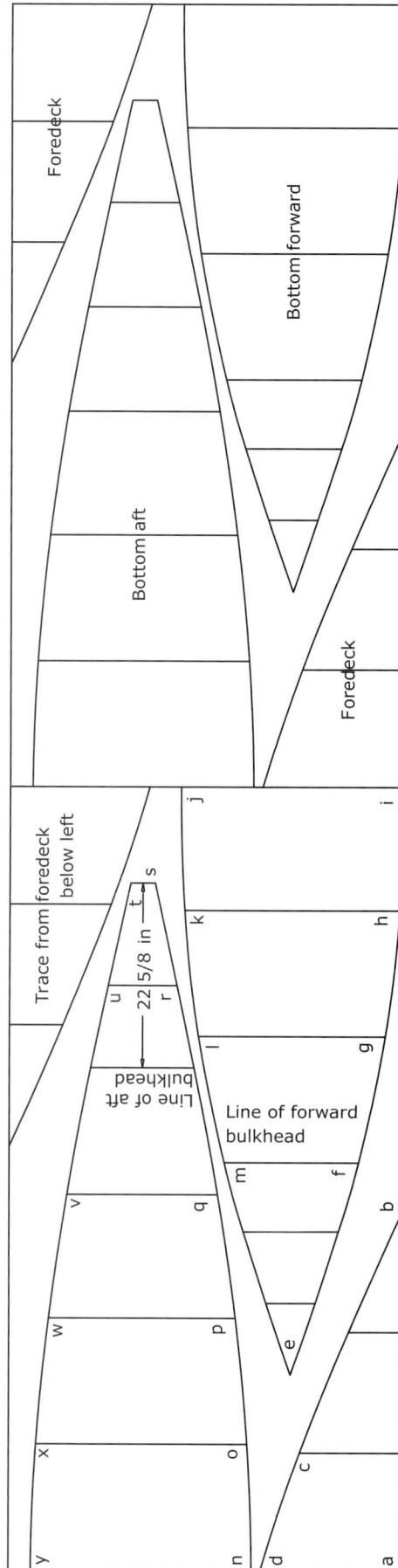

Foredeck

Bottom forward

Bottom aft

Foredeck

Trace from foredeck below left

s

t

22 5/8 in

u

r

Line of aft bulkhead

Line of forward bulkhead

j

i

k

h

l

g

m

f

e

b

v

q

w

p

x

o

c

a

y

n

d

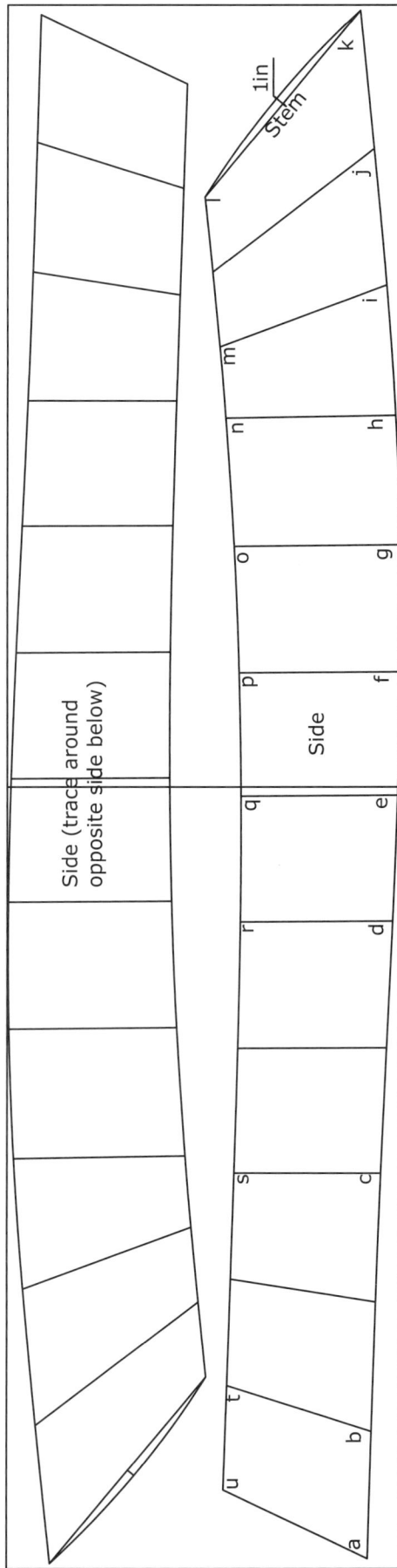

Side (trace around opposite side below)

Side

Stem

1in

a b c d e f g h i j k

u t s r q p o n m l

Doris the Dory Coordinates (in inches), Part 1

	x	y
Foredeck		
a	0	0
b	$44\frac{1}{8}$	0
c	$14\frac{3}{8}$	$12\frac{3}{8}$
d	0	$17\frac{1}{8}$
Bottom panel, forward section		
e	24	$13\frac{1}{2}$
f	50	$5\frac{3}{8}$
g	$65\frac{3}{8}$	$2\frac{1}{8}$
h	$80\frac{7}{8}$	$\frac{1}{2}$
i	96	0
j	96	27
k	$80\frac{7}{8}$	$26\frac{5}{8}$
l	$65\frac{3}{8}$	$24\frac{7}{8}$
m	50	$21\frac{3}{4}$
Bottom panel, aft section		
n	0	$18\frac{1}{4}$
o	$15\frac{3}{8}$	$18\frac{3}{4}$
p	$30\frac{7}{8}$	$20\frac{1}{4}$
q	46	$22\frac{1}{2}$
r	$71\frac{5}{8}$	$27\frac{1}{2}$
s	$84\frac{1}{4}$	$30\frac{1}{4}$
t	$84\frac{1}{4}$	$33\frac{1}{4}$
u	$71\frac{5}{8}$	36
v	46	41
w	$30\frac{7}{8}$	$43\frac{1}{4}$
x	$15\frac{1}{2}$	$44\frac{3}{4}$
y	0	$45\frac{1}{4}$

Part 2

	x	y
Sides		
a	$1\frac{1}{4}$	4
b	$16\frac{7}{8}$	$3\frac{5}{8}$
c	$48\frac{5}{8}$	$2\frac{1}{2}$
d	$79\frac{1}{2}$	$1\frac{1}{8}$
e	95	$\frac{1}{2}$
f	$110\frac{1}{8}$	$\frac{1}{8}$
g	$125\frac{3}{4}$	$\frac{1}{8}$
h	$141\frac{5}{8}$	$\frac{3}{4}$
i	$157\frac{5}{8}$	$1\frac{7}{8}$
j	$174\frac{3}{8}$	$3\frac{1}{2}$
k	$191\frac{3}{8}$	$5\frac{1}{4}$
l	$168\frac{3}{8}$	$24\frac{1}{8}$
m	150	$22\frac{1}{4}$
n	$141\frac{1}{4}$	$21\frac{1}{2}$
o	$125\frac{1}{2}$	$20\frac{1}{2}$
p	110	$19\frac{7}{8}$
q	$94\frac{7}{8}$	$19\frac{5}{8}$
r	$79\frac{1}{2}$	$19\frac{3}{4}$
s	$48\frac{1}{2}$	$20\frac{1}{2}$
t	$22\frac{1}{2}$	$21\frac{3}{8}$
u	$9\frac{3}{4}$	$21\frac{3}{4}$

Part 3

	x	y
Forward frame		
a	0	0
b	$33\frac{3}{4}$	0
c	25	$15\frac{1}{2}$
d	$8\frac{5}{8}$	$15\frac{1}{2}$
Transom		
e	$36\frac{1}{4}$	0
f	$39\frac{1}{4}$	0
g	$45\frac{1}{2}$	$18\frac{5}{8}$
h	30	$18\frac{5}{8}$
Aft deck 1		
i	0	$26\frac{3}{8}$
j	$29\frac{7}{8}$	$26\frac{3}{8}$
k	$29\frac{7}{8}$	$40\frac{1}{8}$
l	0	$33\frac{1}{4}$
Aft frame		
m	$34\frac{5}{8}$	$28\frac{1}{8}$
n	48	$20\frac{3}{4}$
o	48	48
p	$34\frac{5}{8}$	$40\frac{5}{8}$
Aft deck 2		
q	0	$34\frac{1}{4}$
r	$29\frac{7}{8}$	$41\frac{1}{8}$
s	$29\frac{7}{8}$	48
t	0	48

Doris the Dory Coordinates (in millimeters), Part 1

Foredeck

	x	y
a	0	0
b	1080	0
c	353	303
d	0	419

Bottom panel, forward section

	x	y
e	587	332
f	1224	131
g	1603	53
h	1981	11
i	2352	0
j	2352	662
k	1981	651
l	1603	609
m	1224	532

Bottom panel, aft section

	x	y
n	0	447
o	378	461
p	756	497
q	1128	551
r	1756	674
s	2063	740
t	2063	814
u	1756	881
v	1128	1004
w	757	1058
x	378	1095
y	0	1109

Part 2

Sides

	x	y
a	31	99
b	414	88
c	1190	61
d	1949	26
e	2327	11
f	2699	3
g	3080	5
h	3470	19
i	3862	47
j	4272	85
k	4688	128
l	4127	590
m	3678	541
n	3462	528
o	3076	503
p	2696	488
q	2325	482
r	1946	484
s	1191	502
t	552	523
u	237	534

Part 3

Forward frame

	x	y
a	0	0
b	826	0
c	613	380
d	826	380

Transom

	x	y
e	887	0
f	961	0
g	1115	456
h	735	456

Aft deck 1

	x	y
i	0	645
j	733	645
k	733	983
l	0	813

Aft frame

	x	y
m	848	688
n	1176	509
o	1176	1176
p	848	997

Aft deck 2

	x	y
q	0	838
r	733	1008
s	733	1176
t	0	1176

Note: Small discrepancies may exist between the millimeter and inch tables. See page ii.

All dimensions in inches

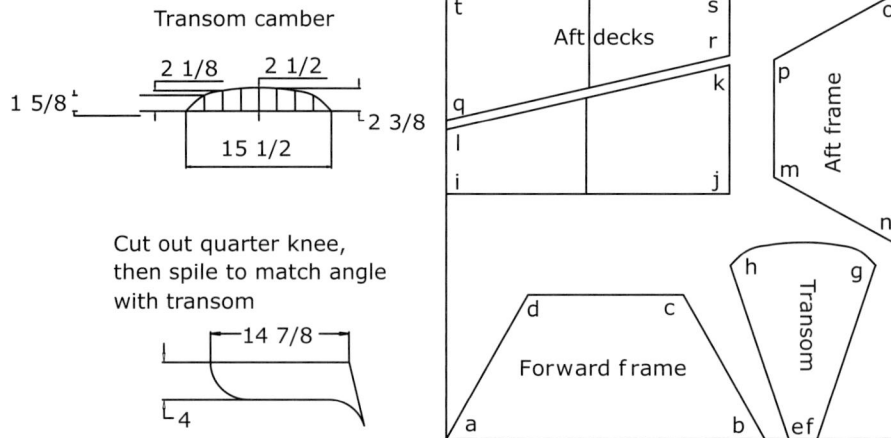

Transom camber

2 1/8   2 1/2

1 5/8

2 3/8

15 1/2

t   s   o

Aft decks   r

q   k   p

l   Aft frame

i   j   m   n

Cut out quarter knee,
then spile to match angle
with transom

14 7/8

d   c   h   g

Forward frame   Transom

4   a   b   e f

The Doris design would make a good hull for a small battery-powered trolling motor. The slippery dory hull form is efficient and offers little resistance to the water, and the improved weight-bearing capacity provided by her wider-than-usual bottom would enable her to carry more battery and human payload than other 16-foot light dories.

As a building project, Doris is very much a standard stitch-and-glue job built from just three and a half sheets of ¼-inch plywood.

In drawing up this project, I decided that the butt joints were more than long enough to be safely made after cutting out the material, so this project begins by squaring off the plywood in 12-inch square blocks and carefully cutting off any excess over 8 feet in length, in a good straight line. Then it's time to plot the panels and the frame positions before driving small nails into each plotted point, bending a flexible batten around them, tracing the curve with a pencil, and cutting the whole thing out in preparation for the stitching and gluing process.

When I built Doris, I butt-jointed the sides and bottom using glass tape and epoxy on both sides of the plywood without difficulty. The weather was unusually sunny and warm for the United Kingdom, and the epoxy hardened nicely by the following afternoon, allowing me to quickly do the stitching with cable ties and the taping of the structure with cloth-backed tape.

The only real complications compared with some of the other boats in this book are due to Doris's greater length and weight. It's difficult to build a boat of this length on a flat concrete surface unless it has a flat bottom with no rocker. Doris is more likely to assume the right shape, therefore, if you build it on a couple of sawhorses placed 9 to 11 feet apart. This will also make it easier on your knees and back. The sawhorses should be solidly built. (It is possible to build this boat using lightweight plastic ones as I did, but they do tend to wriggle as you work.) Because the boat is likely to weigh 110 pounds or so, it's important to tie the sawhorses together with lines or even straps of lumber so that they won't suddenly slide away and drop your boat on your toes.

Once the hull panels are assembled, Doris benefits from a breasthook at the bow and quarter knees at the stern, a solid gunwale, gapped inwales, a couple of blocks to mount the oarlocks, a stretcher to brace your feet, and—if you're brave and experienced enough to venture into sailing a dory—a rig, daggerboard, and rudder.

In this narrow boat, stretchers for your feet can be simply pieces of wood glued and screwed to the side of the boat at the right spot for your leg length: you have to find this by experiment. Alternately, they can be made up using a double thickness of plywood cut into a "comb" and glued and screwed onto the sides of the boat where they meet the chines, as described in Chapter 6.

Because this is a larger project than some of the other boats in this book, covering the

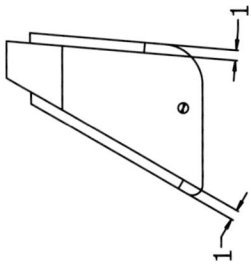

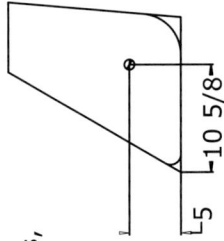

Rudder pivot

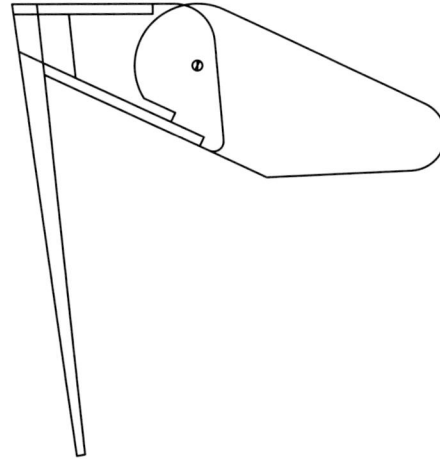

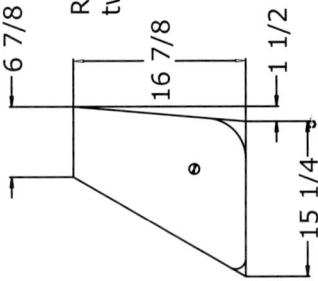

Rudder cheeks,
two-off

All dimensions in inches

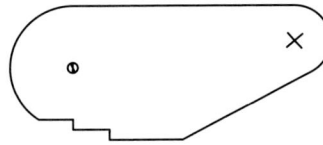

Tiller half, two-off. The halves are screwed and
glued together at 8 and 16 in., and bolted at
21 in. through the head of the rudder

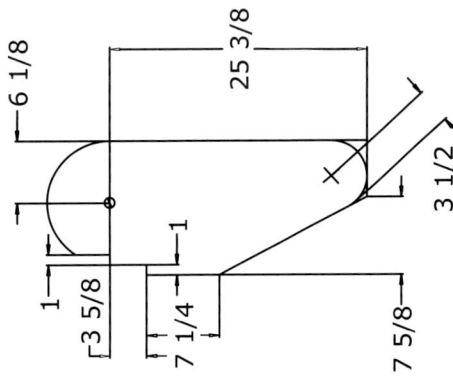

All dimensions in inches

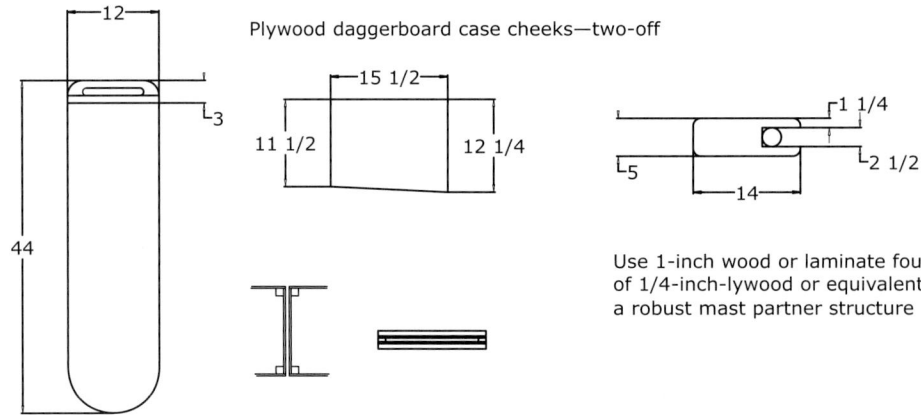

Plywood daggerboard case cheeks—two-off

12

3

44

15 1/2

11 1/2 12 1/4

1 1/4

5 2 1/2

14

Use 1-inch wood or laminate four sheets
of 1/4-inch-lywood or equivalent to create
a robust mast partner structure

bottom with scratch-resistant fiberglass and epoxy is even more advisable, although it is still a matter of personal preference. If you wish to keep the cost or weight down, and are willing to accept a shorter lifespan and the need to take greater care when beaching the boat, you could omit it.

The rig is another sprit rig, very much like the one on the Flying Mouse, although of course it is significantly larger, and its mast does not require the Flying Mouse's forestay and shrouds. The daggerboard and case are very straightforward. The tiller is quite long; it has to be to enable the helmsman to steer the boat from amidships or even forward of that when beating or going about.

Because of its size, it is essential that the tiller can be separated from the rudder so both can be transported in a car. The best way to do this is to attach them using a stainless steel bolt and a couple of nuts.

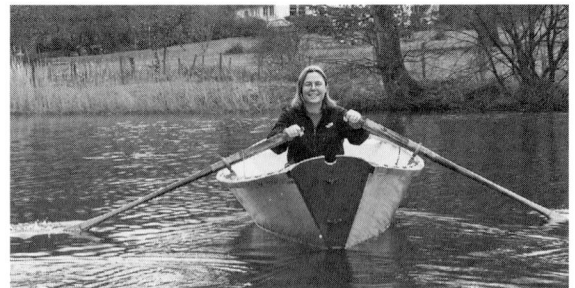

Doris the Dory isn't fast, but she is perfectly happy in conditions like this—probably around Force 5. Some people say that dories don't sail well, but with her wider-than-normal bottom and bit of ballast carried down low, Doris does just fine, even in winds a notch or two stronger than this.

Built much lighter than a traditional Banks dory, Doris is also very competent under oars and will prove an excellent expedition boat.

WHAT SHALL WE BUILD NEXT?

MORE DESIGNS AND THEIR DESIGNERS

The best books are inspirational, and I would be delighted to learn that you have been inspired to build models, full-sized boats, or both from the plans in this one. If you have built boats before, you know how satisfying it is to take to the water in a craft you fashioned with your own hands. The boatbuilding bug will have bitten hard, and once bitten, it's hard to stop building and messing about in all sorts of small boats.

The intense, often dreamlike process of choosing your next boatbuilding project is well recognized, so much so that it's often called boat-dreaming. It can take months or even years. This is normal, and not at all surprising, given that it has much in common with falling in love. As in love, most people will play with a few alternatives before making their final choice. There has to be a certain magic in the selection, and you must convince yourself that you can make a rational justification for your decision. The process itself is mostly enjoyable and life-affirming.

And, as in love, sometimes it turns out all wrong, no matter how hard you try to get it right.

So take your time and enjoy the process of making your design choice as you explore what's available in books and on websites, such as the excellent www.duckworksmagazine.com. Make models. Contact other boat nuts about their experiences of building and using the kind of boat that interests you. Visit them, pay them extensive compliments (as you would someone with a new house or car), learn what you can, and, most important, try out their boats if at all possible.

Work on improving your skills with hand tools, which is more useful than buying expensive power tools. If you go to boatbuilding or woodworking classes, you may well make contacts with people who'll let you use their tools.

Don't lose sight of what it is you enjoyed about building your first boats: if you're a busy, impatient type who likes simple, quick construction in your first boat, that's what you should be looking for in your next. However, if your instinct is to spend hours making every aspect of the job beautiful, you may want to take on some much more difficult challenges and spend years building beautiful lapstrake boats that turn heads every time you're out on the water.

But, above all, avoid getting into a project that's too big or complicated, one that may crush your morale and possibly lead to disagreements at home. Designers will tell you that most plans don't get started, and that most projects that do get started are never finished because they were just too big. Choose a boat that meets your real needs. If you're going to be on your own most of the time—like most boating enthusiasts—you need a boat that you can handle on your own. This often means a smaller boat than you might otherwise be inclined to dream of.

If your makeup includes an interest in theory, reading about boat design will make you more confident about making a good choice. It is useful to read everything you can find by the great designers for home boatbuilding: Philip C. Bolger, Jim Michalak, and John Welsford are some of the obvious candidates, but don't forget Reuel Parker, Thomas Firth Jones, Iain Oughtred, and L. Francis Herreshoff. And there are also some great little books about design, including *How to Design a Boat* by John Teale and *Understanding Boat Design* by Ted Brewer.

Now, if you want to go further than the projects I've covered in this book, let me tell you about some of the boats you might consider from some of my favorite designers.

PHIL BOLGER

To many, Phil Bolger is the most influential designer for home boatbuilders. In fact, his influence has been so great that many labor under the misapprehension that his boxy, utilitarian-looking designs for home boatbuilding represent his entire output. But Bolger's work covers a wide range and includes boats designed for looks almost beyond any other consideration, as well as designs where effectiveness, economy, and ease of building top the list.

He has also written a series of entertaining and inspiring books about his designs that have now become collector's items. Two of the great things about Bolger's writing are that he's so apparently honest and revealing about his thought processes, and that he so clearly communicates many of the issues involved in designing his boats. Third is that he has huge experience of designing and using skiffs and sharpies, both boat types that have long fascinated home boatbuilders, who are instinctively attracted to the apparent simplicity of flat-bottomed boats.

I'm sure that his engaging writing style and original boat designs have been the inspiration of more than a few of the current generation of designers focused on small home-built boats.

Teal: An Instant Boat for Kids

Phil Bolger has drawn up many small boats for home boatbuilding, but I doubt any of them have proved as popular as Teal. This

Contact Info

These plans are available from:

Mr. Philip C. Bolger
P.O. Box 1209
Gloucester, MA 01930
978-282-1349 (fax)

You can also buy the plans from Harold Payson, Phil's long-standing collaborator, at www.instantboats.com.

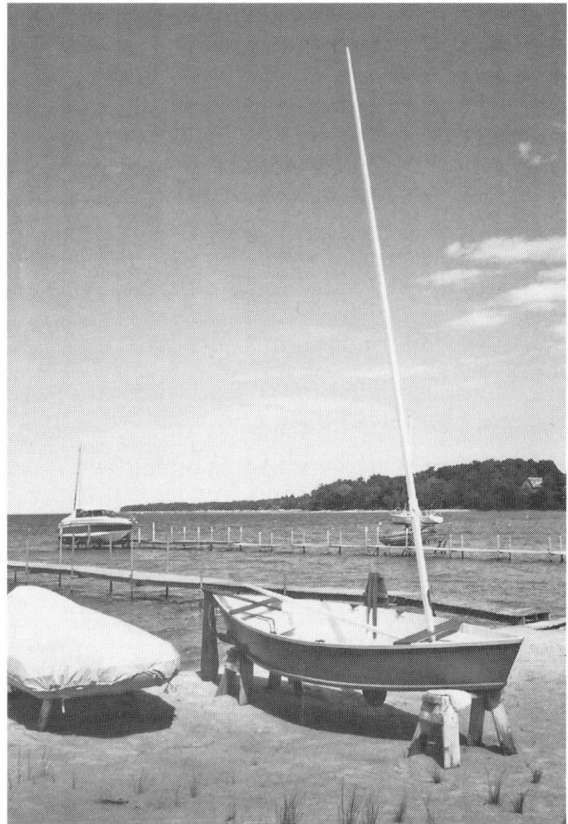

Teal. (Bob Chamberland)

12'-by-3'6" boat can be built from two sheets of plywood.

According to my copy of his book *The Folding Schooner*, published by International Marine, Teal was a long time in gestation as Bolger returned over and over again to the problem of how to get the best double-ended boat possible out of two sheets of material. These days, of course, we'd use computers to help solve the problem, but back in the mid-1970s Bolger did the whole thing on paper, which can't have made it any easier.

Teal is still a contender for anyone considering building their first, second, or third home-built boat: it's simple, rows and sails well, and is a great little boat for a couple of kids or one adult to play with. It won't carry much crew weight, but like the Mouse, it's the kind of small boat that kids love making their own, and I'm quite sure that many families with three kids would not find three Teals were too many.

Teal is actually a little larger than the Cinderella canoe (see page 146), and the fact that she is capable of standing up to

54 square feet of sail is a powerful testament to the stability that comes from having a hard chine rather than a round bottom. Teal is an "instant boat" intended to be built with external chine logs, frames trimmed to the angles of the sides and bottom, and cut and shaped stem and stern pieces, although I'm sure that with a little ingenuity much of the construction could be made to work in stitch-and-glue for those who prefer that approach.

Bolger suggests some options for building buoyancy foam into the ends of this boat, and if you choose to build it I would urge you to do as he suggests. Trying to stick to his two-sheets rule, Bolger couldn't find any room left in his materials allowance to add decks, but I'd suggest that finding some plywood from somewhere to add decks will make this boat stiff and strong, and will protect the foam insulation to boot.

His 'n' Hers Schooner

This is a project for a perfectionist with an extrovert streak. One of the key reasons for building your own boat is that it enables you to own the boat you really want, not just the boat the manufacturers want to make or feel you should buy—and this is a boat no manufacturer would produce.

The His 'n' Hers is a cute, retro-styled micro-schooner that will turn heads wherever it goes—which is why you'd want its finish to be as perfect as possible. It's also a lot of fun because it has enough strings to give everyone on board something interesting to do.

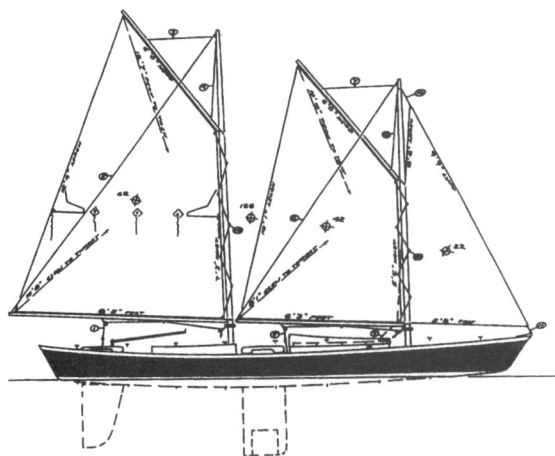

His 'n' Hers Schooner. (Phil Bolger)

Even at 19'6" overall, don't expect the His 'n' Hers to be any kind of rocket ship, particularly on the wind. Schooners start with a disadvantage in sailing to windward compared with boats that have their largest sails forward, and dividing up the sail plan of a small boat into so many low-aspect pieces of sail greatly increases windage and reduces efficiency. However, it does have advantages, not the least of which is that it is very easy to adjust sail area by either raising or taking in sails.

Although the build is standard "instant boat," this is naturally a rather more complex job than Teal and it does require the boatbuilder to master the art of melting a slug of lead into a hole cut in the daggerboard.

Micro: A Tiny Holiday Home on the Sea

Phil Bolger's 15'4" Micro first appeared in print in his book *Boats with an Open Mind*, and since that time it has become a legend.

It's a boat that divides people: some like its bull-doggy and blunt-nosed look and would be delighted to be seen with one in any company, while others see it as simply a particularly foul example of the kind of thing that only a plywood backyard boatbuilder could love, and then only if they'd never seen a real boat.

Both are right. The magic of the Micro is that it is a combination of appealing things: it's a heavy coastal cruiser that two people can use to overnight in reasonable comfort, and it's easy and cheap to build. Add to that the appeal of a rig with the safe qualities of a cat-ketch (let go of the main for any reason and you can be confident that the boat will quietly turn upwind and stop while you sort out whatever's going on), and a clever bow and keel arrangement that allows the crew to step ashore with dry feet, and you have something close to a winning combination.

There are disadvantages to owning a Micro. The boat's heavy enough to make launch and recovery a two-person job; it needs a good outboard and sufficient gas on board to get you home (the Micro's windward performance is not its star attraction); and the flat bottom will often pound in a chop when you're trying to

Micro. (Jay Kammerzell)

sleep while at anchor. But the advantages and the eccentric snub-nosed romance of the Micro outweigh the disadvantages for many, and I think we'll see a great many more Micros built in the coming years.

It's worth noting that there are also the 19-foot Long Micro and the open 11-foot Oldshoe versions to choose from.

JIM MICHALAK

I really like Jim Michalak's designs. They may not always be as attractive as some might wish, but for Jim, function, value for the money, and effort come a long way ahead of looks. The small collection of his boats presented here are all simple to build using the techniques I've discussed in this book. What's different about them is that in general they're larger than the boats we've looked at and have uses that the smaller boats simply can't match.

Contact Info

Jim Michalak's plans are available via his excellent online newsletter:

http://homepages.apci.net/%7Emichalak/#Contents

Or, go to www.duckworksbbs.com.

AF3: A Lake Sailor's Escape

Named after Alison Krauss's fiddle (Jim was an early fan), the AF3 cuddy skiff (Jim likes to call it a sharpie) is 16 feet by 5 feet and weighs about 250 pounds, which makes it easily trailerable. At that weight, I'd expect to be able to launch and recover it single-handedly without a winch. You could use this boat as an overnight camper.

It's just about my favorite among Jim's designs.

The AF3 has a sharpie-style spritsail with its center of effort over the trailing edge of a narrow leeboard, which will seem unusual to many but works well in these flat-bottomed boats with their bows out of the water. Jim has shown external chine logs because he believes they make an easy, quick, and strong bottom attachment, but traditional interior logs can be used instead. Taped epoxy chine joints of a good size would be even better and cleaner, and taped epoxy has the advantage that it eliminates a structural member that has an unfortunate tendency to rot in boats kept where freshwater can reach them.

The AF3 also has a slot-top cuddy, which provides partial shelter and headroom when

AF3. (Peter Simmons)

you need it, and good ventilation. The high sides and side decks that make up the cuddy are designed to prevent the boat from swamping in a knock-down, but I fear that the slot may be wide enough to take in water when the boat is on its side in waves of any size. If I owned one of these boats, therefore, I'd probably prefer to sail with a watertight cover in place in anything but light winds.

Then again, I also see some big advantages in the slot-top, for together with some useful shelter from the weather and sun, it allows easy access to the mast and bow, excellent ventilation (on hot days it's said to make good use of down-draught from the sail), and only the very hard-hearted would find the prospect of sleeping under the stars on a warm, still night unappealing.

I should add that a lightweight flat-bottomed skiff or sharpie of this size and form is very much a fair weather, sheltered water sailer, and not really suitable for windy and rough waters such as those found along the coasts of the United Kingdom and New Zealand. What's more, like Bolger's Micro it will need to be dragged out of the water or moored in a very sheltered spot for sleeping, as the pounding of even small waves on the AF3's bottom will be noisy, even when they are too small to be any threat to the boat.

However, an AF3 equipped with a small outboard could be a great deal of fun in the right conditions on sheltered estuaries, and in the hands of an experienced sailor with an extra sail—such as a foresail used for reaching—it could be a hoot on networks of small inland waters such as our Norfolk Broads.

The first AF3 builder, Herb McLeod, and a string of subsequent builders have reported that the AF3 sails well and can be easily righted by a single-handed skipper in a capsize. Herb also reported that he completed the boat in 70 hours of work using spruce exterior plywood—it uses only six sheets of ¼-inch and two sheets of ½-inch plywood, which I guess accounts for its relatively low weight, and which in turn contributes to its performance.

213

AF4: A Low Horsepower Motor Cruiser

The AF4 is a lightweight, flat-bottomed cuddy power skiff, 18 feet by 5 feet, that planes easily and economically with an outboard engine of 10 to 15 hp. The AF4 is a little larger than the AF3 and offers more internal space, but the family likeness couldn't be more obvious.

At around 350 pounds when empty, it's still very trailerable, but you'll probably need two people to launch and recover. (The smaller 15-foot AF4 Breve, also from Jim's catalog, is better suited for regular single-handers.)

The flat bottom that gives the AF3 the ability to stand up to its sails (albeit at the cost of a large wetted surface area) is also the AF4's secret weapon, for it is what enables it to plane efficiently and travel so cheaply. Like the AF3, this is very much a sheltered water boat, but for the low budget boater, I'd say that limitation is well worth living with in return for this boat's convenience, value for money, and fun potential in sheltered areas such as estuaries, rivers, lakes, and the occasional sheltered-sea trip when wave conditions and the weather forecast permit.

The large and unimpeded cockpit is 6 feet long and 4 feet wide, and aft of that there is a draining motor well with space for a good-sized fuel tank. Jim says that a 15 hp engine is a good maximum, and I think he's right to discourage over-large engines, which are dangerous in a boat not capable of handling them and encourage irresponsible speeds. Going smaller in terms of power is also better for the environment and easier on your wallet.

AF4. (Max Wawrzyniak)

Ladybug. (Chuck and Sandra Leinweber)

Ladybug: A Simple Open-Boat Camp-Cruiser

Ladybug is Jim's all-purpose cruising dinghy. At 14 feet long, it has a big cockpit and lots of storage in buoyancy boxes at each end. There's a single balanced lug of 75 square feet. This is not a big sail, but this is meant to be a safe, spacious little cruiser capable of taking a family for a day's outing, not a racing boat, and it's light enough to be launched and recovered by most single-handed sailors. In fact, I'd say that it should be a serious consideration for anyone interested in open-boat cruises that include camping overnight on board under a boom tent.

It has a good slippery hull, and the wooden spars will prevent the boat from turning turtle in a capsize. After you get the boat back on its feet, climbing back on board is easy since Jim's plans include a slot cut into the rudder for use as a step.

Harmonica: Home Away from Home

Ask any group of people what kind of boat they'd want to have, and quite a few will say that they'd go for a shanty boat like Harmonica, a little 12-by-5-foot rectangular floating box that includes a small balcony area and a place

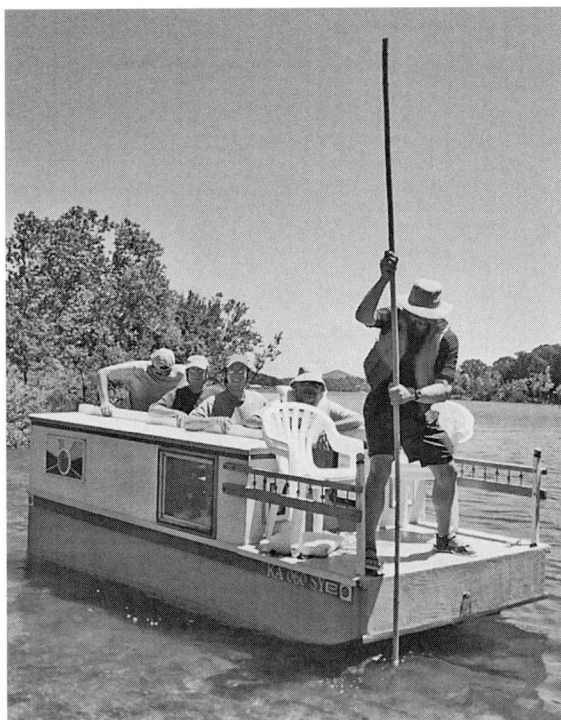

Harmonica. (Chris Crandall)

to clamp a tiny outboard. What makes this boat so attractive as a low-cost building project is that it uses so little material: four sheets of ⅜-inch plywood and six sheets of ¼-inch plywood banged together with glue and nails and hardly a bevel cut in sight. Really, it couldn't be easier.

Sitting on a small trailer, Harmonica would look just like a tiny recreational vehicle (caravan) and I imagine that it might be used in both roles.

Harmonica's limitations are the ones you would expect for a flat-bottomed boat of this kind: it can't be used anywhere there may be sizeable waves, and the windage is so great that I doubt it would be possible to steer it in windy conditions.

MURRAY ISLES

Murray Isles lives and works in Hobart on the large island of Tasmania. It's Australia's southernmost city and famous for being the destination of one of the world's toughest sailing races, the Rolex Sidney Hobart Yacht Race. The sea has a large influence on Australian life, not least because so much of that huge country's population lives near the coast, and

Contact Info

Murray Isles' plans are available at:

www.islesdesign.com

the influence and economic importance of the sea in Tasmania is as strong as anywhere in Australia.

Murray's work as a designer reflects all this economic activity. Much of his design work is specialist and commercial. It includes workboats, tugs, ferries, charter passenger vessels, small cargo vessels, site barges, conversions, and a range of pens, tanks, and transfer systems used in fish farming. Luckily for us, however, somewhere along the way he has found time to design a range of practical and interesting small craft, some of which can be built by the simple methods covered in this book.

Swallowdale 15: A Favorite All-Purpose Dory Skiff

Jim Michalak's Ladybug represents one good approach to the cruising dinghy, but this 15-foot dory-skiff represents an alternative that many might prefer. By dint of its narrow beam, Swallowdale is a significantly smaller boat than Jim's, despite its slightly longer length.

A dory-skiff is somewhere between the flat-bottomed skiff form we saw in David Beede's Summer Breeze and a genuine dory form. Its bottom is broader at the stern than a traditional dory, but the boat is still proportionately narrow compared with conventional sailing dinghies.

One consequence of this is that it's a slippery boat despite its relatively simple hull form, and will row very well for a boat that's also capable of sailing. More, it can be used with a 2½ hp outboard, which makes it unusually versatile. Most small boats either row well, sail well, or motor well, but Swallowdale achieves all three in an easily built hull. This makes it a good boat for many beginning boatbuilders, particularly if they are not yet experienced in handling small boats and would like a boat that can be used in a variety of ways.

- SWALLOWDALE 15 -

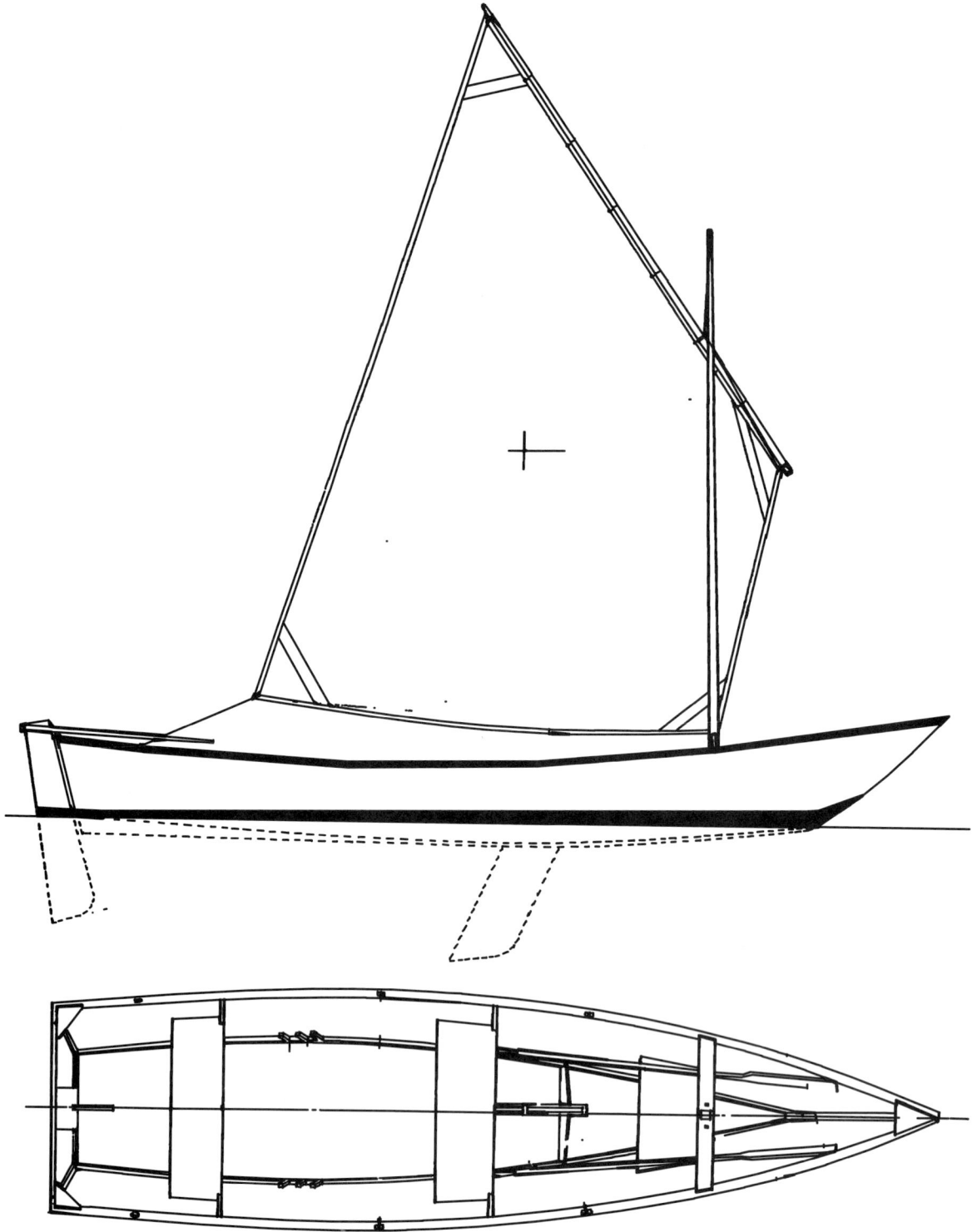

Swallowdale. (Murray Isles)

Another consequence of this particular hull form is that it will be difficult, if not impossible, to find space for two to sleep in it under a tent, as compared with, say, Ladybug. Does that matter? In practice, it won't matter to many, though it may to some where camping on privately owned land is not allowed.

Swallowdale 15 was originally designed with the elegantly high-peaked single standing lugsail shown in the drawing. It is quick and easy to set up, reefs well, and is cost-effective, but demand from builders has led to an alternative cat-ketch rig. The plans include full details for both rigs.

Swan Bay: A Spacious but Light-Sailing Semi-Dory

At 17 feet 10 inches, Murray's Swan Bay has a similar hull form to Swallowdale and many of the same virtues (see next page). In fact, Swan Bay could be called its big sister. Like the Swallowdale, Swan Bay can be rigged as a cat-ketch and makes a good cruising dinghy, but with a proportionately broader bottom and a five-panel form rather than slab sides, it is much more of a cruising boat for two. And yet at less than 200 pounds of hull weight, it's still in the weight range where a couple of reasonably fit people can pull it up or down a beach.

Murray has also drawn up a gunter-rigged sloop rig for this design. While simple one-sail and cat-ketch rigs have much to recommend them, one of the advantages of the sloop is that it gives the crew something interesting to do, and that can be particularly important if that other person is an easily bored child. Enthusiasts could rig the boat both ways.

Pepper Gal: A Catboat for the 21st Century

This design is for boatbuilders who are both ambitious and perfectionists at heart because building it will demand a bigger investment in effort and materials than any boat in this book, even if its build does not require any techniques you haven't already learned here. Still, I really like the design (see next page).

Pepper Gal is a little catboat of 15 feet 3 inches for daysailing, but with a cuddy just large enough for one or two people to overnight aboard. It's built using the stitch-and-glue method.

Pepper Gal is not really intended for coastal cruising, but it is certainly seaworthy enough for most semi-sheltered waters, and it's small enough to be built in your garage.

Catboats are traditionally wide boats, and with a beam of 7 feet 2 inches, Pepper Gal is no exception. It's as beamy as many 19-footers, which makes it a "large small boat," and in sheltered waters it'll have a stable, broad-shouldered feel and will stand up to her good-sized but simple rig of 142 square feet. With that much sail, Pepper Gal will move smartly in a good breeze.

However, the price you pay for having a big boat in a small envelope is that in waves you may have a rougher ride than you'd experience in a less beamy boat, as the hull tries to follow the moving surface of the water. Also, because Pepper Gal's broad shoulders have to push a good amount of water aside, I'd also expect her to make slow progress when beating against waves in light winds—and when that happens it may be time to start the engine.

All the controls for the single sail are available to the skipper without leaving the cockpit, and Murray has found a clever way to enable a skipper working on his own to step the mast without difficulty: the unstayed mast is mounted in a tabernacle and pivots through the forward hatch.

Below the foredeck, the cuddy has space for two generous canvas quarter berths and a galley area forward, and by placing the centerboard case slightly to one side of the hull centerline it is also possible to slide a Porta Potti under the cockpit well.

JOHN WELSFORD

The designs of New Zealander John Welsford reflect the coastal conditions where he lives. New Zealand is so far south that it picks up weather systems associated with the legendary Roaring Forties, and it comes as no surprise that most of his boats are strong, with round plywood lapstrake bottoms, handsome sheerlines, and a good amount of freeboard.

BOUYANCY TANK
UNDER FOREDECK

MAST
STEP

RF134A

THWART

CENTREBOARD
CASE

ROWING THWART

39

BOUYANCY TANK UNDER

THWARTS (P&S)

RONSTAN RF420 GOOSENECK

64x46

MAST

AFT EDGE STRAIGHT

1030 | 1770 | 2770 | 3770 | 4770
64x46 | 60x43 | 55x39 | 50x36

BOOM

TOP EDGE STRAIGHT

3100 | 1800 | 1000
40x25 | 50x33 | 45x33

RONSTAN RF180 & RF895

SHAPED 20x45x70 CHEEKS GLUED & SCREWED

2800 | 2400 | 1800 | 1200
25 DIA | 33 DIA | 40 DIA | 45 DIA

YARD

10 DIA HOLE

FORWARD EDGE STRAIGHT

45 DIA

FORE STAY – 3mm DIA
1x19 STAINLESS STEEL
WIRE LACED TO
CHAINPLATE

WORK EYE IN
FRONT OF YARD
SOCK FOR
HALYARD

RF469

300

LUFF LACING
EYELETS ON 400
SPACING

SHROUDS – 2.5mm
DIA 1x19 STAINLESS
STEEL WIRE WITH
RF445 ADJUSTER

3.50 SQ MTRES

8.99 SQ MTRES

12.49 SQ MTRES

HEADSAIL DIMENSIONS
3834 LUFF
3459 LEECH
2264 FOOT

CE

CLR

MAINSAIL DIMENSIONS
2491 LUFF
2700 HEAD
5598 LEECH
3000 FOOT
3691 CLEW–THROAT DIAGONAL
800 REEF SPACINGS
300 LEECH ROACH

DRILL FOR 20x6 BOLT TO BE SET
IN 10 DIA HOLE WITH THICKENED
EPOXY

FORWARD TANG SIMILA
TO SHROUD TANGS,
BENT 67°

6 RAD

65

10

18

DRILL FOR 8mm PIVOT
BOLT

RF429

65

12

6 DIA

TANG BENT 10°

MASTHEAD FITTING DETAIL
– FROM 1.5mm STAINLESS
STEEL
– SHOWN HALF SIZE

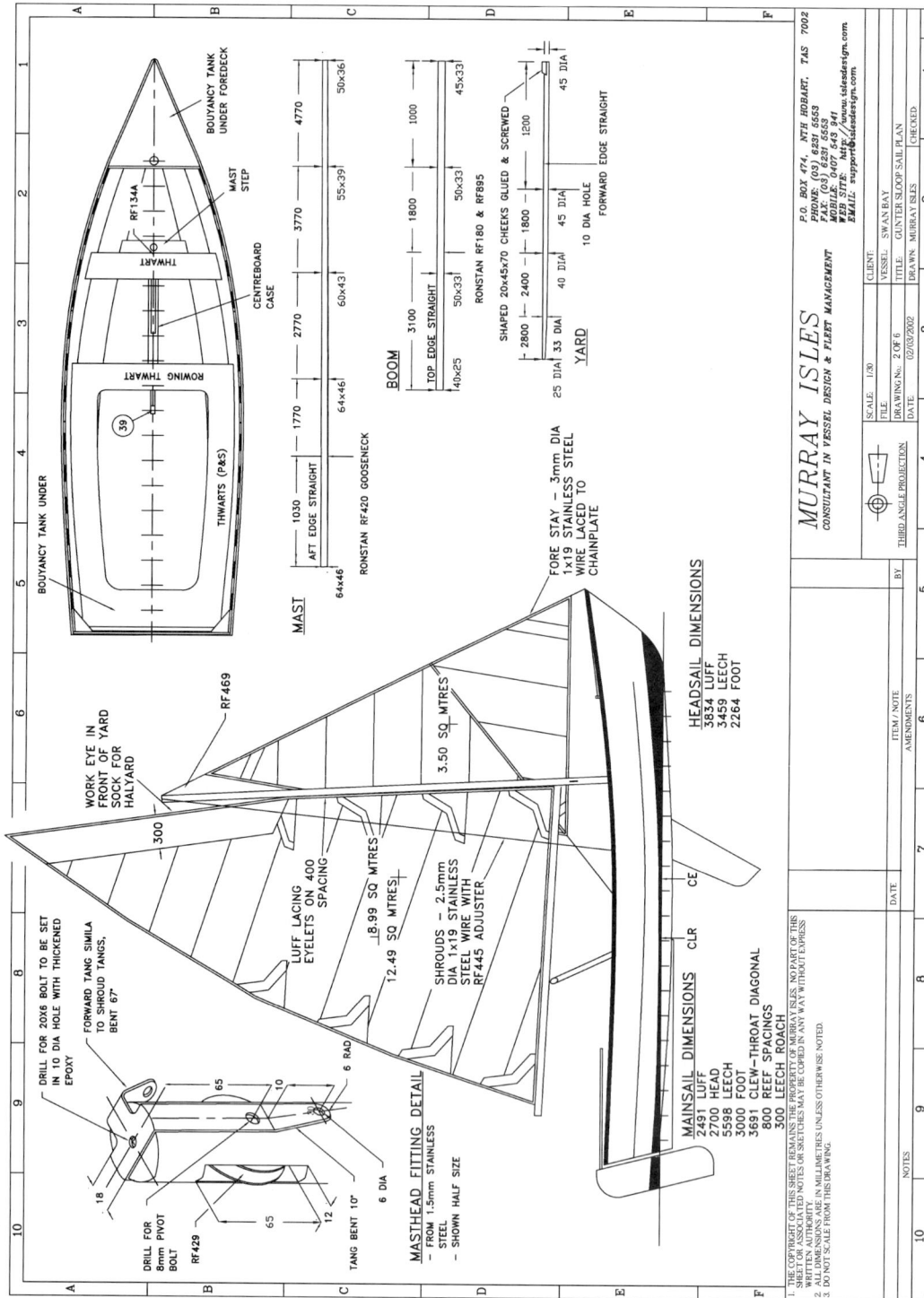

MURRAY ISLES
CONSULTANT IN VESSEL DESIGN & FLEET MANAGEMENT

P.O. BOX 474, NTH HOBART, TAS 7002
PHONE: (03) 6231 5553
FAX: (03) 6231 5553
MOBILE: 0407 543 941
WEB SITE: http://www.islesdesign.com
EMAIL: support@islesdesign.com

CLIENT:
VESSEL: SWAN BAY
TITLE: GUNTER SLOOP SAIL PLAN
DRAWN: MURRAY ISLES | CHECKED:

SCALE: 1/30
FILE:
DRAWING No: 2 OF 6
DATE: 02/03/2002

THIRD ANGLE PROJECTION

NOTES | AMENDMENTS | ITEM / NOTE | DATE | BY

Swan Bay. (Murray Isles)

Pepper Gal. (Murray Isles)

Contact Info

John Welsford's plans are available at:

www.jwboatdesigns.co.nz

That said, he has also designed a number of boats that are perfect for those who lack confidence in their woodworking skills and are attracted to simpler boats.

Houdini: Classy Dinghy Cruising for One or Two

The 13-foot 2-inch Houdini is an excellent general-purpose and cruising dinghy that John calls his "escape machine." It's like a catboat with its large single sail of 115 square feet and a generous beam of 5 feet 10 inches, a centerboard positioned well forward, and a good-sized skeg and rudder. But there the similarity to catboats ends, for despite its beam Houdini has a relatively narrow waterline. This combination is achieved by a generous flare and is there to provide the leverage to allow a skipper to keep the boat on its feet in strong winds. Another unusual feature is that Houdini has a self-draining floor that,

with the forward-placed centerboard, provides a long and wide platform for camping, lounging in a deckchair, and so on.

The boat came into existence some years ago when John found himself without a boat to cruise in and needed a new one he could build in a hurry—which is perhaps the reason she's a stitch-and-glue chine boat that's definitely suitable for a second or third boatbuilding project. The design reflects John's desire to be able to sleep two on board under a tent, and that it not be too large for his wife and 12-year-old daughter to sail alone.

The sprit-boomed standing lugsail rig is a favorite of John's. It offers a lot of sail area for little money, it has a low center of effort, and it is incredibly powerful when reaching and running. The spars are short enough to stow securely in the boat, which is particularly useful when it's being trailered.

Tread Lightly: Serious Cruising for One

A host of factors came together to prompt the development of this surprising little solo

Houdini. (Burton Blais)

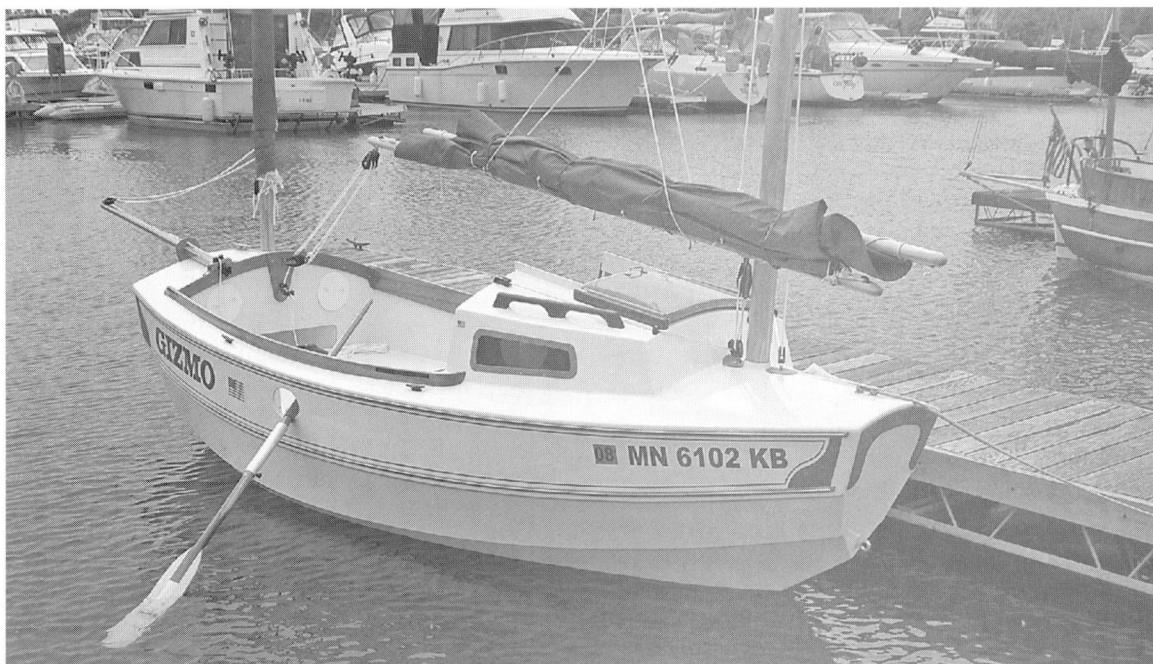

Tread Lightly. (Bob Trygg)

cabin sailing cruiser, which is based on a scaling up of John's best-known tender, the Tender Behind. I've often thought that the kind of pram hull form that makes a good tender could also be developed to make a good one-man microcruiser, and Tread Lightly is the proof.

A lot of questioning went into this design. How much room does it take to lie down? How much space does a month's worth of provisions take if the boat can be resupplied with water once a week? What sort of hull form could carry a really decent weight on a length that would fit into a standard garage? How can a boat that would suit all of the other parameters be made to have the smallest possible range of stability if inverted? How could you make it capable of being reefed or hove-to in a seaway? Where would the cooking stove go? Where would you store dry clothes? How much space is needed for sitting headroom? How much natural light is needed?

John's answer was a flat-bottomed, round-sided pram dinghy 13 feet in length with a 5-foot beam, a sail area of 123 square feet, and a displacement of 661 pounds. A pram hull can cope with all the gear for cruising, as well as some ballast for stability under sail, all in a short length.

The rig has a balanced lugsail main and sharpie-style spritsail mizzen, which allows the masts to be placed close to the ends of the boat. Balanced lugsails are rare on cruising boats designed for the sea, but this is a simple and economical rig, and because its sail area is well spread out and low the boat will often be able to self-steer. Overall, it's difficult to imagine a better design for a garage-built solo cruiser.

Trover: Economical Outboard Boat

John designed Trover for an aid organization that wanted a simple inshore and estuary fishing boat for the people of East Timor. These villagers use boats for just about everything:

Trover. (Burton Blais)

not just fishing and trucking material around, but also as the local ambulance, taxi, hearse, and wedding carriage.

The problem was a terrible one that few boat designers ever need to think about. A disastrous war had left the Timorese without boats, and the country needed several thousand to recover, yet people were starving, money was short, there were few skilled boatbuilders, and local materials were almost unobtainable. Time was at a premium, but they did have stocks of plywood and Yamaha 15 hp engines supplied by the United Nations.

To cut a long story short, John devised a simple and effective 14-foot 5-inch by 4-foot 7-inch boat that was a significant contribution to solving the country's problems.

Fast, stable, roomy, and dry, it is also very well suited to the needs of people who want to build their own small and economical outboard boat for fishing and messing about. It has a narrow, flat bottom panel and a fine entry; the entry ensures that the boat handles waves well while a planing shoe enables it to plane a heavy load with a relatively small motor. The chine panels incorporate sufficient V to ease the motion in a seaway, and the topsides are flared enough to keep the boat dry inside. In short, it's ideal for the short sloppy waves of lakes and estuaries.

JACQUES MERTENS-GOOSENS

A Frenchman who lives and works in the United States, Jacques Mertens-Goosens is a designer whose collection of boat designs neatly combines both American and European influences: while he's drawn a series of classic looking little plywood skiffs and power dories, for example, his pocket cruisers and V-bottomed dinghies are clearly inspired by boats often seen in European waters. Looking at his designs, it's obvious that he's a great draftsman and has the knack of making even simple little boats look good.

Contact Info

Jacques Mertens-Goosens's plans are available at:

www.bateau.com

He's also the man behind the D4 8-foot V-bottomed dinghy design, which is available as a free download from his website. This useful small boat was one of the very first free boat designs to be made available online and has been built many times. Along the way it must have introduced large numbers of people to stitch-and-glue boatbuilding.

Cat Ketch 17: Big, Comfortable Two-Sail Skiff

One of Jacques's American-style boats, Cat Ketch 17 is a large general-purpose cat-ketch-rigged skiff that's 17 feet 6 inches long with a 6-foot beam. It has a lot of storage space, a flat gently-rockered bottom for ease of building and for camping, and an easily adjusted and distinctive two-stick rig that can be reefed right down whenever required. I think a well-built and nicely finished boat to this design would attract attention almost everywhere.

Jacques comments that it would be a great boat for the light winds and shoal waters of the Florida Keys, and he's right. It has a clever daggerboard designed to kick up when it hits something, making the boat particularly suitable for shoal waters. It's also big enough and narrow enough in beam to be able to cope with waves better than shorter, relatively fatter skiffs.

However, I'd suggest that even at these dimensions an unballasted flat-bottomed boat will be difficult to manage in one of the windy and rough corners of the world, and that flat bottom will be noisy at times, particularly if you do decide to follow his recommendation and use her for camping.

Still, the CK 17 is a great boat. It would be a perfect summer picnic boat for a family, but perhaps its most obvious use is sailing with groups of kids and for fishing parties of three or even four adults who like to fit in a little sailing on their day out.

Otter 16: Row-Out, Sail-Home Daysailer

This is a much more sophisticated boat than Doris, yet like the sailing light dory I've presented in this book it's intended to be a rowboat

Cat Ketch 17. (Jacques Mertens-Goosens)

first and a sailer second. The difference is that Otter's more sophisticated hull makes it a better sailer, albeit with more complicated construction. Jacques designed it for his own purposes, which is always a good sign.

Jacques wanted a boat that could be rowed from his dock to the local fishing spots between one and three miles away, and then sailed home downwind. The rowing had to be pleasant, because he loves rowing and can't stand a sluggish rowboat, so it should have a narrow waterline. The hull would widen as it goes up, to enable it to stand up to its sails. The end result is a hull cross section amidships that is a five-panel shape not unlike that of a Swampscott dory.

223

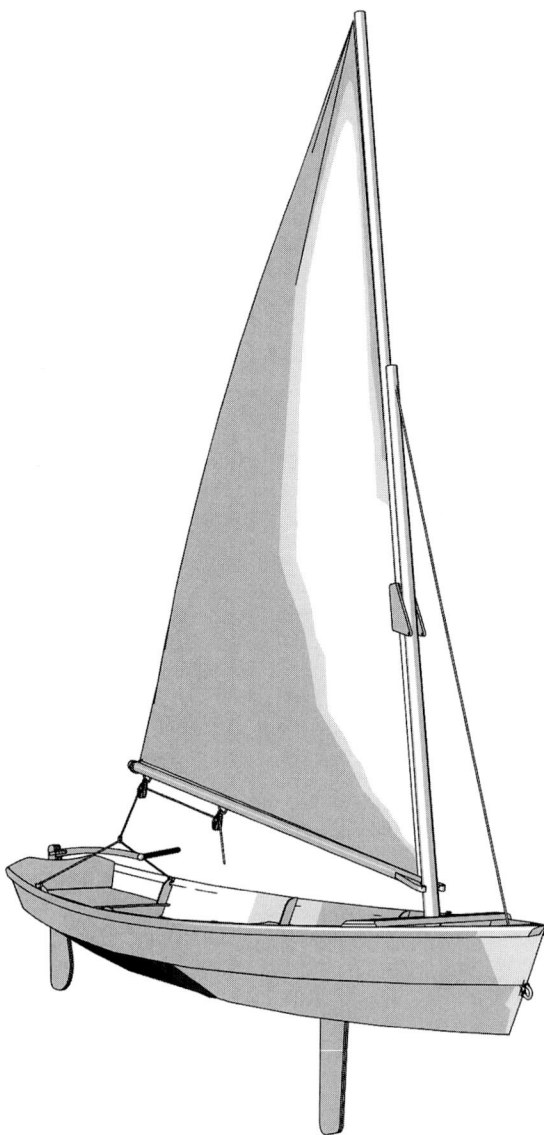

Otter 16. (Jacques Mertens-Goosens)

The rig did not have to be very large because the sailing was intended to be lazy and undemanding. In fact, the most important aspect of the rig was that the whole thing had to fit inside the boat and be out of the way while fishing. The rudder also had to be easy to install and unobtrusive, so as not to interfere with fishing lines when trolling them astern.

Jacques concluded that a design from the early days of stitch-and-glue plywood boats called the Dobler 16 came closest to the boat he felt he needed, and so he designed this 15-foot 6-inch boat for himself along the same lines. He says he hopes to find time to build her one day!

Indian River Skiff: The Poor Man's Power Dory

This is another of our Frenchman's distinctively American-style designs, a small 15-foot 4-inch by 5-foot power dory.

We don't see many boats like this in Europe. They typically have a bow section not unlike that of a conventional rowing or sailing dory, but with a central section that broadens out to make a straight run. It is this shape that gives them their characteristic combination of seaworthiness and performance, and it's also a relatively straightforward shape to build. I'd expect this boat to have some of the seaworthiness of the larger power dories, but with its slim proportions, light weight, and narrow flat bottom leading to a strong stern, I'd expect it to perform really well with a 25 hp engine.

The Indian River plans come with three different internal layouts for different purposes, including one that is largely decked-over and is specifically designed for fishing. Jacques says it'll carry three to five adults, depending on the conditions.

CONRAD NATZIO

Like many of us involved in small boat design, Conrad Natzio came from another discipline entirely: he spent 20 years as a probation officer. At the end of that long stint, he needed a more physically demanding occupation and retrained as a boatbuilder at the International Boatbuilding College.

In the United Kingdom, we have almost no tradition of boats made from wide, flat panels, but Conrad is nevertheless fascinated by traditional North American boat styles that

Contact Info

Conrad Natzio's plans are available from:

The Old School

Brundish Road, Raveningham

Norwich NR14 6NT UK

+44 1508 548675

Or, go to: www.broadlyboats.com.

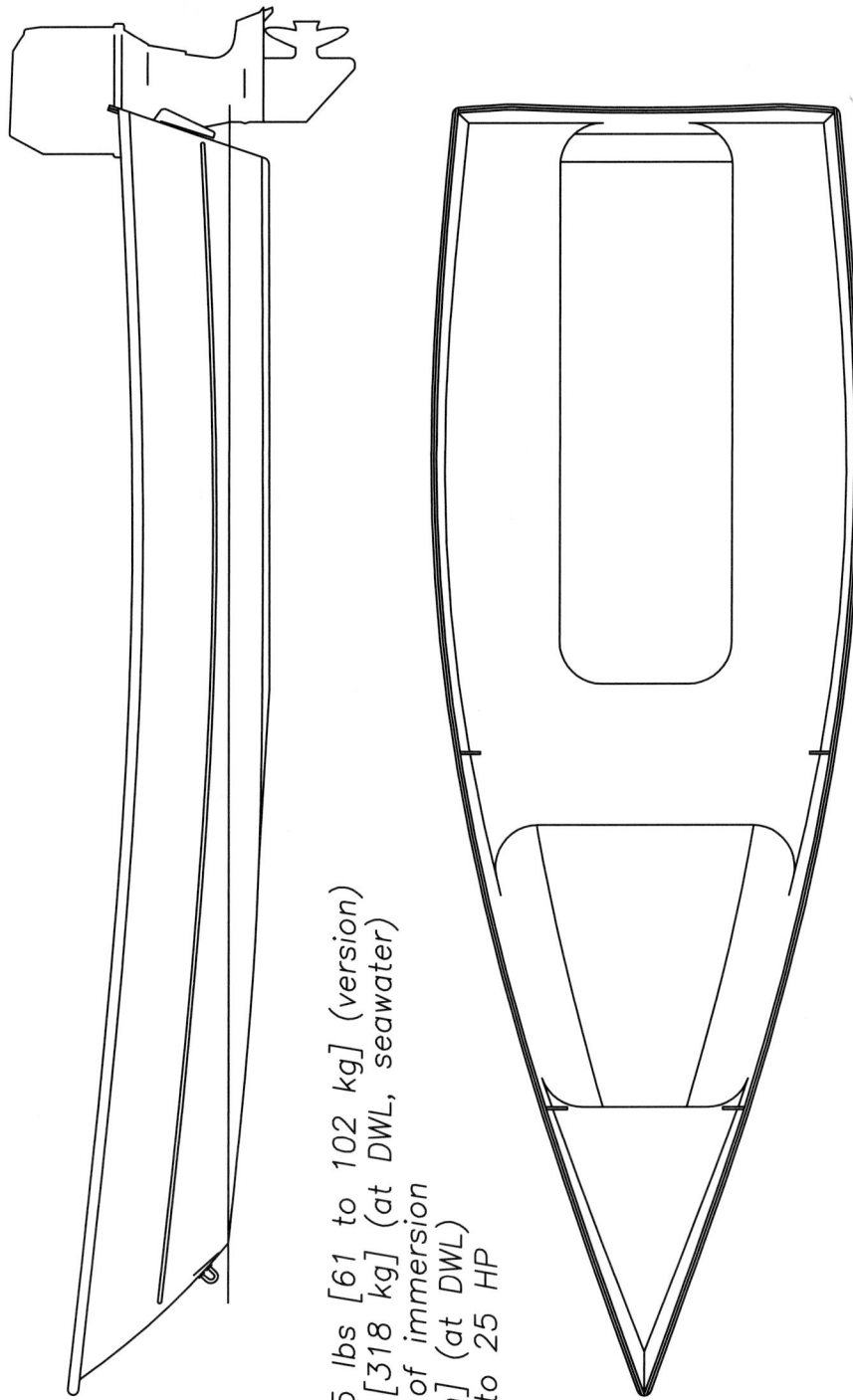

D15
LOA: 15'–4" [4,68 m]
Beam: 5' [1,53 m]
Hull weight: 135 to 225 lbs [61 to 102 kg] (version)
Displacement: 700 lbs. [318 kg] (at DWL, seawater)
PPI: 93.4 lbs per inch of immersion
Hull draft: 5" [127 mm] (at DWL)
Recommended HP: 10 to 25 HP

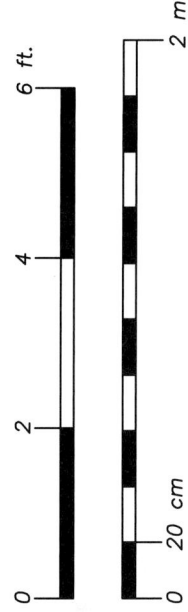

6 ft.
4
2
0

2 m
20 cm
0

Indian River Skiff. (Jacques Mertens-Goosens)

225

use wide, flat areas of lumber or plywood, and also by Phil Bolger's bold and simple designs for home boatbuilders.

Having built several from Bolger's designs, around 10 years ago he decided to develop and sell plans of his own that would allow boatbuilders to construct boats quickly and confidently, and which would be useful for a wide range of purposes.

Conrad's boats have a more traditional look than many of Bolger's, and I strongly suspect he has also been very much influenced by the shallow waters of the Norfolk Broads, an area of rivers and flooded medieval peat workings near his home. Several of his designs have unusual twin shallow bilge keels. Easy to build, they also strengthen and stiffen the hull, and open up the space inside the boat.

Another shallow water feature favored by Conrad but which hasn't seen much use in small boats in the United Kingdom is the end-plate rudder developed some years ago by Phil Bolger. This is a short rudder with a horizontal plate at its lower edge to create an inverted T-shaped structure that holds on to the water as well as a deeper conventional rudder. Because it does not need to be lifted when the water is shallow or when beaching, it's extremely easy to make and requires a minimum of hardware.

Sandpiper: A Skiff for a Summer Day

Originally designed to be built at a boat show over a four-day weekend and subsequently used on sheltered waters, Sandpiper is a 13-foot 7-inch by 4-foot 8-inch flat-bottomed skiff with Conrad's trademark long bilge keels.

Sandpiper's open interior makes it a comfortable daysailer for a couple of adults, and with a tent it would also be a good camp cruiser for one on inland waters. As you might expect in a boat of this size meant to be built in just a few days, Conrad has kept everything as simple as can be: the side panel edges are parallel to each other, and the

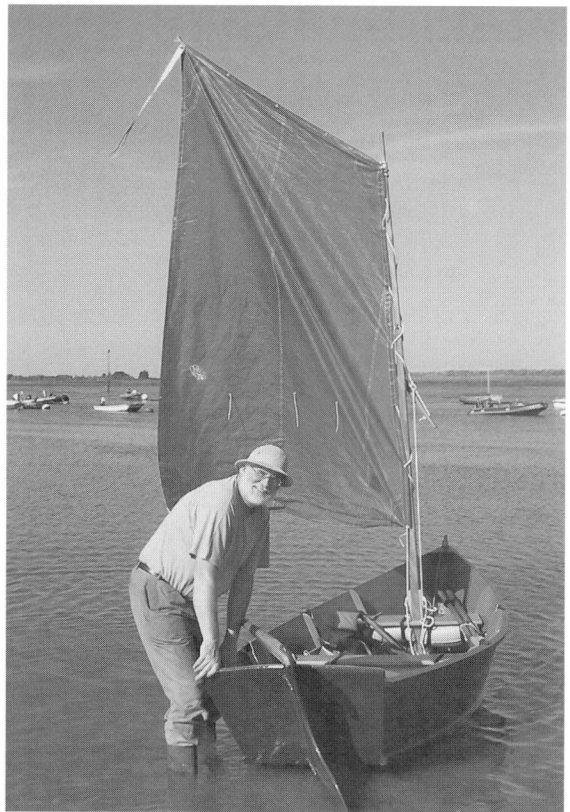

Sandpiper. (Chris Partridge)

sheerline and bottom profile are created by bending them around internal frames. Just three sheets of ¼-inch plywood are required, but the long keels make the bottom of the boat extremely stiff.

Buoyancy bags are recommended to enable the skipper to right the boat in a capsize, and the plans include two snug rigs: a spritsail with a jib and a balanced lug.

Oystercatcher: A Skiff That Lets Everyone Have a Go

Oystercatcher is a bigger, heavier, flat-bottomed skiff that at 15 feet 3 inches by 5 feet is big enough to allow a small family to sail together. The design is based on New England skiffs, which means that it has a jaunty near-vertical stem and a well-rockered (curved fore-and-aft) bottom that enables it to carry a large load while retaining the ability to go about quickly.

Oystercatcher can be rigged with a balanced lug main and a leg-of-mutton mizzen in

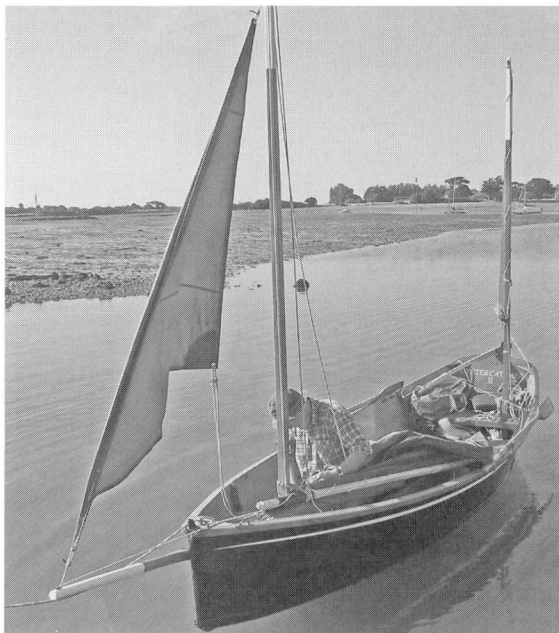

Oystercatcher. (Conrad Natzio)

either of two sizes. This rig looks very salty indeed. However, it can also be set up as a schooner-rigged lugger, which will suit many children. Most kids prefer to take part in sailing rather than simply watching their parents do all the work, and the schooner rig provides the maximum number of interesting strings to pull.

A little larger than the Sandpiper, Oystercatcher is very suitable for camping with two adults.

SUPPLIERS OF BOATBUILDING MATERIALS

POLYURETHANE GLUES

The Chemical Company Ltd. (Balcotan polyurethane glue)
22 Highgate
Cherry Burton
Beverley, East Yorkshire, HU17 7RR
United Kingdom
+44 (0) 1482 874059
www.holdich.demon.co.uk/chemical/balcotan.html

Henkel Consumer Adhesives (PL Premium polyurethane construction adhesive, available at most builders' supply stores)
7405 Production Drive
Mentor, OH 44060
800-999-8920
www.stickwithpl.com/contact.asp

EPOXIES

Chinawind Yachts
The Old Workshop
Barrs Lane
Benington
Boston, PE22 0ED
United Kingdom
+44 (0) 1205 760662
E-mail: PeterJchinawind@aol.com
www.masepoxies.co.uk

Fyne Boat Kits
Unit 5
Station Yard
Burneside
Kendal
Cumbria LA9 6QZ
United Kingdom
+44 (0) 01539 721 770
info@fyneboatkits.com
www.fyneboatkits.com

MAS Epoxies
2615 River Road #3A
Cinnaminson, NJ 08077
888-627-3769
www.masepoxies.com

System Three Resins, Inc.
3500 West Valley Highway North
Suite 105
Auburn, WA 98001
800-333-5514
www.systemthree.com

UK Epoxy Resins
3 Square Lane
Burscough
Lancashire
L40 7RG
United Kingdom
+44 (0)1704 892364
Fax: +44 (0)1704 892364
enquiry@epoxy-resins.co.uk
www.epoxy-resins.co.uk

West System Inc.
102 Patterson Ave.
P.O. Box 665
Bay City, MI 48707-0665
866-937-8797
www.westsystem.com

FIBERGLASS SUPPLIES, EPOXY, POLYESTER RESINS

Fiberglass Supply
314 West Depot
P. O. Box 345
Bingen, WA 98605-0345
509-493-3464
www.fiberglasssupply.com

BOAT HARDWARE AND SUPPLIES

Duckworks Boat Builders Supply
608 Gammenthaler
Harper, TX 78631
www.duckworksbbs.com

Hamilton Marine
155 E. Main St
Searsport, ME 04974
800-639-2715
www.hamiltonmarine.com

Jamestown Distributors
17 Peckham Drive
Bristol, RI 02809
800-497-0010
www.JamestownDistributors.com

Jeckells the Sailmakers
Jeckells of Wroxham Ltd.
Station Road
Wroxham
Norfolk
England, NR12 8UT
United Kingdom
www.jeckells.co.uk

Trident UK
Trident Quay
South Shore Road
Gateshead
Tyne and Wear NE8 3AE
United Kingdom
+44 (0)191 490 1736
Fax: +44(0)191 478 2122
enquiries@trident-uk.com
www.trident-uk.com

MAGAZINES

Duckworks Magazine
608 Gammenthaler
Harper, TX 78631
chuck@duckworksmagazine.com
www.duckworksmagazine.com

Messing About in Boats
29 Burley St
Wenham, MA 01984-1943
www.messingaboutinboats.com

Watercraft
Bridge Shop
Gweek, Helston
Cornwall TR12 6UD
United Kingdom
+44 (0)1326 221424
Fax: +44 (0)1326 221728
subs@watercraft-magazine.com
www.watercraft.co.uk

INDEX

230